optimizing materials for nuclear applications

Edited by

F. A. Garner, D. S. Gelles

F. W. Wiffen

Cover Credit:

The cover shows that TiC precipitates in titanium-modified type 316 stainless steel are effective collectors of helium. Irradiation at 630°C generated the 880 parts-per-million helium contained in the bubbles. This micrograph, with a magnification of 337,500x, was produced by P.J. Maziasz of the Oak Ridge National Laboratory.

optimizing materials for nuclear applications

Proceedings of a symposium sponsored by the Nuclear Metallurgy Committee at the Winter Meeting of The Metallurgical Society, Los Angeles, California, February 27-29, 1984.

Edited by

F. A. Garner
Hanford Engineering Development Laboratory
Richland, Washington

D. S. Gelles
Hanford Engineering Development Laboratory
Richland, Washington

F. W. Wiffen
Oak Ridge National Laboratory
Oak Ridge, Tennessee

A Publication of The Metallurgical Society, Inc.

Library of Congress Cataloging-in-Publication Data
Main entry under title:

Optimizing materials for nuclear applications.
"The Symposium on Optimizing Materials for Nuclear Applications was held during the winter meeting of The Metallurgical Society of AIME at Los Angeles, California, in January 1984"--CIP pref.
Includes indexes.
1. Nuclear reactors--Materials--Congresses.
I. Garner, F. A. **II.** Gelles, D. S. **III.** Wiffen, F. W.
IV. Symposium on Optimizing Materials for Nuclear Applications (1984 : Los Angeles, Calif.)
V. Metallurgical Society of AIME. Nuclear Metallurgy Committee.
TK9185.068 1985 621.48'33 85-18759
ISBN 0-87339-002-4

A Publication of The Metallurgical Society, Inc.
420 Commonwealth Drive
Warrendale, Pennsylvania 15086
(412) 776-9000

Printed in the United States of America.
Library of Congress Catalogue Number 85-18759
ISBN NUMBER 0-87339-002-4

preface

The symposium on Optimizing Materials for Nuclear Applications was held during the Winter Meeting of the Metallurgical Society of AIME at Los Angeles, California in January 1984. While it was a relatively small meeting, the material presented here and in other oral presentations addressed the important issue of how to pursue material optimization for purposes of extended component life or public safety.

Optimization efforts for materials selected for use in nuclear applications are often quite complex. Such materials usually must serve in exceptionally demanding environments involving applied stresses, high temperature and exposure to the corrosive influence of coolants, fuel and fission products. Even more important, however, structural materials must resist the often deleterious influence of atomic displacements created within the material by neutron or charged-particle irradiation. Displacive irradiation can cause extensive changes in mechanical properties, component dimensions, phase equilibria and corrosion resistance.

In each nuclear environment the alloy designer must identify the life-limiting degradation mechanisms or safety concerns and optimize the material to meet these concerns via changes in composition, fabrication or environment. The optimization process often involves a considerable number of trade-offs between two or more properties. An example is found in the optimization of Fe-Cr-Ni alloys used for breeder reactor service, in which two of the simplest means of improving the swelling resistance lead to a decrease in corrosion resistance or large reductions in post-irradiation ductility.

In these proceedings a wide variety of service environments, reactor concepts, structural materials, safety concerns and damage mechanisms are investigated, most by experiment and some by theoretical analysis. While many of the papers are directed toward optimization of components in currently operating reactor concepts, some of the papers involve problems expected in anticipated fusion environments. For example, three papers discuss the development of alloys that will exhibit a reduced level of long-term radioactivity and thus mitigate future waste disposal problems for fusion reactors.

Another three papers address safety concerns such as the embrittlement of pressure vessel steels, radiation exposure to plant maintenance personnel and the protection of the general public from the consequences of deep space probes which might fail during launch and fall to earth. Five papers in this volume are directed toward reducing the dimensional instabilities that result from swelling, irradiation creep and irradiation growth.

The editors are indebted not only to the authors for their diligence in meeting the standards of the Metals Society of AIME but also to the volunteer experts who assisted in the reviewing procedure.

F. A. Garner and D. S. Gelles
Hanford Engineering Development Laboratory
Richland, Washington

F. W. Wiffen
Oak Ridge National Laboratory
Oak Ridge, Tennessee

Table of contents

Preface v

OPTIMIZATION OF MATERIALS FOR NUCLEAR SERVICE

Liquid Metal Corrosion Considerations in Alloy Development 3
P. F. Tortorelli and J. H. DeVan

Tailoring and Evaluating Wear-Resistant Alloys for Nuclear Applications 15
H. Ocken

Factors Affecting In-Core Dimensional Stability of Zircaloy-2 Calandria Tubes 35
V. Fidleris, A. R. Causey, and R. A. Holt

Control of Microstructure in Brazed Zone of Zircaloy-4 Nuclear Fuel Sheathing by Optimization of Σ(C+P+Si) Contents and Cooling Schedules 51
V. Quach and D. O. Northwood

Optimization of Martensitic Stainless Steels for Nuclear Reactor Applications 63
D. S. Gelles

The Development of Austenitic Steels for Fast Induced-Radioactivity Decay for Fusion Reactor Applications 73
R. L. Klueh and E. E. Bloom

Dependence of Neutron-Induced Swelling on Composition in Iron-Based Austenitic Alloys 87
F. A. Garner and H. R. Brager

Overview of the Swelling Behavior of 316 Stainless Steel 111
F. A. Garner

The Origin of the Large Resistance to Void Swelling Observed in Fe-35.5Ni-7.5Cr 141
H. R. Brager and F. A. Garner

An Evaluation of Steels and Welds Produced Overseas for 288°C Radiation Embrittlement Resistance 167
J. R. Hawthorne

Ir/PuO_2 Compatibility; Transfer of Impurities from Plutonium Dioxide to Iridium Metal During High Temperature Aging 201
D. H. Taylor, W. H. Christie, and D. Pavone

THEORETICAL SUPPORT STUDIES

Calculated Alloying Effects on Irradiation Hardening of Ferritic Steels 225
E. P. Simonen

Numerical Simulation of Void Nucleation in Metals Under Cyclic Fusion Irradiation Conditions 239
S. J. Primeau and K. C. Russell

Irradiation-Induced Cavity Nucleation on Dislocation Lines 257
B. Bendriem and K. C. Russell

Subject Index 269

Author Index 273

optimization of materials for nuclear service

LIQUID METAL CORROSION CONSIDERATIONS IN ALLOY DEVELOPMENT*

P. F. Tortorelli and J. H. DeVan

Metals and Ceramics Division
Oak Ridge National Laboratory
Oak Ridge, Tennessee 37831
USA

Abstract

Liquid metal corrosion can be an important consideration in developing alloys for fusion and fast breeder reactors and other applications. Because of the many different forms of liquid metal corrosion (dissolution, alloying, carbon transfer, etc.), alloy optimization based on corrosion resistance depends on a number of factors such as the application temperatures, the particular liquid metal, and the level and nature of impurities in the liquid and solid metals. The present paper reviews the various forms of corrosion by lithium, lead, and sodium and indicates how such corrosion reactions can influence alloy optimization.

---

*Research sponsored by the Office of Fusion Energy, U.S. Department of Energy under contract DE-AC05-84OR21400 with the Martin Marietta Energy Systems, Inc.

Introduction

Due to their excellent heat transfer properties, liquid metals have been used, or are being proposed, as heat transfer fluids in a variety of nuclear and non-nuclear power systems. Examples of such uses include molten sodium for liquid metal fast breeder reactors and central receiver solar stations, and liquid lithium for fusion and space nuclear reactors. Furthermore, since requirements for tritium breeding in deuterium-tritium fusion reactors necessitate the exposure of lithium atoms to fusion neutrons, liquid metal breeding fluids of lithium or Pb-17 at. % Li have been considered for this function while molten lead or bismuth can serve as neutron multipliers to effectively raise the tritium breeding ratio if other types of lithium-containing breeding material are used. Also, liquid metals can be used as two-phase working fluids in Rankine cycle power conversion devices (potassium) and in heat pipes (potassium, lithium, sodium, NaK). NaK has even been employed as a static heat sink in automotive and aircraft valves due to its high thermal conductivity.

Whenever liquid metals are used, whether in specific applications or in handling of melts during processing, a compatible containment material must be selected. In some cases, liquid metal corrosion is not important, but for many of the applications cited above, corrosion considerations can play a significant role in choosing the appropriate structural material and operating conditions. Thus, liquid metal corrosion studies in support of aircraft, space, and fast breeder reactor programs and heat pipe technology date back many years. More recently, such research has also been done as part of the fusion energy technology program. In the present paper, we review our understanding of liquid metal corrosion gained from such studies through a discussion of the principal corrosion reactions and the important parameters that control these processes, particularly with regard to how alloy modification can be used to reduce the deleterious effects of this type of corrosion.

The subject of this paper will be restricted to corrosion phenomena; liquid metal embrittlement factors and environmental effects on mechanical properties will not be considered in developing alloy development principles. Furthermore, the discussion will be limited to corrosion under single-phase (liquid) conditions, which is the primary area of interest for most nuclear and solar applications. The various forms of liquid metal corrosion will first be described and then the influence of alloy composition and microstructure on the underlying corrosion reactions will be discussed.

Forms of Liquid Metal Corrosion

Liquid metal corrosion may manifest itself in a variety of forms. The different types can be broadly classified as

1. dissolution,
2. alloying,
3. intergranular penetration,
4. impurity and interstitial reactions, and
5. mass transfer.

There is overlap of these categories; the various forms of corrosion are not necessarily independent of each other. For example, intergranular penetration can be related to impurity reactions while dissolution is one step in the mass transfer process. However, such a classification scheme is adequate for the purposes of this paper; it establishes a base on which to discuss the corrosion reactions of importance to alloy development.

Dissolution

This form of liquid metal corrosion is the most obvious — the simple solution of the atoms of the containment material in the liquid metal. Dissolution is governed by the elemental solubilities in the liquid metal and the kinetics of the dissolution reaction(s). Simple dissolution is the principal form of corrosion that occurs in isothermal, single alloy, static systems of liquid metals containing very low levels of impurities in both the alloy and melt.

Alloying

Reactions between atoms of the liquid metal and those of the structural alloy can occur such that a stable product forms without the participation of impurity or interstitial elements. This is not a common form of liquid metal corrosion (particularly with the alkali metals) and will not be discussed further in this paper. Alloying reactions, however, can be used to inhibit corrosion by adding an element to the liquid metal that will react with the containment material to form a corrosion resistant layer. Examples of this can be found in lithium and lead systems (1).

Intergranular Penetration

Under certain impurity and/or microstructural conditions, localized attack of grain boundaries can occur upon exposure to a liquid metal. (An example is given in Figure 1.) The tendency for this type of corrosion can generally be related to the instability of grain boundary precipitates relative to the liquid metal or a susceptible band of altered composition near grain boundaries.

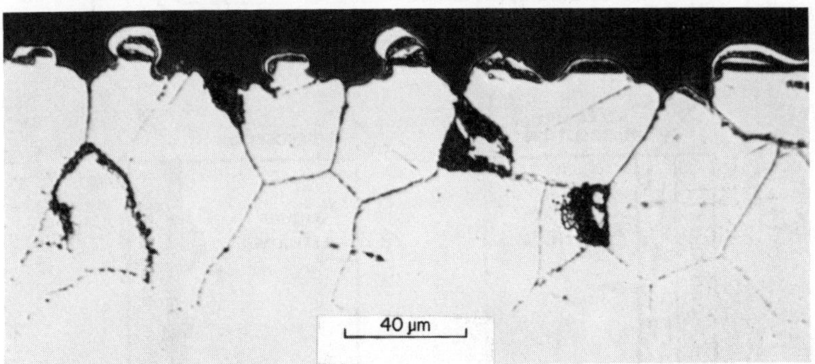

Figure 1 — Polished cross-section of type 316 stainless steel exposed to nitrogen contaminated lithium for 2000 h at 700°C (from reference 2).

Impurity and Interstitial Reactions

Examples of impurity and interstitial reactions include the decarburization of steel in sodium or lithium and the oxidation of steel in sodium or lead of high oxygen activity. In many cases when the principal elements of the containment material have low solubilities in liquid metals (for example, refractory metals in sodium and lithium), reactions involving light elements such as oxygen, carbon, and nitrogen dominate the corrosion process.

Mass Transfer

In this form of corrosion, there is a net movement of material in response to a gradient in temperature or composition across a liquid metal system. Thermal gradient mass transfer is shown schematically in Figure 2. Since the liquid metal is the means by which the material is transported, the dissolution and deposition reactions are of prime concern. In cases where liquid metals are used under nonisothermal conditions, nonuniform deposition of material in the cold zone of the circuit can be a more severe problem than dissolution in the hot region due to flow restrictions and, in nuclear applications, excessive radiation levels outside the core caused by radionuclide deposition. Concentration gradient mass transfer can occur in the absence of a thermal gradient or add to the severity of the mass transport problem in nonisothermal systems due to elemental activity differences across the liquid metal arising from the simultaneous exposure of containment materials of different composition. An example of concentration gradient mass transfer would be carbon transfer from a steel to a refractory metal in a bimetallic liquid metal system.

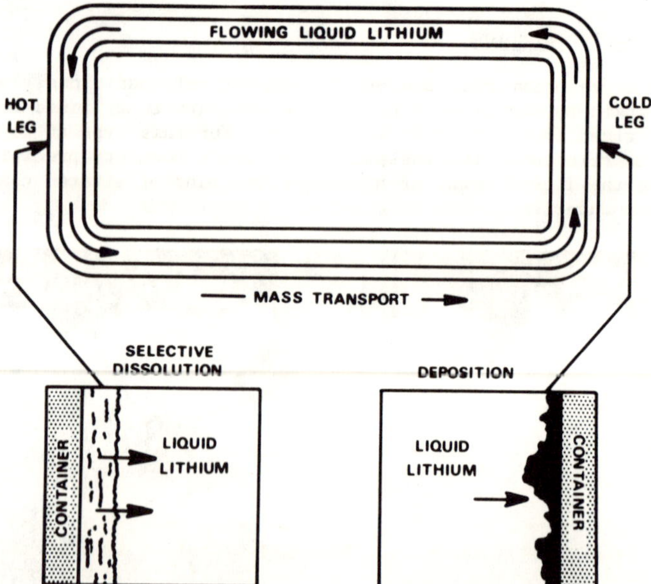

Figure 2 — Schematic representation of thermal gradient mass transfer in a lithium-austenitic stainless steel system (from reference 3).

Corrosion Reactions — Influence of Alloy Composition and Microstructure

Direct Dissolution Reactions

Direct dissolution reactions are those that involve the simple solution of an element from the containment material in the liquid metal in the absence of any impurity effects. An example of this reaction would be

$$\text{Fe (in solid)} = \text{Fe (in liquid)} \ .$$

This type of reaction is governed by the solubility of the particular element of the solid in the liquid metal and the kinetics of the dissolution process. For pure metals, a straightforward approach to the selection of a containment material would therefore be to use an element that has a very low solubility in the liquid metal. For example, refractory metals have low solubilities in sodium, lithium, and lead and are essentially inert to these molten metals aside from some impurity-induced effects (see below). For the more practical situation where an alloy is required, it is sometimes possible to develop a dissolution-resistant alloy based on reducing or eliminating an element that exhibits a high solubility in the liquid metal of interest and increasing the relative enrichment in elements that have lower solubilities. Such behavior has been found in sodium, lead, and lithium systems, where nickel exhibits significant solubilities: decreasing the concentration of nickel in Fe-Cr-Ni alloys decreased the extent of dissolution of these alloys (4–6). An alternative to reducing an "active" elemental concentration in an alloy is to reduce its activity. It has been shown that a high nickel alloy can have a significantly lower dissolution rate in lead than alloys with lower nickel concentrations if the former material consists of an intermetallic compound (7).

It should be noted that while alloy development can often be guided by solubilities in the liquid metal, this "rule of thumb" can have exceptions. Indeed, as in many other processes, the kinetics of the dissolution (or deposition) reaction may control the overall rate of weight loss such that solubility does not determine the ultimate amount of dissolution, particularly for short-time exposures and for nonisothermal conditions in which deposition is simultaneously occurring.

Corrosion Product Formation

Corrosion product formation is one type of what was classified as an interstitial or impurity reaction. The general form of such a reaction is

$$xL + yM + zI = L_xM_yI_z \ ,$$

where L is the chemical symbol for a liquid metal atom, M is one for the containment alloy, and I represents an interstitial or impurity atom in the solid or liquid ($x, y, z \geqslant 0$). The $L_xM_yI_z$ corrosion product that forms may be soluble or insoluble in the liquid metal. If soluble, the effect of the "I" atom could be to cause higher measured weight losses and result in an apparently higher solubility of M in L. (This is a frequent cause of erroneous solubility measurements.) Furthermore, if a soluble corrosion product forms at selected sites on the surface, localized attack will be observed. If the corrosion product is insoluble, then a surface layer will develop. However, this does not necessarily mean that it can be directly observed: the product may dissolve during the process of removing the residual liquid metal from the exposed containment alloy.

A good example of the importance of impurity or interstitial reactions that form corrosion products can be found in the sodium-steel-oxygen system. It is thought that the reaction

$$3\ Na_2O(l) + Fe(s) = (Na_2O)_2 \cdot FeO + 2\ Na(l)$$

increases the apparent solubility of iron in sodium at higher oxygen activities while the interaction of oxygen, sodium, and chromium can lead to the formation of a corrosion product film

$$2\ Na_2O(l) + Cr(s) = NaCrO_2 + 3\ Na(l)\ .$$

This second reaction is one of primary importance in sodium corrosion of chromium-containing steels. It can be controlled by reduction of the oxygen concentration of the sodium to less than about 3 wt ppm and/or by composition modification of the alloy, that is, by reducing the chromium concentration of the steel.

An example of such corrosion product reactions can also be found in lithium-steel systems, where nitrogen has been observed to increase corrosion (2,8,9). In particular, the reaction

$$5\ Li_3N + Cr(s) = Li_9CrN_5 + 6\ Li$$

has been observed (10) and this corrosion product has been found to be localized at grain boundaries of exposed steels (11,12). In addition, nitrogen is thought to increase the dissolution of nickel in molten lithium (10) and is effective at increasing corrosion whether it is in the lithium or in the steel (12,13). Therefore, reducing the nickel, chromium, and nitrogen concentrations of steels exposed to lithium could improve their corrosion resistance.

A final example of a corrosion product reaction that can occur in a liquid-metal system is that of oxidation of a metal or alloy exposed to molten lead. In some cases, this reaction may actually be beneficial by providing a protective barrier against the highly aggressive lead (6). [Pure lead is more corrosive toward ferrous alloys than pure lithium or sodium (1,4,6)]. However, this surface product will form and then heal only when the oxygen activity of the melt is maintained at a high level or when oxide formers such as aluminum or silicon have been added to the containment alloy to promote this form of protection (6). Therefore, alloy development considerations based on the oxidation reaction in lead-steel systems would favor the use of alloying additions that are strong oxide formers to help protect the steel from dissolution.

Impurity Transfer

The second general type of impurity or interstitial reaction is that of elemental transfer. Such a reaction is of crucial importance for oxygen-metal-lithium systems since lithium is such a strong sink for oxygen. In fact, despite their low solubilities in lithium, niobium and tantalum can be severely attacked when exposed to lithium if the oxygen activity of these metals is not minimized. Their oxides are rapidly attacked with concomitant transfer of the oxygen into the molten lithium. The result is localized penetration along grain boundaries and selected crystallographic planes (14,15). This form of corrosion can be eliminated, however, by minimizing the oxygen concentration of these refractory metals and by using alloying additions that form oxides that are not reduced by lithium (1–2 at. % Zr in Nb and Hf in Ta) (14).

Another important example of impurity transfer is that of decarburization/carburization. Carbon transfer to or from the liquid metal can cause decarburization of Fe-Cr-Mo steels (particularly lower chromium steels) and carburization of refractory metals and higher chromium alloys. There have been many studies of such reactions for sodium-steel systems (see, for example, references 16–18) while less work has been done in the area of lithium-steel carbon transfer (18,19). However, the same type of considerations apply in both cases and as an example, results from two lithium studies (19,20) will be briefly discussed. The equilibrium partitioning of the carbon between the Fe-Cr-Mo steel and the lithium can be described as (19)

$$\frac{C_{\underline{C}(s)}}{C_{\underline{C}(Li)}} = (a_{Cr})^{x/y} \frac{C^0_{\underline{C}(s)}}{C^0_{\underline{C}(Li)}} \exp \frac{\Delta F^0_{\frac{1}{2}Li_2C_2}}{RT} \exp \frac{-\Delta F^0_{(1/y)Cr_xC_y}}{RT}$$

where

$C_{\underline{C}(s)}$, $C_{\underline{C}(Li)}$ = concentration of carbon in the steel and lithium, respectively,

a_{Cr} = chromium activity of the steel,

$C^0_{\underline{C}(s)}$, $C^0_{\underline{C}(Li)}$ = solubilities of carbon in the steel and lithium, respectively,

ΔF^0 = free energies of formation of the indicated compounds,

and x,y are defined by the stoichiometry of the chromium carbide. This equation indicates that to decrease the tendency for decarburization of an alloy [i.e., to increase the "partitioning coefficient," $C_{\underline{C}(s)}/C_{\underline{C}(Li)}$], the chromium activity of the alloy must be increased or the free energy of formation of the matrix carbide must be lowered (made more negative) by alloy manipulation to form a more stable carbide. Experiments in lithium and sodium have shown these factors indeed have the desired effect. Tempering of Fe-Cr-Mo steels to yield more stable starting carbides can significantly reduce decarburization by these two liquid metals (16,18,19). Furthermore, alloying additions such as niobium form very stable carbides and can dramatically reduce decarburization (20). [With very unstable microstructures, the carbides are rapidly dissolved (20) in a manner similar to that described above for oxygen-containing refractory metals in lithium.]

In addition, as shown by the above partitioning equation, increasing the chromium level of a steel effectively decreases the tendency for carbon loss in sodium (18) and lithium (19). With higher chromium steels (for example, (austenitic stainless steels), carburization can then become a problem at higher temperatures.

Based on thermodynamic predictions and experimental results, it is apparent that carbon transfer in sodium and lithium systems containing steels can be minimized by optimizing chromium levels and increasing the stability of the carbide precipitate structure by proper heat treatment and/or addition of carbide stabilizers.

Mass Transfer

Alloy modification may also affect mass transfer. In a trivial sense, concentration gradient mass transfer may be eliminated by not using more than one alloy composition in contact with the same liquid metal. Otherwise, the principal way to decrease mass transfer via alloy development is to reduce the amount of dissolution, corrosion product formation, and impurity transfer by the ways already discussed above.

Another method to affect mass transfer is to control deposition. This can be done by tailoring the temperature distribution around a loop, but, with respect to alloy considerations, the effects that can be influenced are rather special cases. For example, if the liquid metal has to flow through a magnetic field, as it would in certain designs of magnetic fusion reactors, preferential deposition of magnetic elements can occur. A reduction in the amount of material available for magnetic trapping (Fe, Ni, Co) may then be necessary. In another case, preferential deposition of chromium was observed in some lithium systems during the early stages of mass transfer (21) and was sufficiently severe to cause plugging in one instance. Reducing the amount of starting chromium in the containment material would decrease this tendency. Finally, lowering the nitrogen content of some alloys may eliminate some deposits in the cold regions of lithium loops — nitrogen may have played an intermediary role in the chromium deposition described above (21) and some HfN deposits were observed (as the only mass transfer products) in the cold zones of two Ta-10 W-0.8 Hf (wt %) lithium loops (22).

Alloy Development Example: Fe-Cr-Mo Steels Exposed to Lithium

A specific example of how liquid metal corrosion considerations can influence development of Fe-Cr-Mo steels for lithium service will be briefly described. For these steels, the possible reactions with chromium are of prime importance. As discussed above, carbon transfer in liquid metal systems can be controlled by increasing the chromium concentration and/or stabilizing the microstructure. A concentration of about 8 to 10 wt % Cr in Fe-Cr-Mo alloys would result in a minimization of carburization-decarburization at 500°C. However, such a chromium concentration is not optimal from the viewpoint of direct dissolution, deposition, and nitrogen-induced corrosion reactions, which are minimized by reducing the chromium activity of the alloy (for a fixed nitrogen concentration). Alloy development based on liquid lithium corrosion considerations would then result in a lower chromium steel (for example, Fe-2 1/4 Cr-1 Mo steel) that has been stabilized by use of alloying additions such as niobium or vanadium to minimize decarburization. Furthermore, such a steel should be produced so that certain minor elements are minimized. For example, elements that dissolve readily in lithium, such as nickel and manganese, should be maintained at low levels; concentrations of 1 wt % could lead to significant mass transfer (23). In addition, the nitrogen level of the steel should be as low as possible to avoid the corrosion reactions involving nitrogen.

Summary

The present paper shows what type of factors have to be considered in alloy selection for liquid metal containment and describes how alloy development can be used to minimize particular types of liquid metal corrosion reactions. A general summary of the types of the most common corrosion reactions and their consequences for alloy development is given in Table I, which also includes specific examples in each category from among those discussed in the text. Since two or more concurrent corrosion reactions are possible in a liquid metal system, consideration of all the applicable alloy development consequences may lead, in some cases, to opposite alloy strategies. (For example, chromium in Fe-Cr-Mo steel-lithium systems should be minimized on the basis of mass transfer, while a higher chromium is preferred for minimizing carbon transfer.) In such instances, optimization of concentrations and/or microstructures would be necessary.

Table I. Consequences for Alloy Development Based on Liquid Metal Corrosion Reactions

Corrosion reaction	Consequence for alloy development	Example
Direct dissolution	Lower activity of key elements	Reduce Ni in Li, Pb systems
Corrosion product formation	1. Lower activity of reacting elements	1. Reduce Cr, N, and Ni in Li systems
	2. In case of protective layer, add elements to promote formation	2. Add Al or Si to steel exposed to Pb
Impurity and interstitial transfer	1. Increase (or add) elements to decrease transfer tendency	1. Increase Cr in steels; add Zr to Nb exposed to Li
	2. Minimize element being transferred	2. Reduce O in Nb and Ta exposed to Li
Mass transfer	1. Reduce extent of other corrosion reactions	1. See above
	2. Reduce concentration of elements that tend to deposit nonuniformly in cold zone	2. Reduce Cr in Li systems
	3. Reduce amount of material available for magnetic trapping (magnetic devices only)	3. Ni and Fe in Pb or Li fusion systems

References

1. P. F. Tortorelli and O. K. Chopra, "Corrosion and Compatibility Considerations of Liquid Metals for Fusion Reactor Applications," Journal of Nuclear Materials, 103&104 (1981) pp. 621–632.

2. P. F. Tortorelli, J. H. DeVan, and J. E. Selle, "Effects of Nitrogen and Nitrogen Getters in Lithium on the Corrosion of Type 316 Stainless Steel," National Association of Corrosion Engineers Corrosion/79 (preprint 115) (March 1979).

3. J. E. Selle and D. L. Olson, "Lithium Compatibility Research — Status and Requirements for Ferrous Materials," pp. 15–22 in Materials Considerations in Liquid Metal Systems in Power Generation, G. A. Whitlow and N. J. Hoffman, eds.; National Association of Corrosion Engineers, Houston, Texas, 1978.

4. J. R. Weeks and H. S. Isaacs, "Corrosion and Deposition of Steels and Nickel-Base Alloys in Liquid Sodium," Advances in Corrosion Science and Technology, 3 (1973) pp. 1–66.

5. P. F. Tortorelli and J. H. DeVan, "Effect of Nickel Concentration on the Mass Transfer of Fe-Ni-Cr Alloys in Lithium," Journal of Nuclear Materials, 103&104 (1981) pp. 633–638.

6. R. C. Asher, D. Davis, and S. A. Beetham, "Some Observations on the Compatibility of Structural Materials with Molten Lead," Corrosion Science, 17 (1977) pp. 545–557.

7. J. V. Cathcart and W. D. Manly, The Mass Transfer Properties of Various Metals and Alloys in Liquid Lead, ORNL-2008, Oak Ridge National Laborabory, Oak Ridge, Tenn., October 1956.

8. D. L. Olson, G. N. Reser, and D. K. Matlock, "Liquid Lithium Corrosion Rate Expressions for Type 304L Stainless Steel," Corrosion (Houston), 36 (1980) pp. 140–144.

9. O. K. Chopra and D. L. Smith, ADIP Semiannual Progress Report for Period Ending March 31, 1983, DOE/ER-0045/11, U.S. Department of Energy, Office of Fusion Energy, Oak Ridge, Tenn., pp. 195–200.

10. R. J. Pulham and P. Hubberstey, "Comparison of Chemical Reactions in Liquid Lithium with Those in Liquid Sodium," Journal of Nuclear Materials, 115 (1983) pp. 239–250.

11. M. G. Barker, P. Hubberstey, A. T. Dadd, and S. A. Frankham, "The Interaction of Chromium with Nitrogen Dissolved in Liquid Lithium," Journal of Nuclear Materials, 114 (1983) pp. 143–149.

12. E. Ruedl and T. Sasaki, "Effect of Lithium on Grain-Boundary Precipitation in a Cr-Mn Austenitic Steel," Journal of Nuclear Materials, 116 (1983) pp. 112–122.

13. E. Ruedl, V. Coen, T. Sasaki, and H. Kolbe, "Intergranular Lithium Penetration of Low-Ni, Cr-Mn Austenitic Stainless Steels," Journal of Nuclear Materials, 110 (1982) pp. 28–36.

14. J. R. DiStefano and E. E. Hoffman, Corrosion Mechanisms in Refractory Metal-Alkali Metal Systems, ORNL-3424, Oak Ridge National Laboratory, Oak Ridge, Tenn., August 1963.

15. R. L. Klueh, "Oxygen Effects on the Corrosion of Niobium and Tantalum by Liquid Lithium," Metallurgical Transactions A, 5 (1974) pp. 875–879.

16. K. Natesan, O. K. Chopra, and T. F. Kassner, "Compatibility of Fe-2 1/4 wt % Cr-1 wt % Mo Steel in a Sodium Environment," Nuclear Technology, 28 (1976) pp. 441–451.

17. K. Matsumoto et al., "Carbon Transfer Behavior of Materials for Liquid-Metal Fast Breeder Reactor Steam Generators," Nuclear Technology, 28 (1976) pp. 452–470.

18. A. Saltelli, O. K. Chopra, and K. Natesan, "An Assessment of Carburization-Decarburization Behavior of Fe-9 Cr-Mo Steels in a Sodium Environment," Journal of Nuclear Materials, 110 (1982) pp. 1–10.

19. P. F. Tortorelli, J. H. DeVan, and R. M. Yonco, "Compatibility of Fe-Cr-Mo Alloys with Static Lithium," Journal of Materials for Energy Systems, 2 (1981) pp. 5–15.

20. T. L. Anderson and G. R. Edwards, "The Corrosion Susceptibility of 2 1/4 Cr-1 Mo Steel in a Lithium-17.6 Wt Pct Lead Liquid," Journal of Materials for Energy Systems, 2 (1981) pp. 16–25.

21. P. F. Tortorelli and J. H. DeVan, "Mass Transfer Deposits in Lithium-Type 316 Stainless Steel Thermal-Convection Loops," pp. 13-55–13-63 in Proceedings of Second International Conference on Liquid Metal Technology in Energy Production, CONF-800401-P2, U.S. Department of Energy, Office of Fusion Energy, Oak Ridge, Tenn., 1980.

22. J. H. DeVan and E. L. Long, Jr., Evaluation of T-111 Forced-Convection Loop Tested with Lithium at 1370°C, ORNL/TM-4775, Oak Ridge National Laboratory, Oak Ridge, Tenn., April 1975.

23. P. F. Tortorelli and J. H. DeVan, "Corrosion of an Fe-12 Cr-1 Mo VW Steel in Thermally Convective Lithium," pp. 215-221 in Proceedings of Topical Conference on Ferritic Alloys for Use in Nuclear Energy Technologies, J. W. Davis and D. J. Michel, eds; AIME, New York, NY, 1984.

TAILORING AND EVALUATING WEAR-RESISTANT ALLOYS

FOR NUCLEAR APPLICATIONS

H. Ocken

Electric Power Research Institute
3412 Hillview Avenue
P. O. Box 10412
Palo Alto, California 94303
USA

Summary

Efforts to limit radiation field buildup in U.S. nuclear plants by reducing the cobalt content of wear-resistant alloys used in the primary circuit of LWRs are reviewed. The earliest work was motivated by the need to identify the key sources of cobalt. These projects lacked accurate wear data from specific plant components, and the results of such measurements are discussed. Cobalt releases resulting from valve maintenance operations have been estimated. Progress in identifying cobalt-free alloys for use in specific components and in developing a substitute for the current cobalt-base hardfacing alloys also is reviewed.

Introduction

Cobalt is an essential constituent in alloys that exhibit superior wear resistance, in the superalloys that are used in high temperature applications, in magnets, and in cemented carbides. The United States relies on foreign sources for its supply of cobalt and in 1980 imported some 15.3 million pounds. The supply of cobalt was interrupted in 1978, which led to a sharp increase in cobalt prices. Our reliance on foreign sources for this strategic element has sparked support of cobalt replacement programs whose goal is to develop low-cobalt or cobalt-free alloys with attributes similar to those of the cobalt-base alloys.

The cobalt released annually by the wear and corrosion of plant materials to the primary circuit in light water reactors amounts to a few hundred grams. Yet, strong incentives exist to reduce such releases to still lower levels, to reduce the specifications for impurity levels of cobalt in construction materials, and to develop substitutes for cobalt-base alloys. This motivation is provided by the observation that over the last decade occupational radiation exposure generally has increased at U.S. nuclear power plants (Figure 1). Occupational radiation exposure results when plant maintenance personnel inspect, repair, or replace out-of-core components upon which radiation fields have built up.

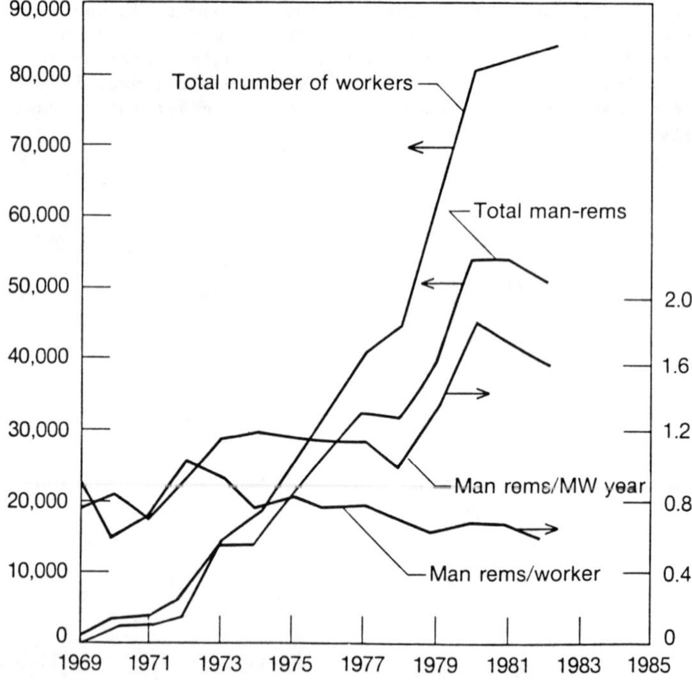

Figure 1 Occupational radiation exposure at U.S. LWRs.

The primary sources of radiation fields are isotopes of cobalt. Cobalt-60, with its penetrating gamma rays and long-5.3 year half-life, is the primary offender, while cobalt-58 contributes substantially to fields in PWRs early in life. Cobalt-60 is produced by the bombardment of its precursor, cobalt-59, with thermal neutrons: $^{59}Co(n,\gamma)^{60}Co$. Cobalt-59 is the naturally occurring 100% abundant isotope of cobalt. Cobalt-58 is produced from nickel-58 by the fast neutron reaction $^{58}Ni(n,p)^{58}Co$.

Cobalt is present in nuclear reactors as a low-level impurity (~0.2%) in construction materials and at high levels in the hardfacing alloys used in applications requiring resistance to mechanical wear. Cobalt is found as an impurity in the Inconel 600 used as steam generator tubing, in the Inconel 718 used as grid spacers in PWR fuel assemblies, in the type-304 stainless steel tubing and sheathing used in control blades and in recirculation lines in BWRs, and in the type-304 stainless steel and the Inconel 625 used in PWR control rod cladding. The cobalt-base alloys are used in applications requiring superior wear resistance, such as in bearing surfaces, valve trim, and reactivity control mechanisms.

These cobalt-base alloys are typically deposited as hardfacings on surfaces subjected to severe service. The most popular hardfacing alloys, which contain 50 to 60 w/o cobalt, are generically referred to as Stellite (a trademark of the Cabot Corp.) These alloys were developed in the early 1900s. Service experience over the years indicates that these are premier alloys for wear-resistant and corrosion-resistant applications; they are the standard against which other wear-resistant hardfacing alloys are compared.

Given the need for reliable performance of critical components in key safety-related systems over the projected forty-year lifetime of the plant, it is no surprise that the Stellites gained wide use and acceptance in the nuclear industry. Apart from routine maintenance, field experience in nuclear plants has shown that hardfaced components usually have performed admirably. However, the increase in radiation fields and in occupational radiation exposure resulting from the repair and replacement of other activated components and the recognition of the role of cobalt-60 in radiation field buildup has provided a strong incentive to develop low-cobalt alloys to replace the Stellites.

These introductory remarks suggest that one way to minimize the buildup of radiation fields is to decrease the amount of cobalt that can enter the primary circuit of the reactor. This paper reviews those elements of the EPRI program on radiation control that focus on replacing cobalt-base wear-resistant alloys. The following topics are discussed: (a) cobalt release mechanisms, (b) cobalt inventory calculations, (c) wear measurements on nuclear components, (d) qualification of commercially available low-cobalt alloys for nuclear applications, and (e) projects addressing the development of new cobalt-free hardfacing alloys for nuclear applications.

Cobalt Release Mechanisms

The release of alloy constituents such as iron, nickel, chromium, and cobalt from construction materials into the coolant is responsible for the growth of oxide films and radiation field buildup on out-of-core surfaces. Corrosion products are released as soluble, colloidal, or particulate species. These corrosion products are carried around the system, are deposited on the Zircaloy fuel rod cladding in the reactor core, and become activated by the intense neutron flux in the reactor core. These activated deposits are subsequently released by dissolution or erosion after residence times on the fuel on the order of a few weeks and are then transported by the coolant to surfaces outside the core. Deposition of these activated corrosion products on out-of-core surfaces renders the latter radioactive. Wear and corrosion of the high-cobalt hardfacing alloys during normal operation also introduces cobalt into the coolant, which also contribute to radiation field buildup as described above. Recent work by Heard and Freeman (1) showed that valve maintenance also releases cobalt to the primary circuit.

Cobalt Inventory Calculations

This brief description of the activation process indicates that a cobalt replacement program must identify both the components and the mechanisms responsible for the majority of cobalt release. Is the primary contribution from the large surface area of PWR steam generators containing trace amounts of cobalt, or is it from wear or corrosion of the small surface area of components hardfaced with the cobalt-base alloys? What are the respective contributions from corrosion, wear, and valve maintenance?

Cobalt sources in PWRs were identified by Young et al. (2) for Combustion Engineering, Inc., plants and by Bergmann et al. (3) for Westinghouse Electric Corp. plants. Falk (4) reviewed the situation for General Electric Co. BWR plants. These inventory calculations led to the following calculated annual cobalt releases: 60 gm/yr for the three-loop (Beaver Valley) PWR and 71 gm/yr for the four-loop (Trojan 2) PWR designed by Westinghouse Electric, 99.8 gm/yr for the Millstone 2 plant designed by Combustion Engineering, and 243.6 gm/yr for the Brunswick 2 BWR/4 plant designed by General Electric. The calculations showed that valve wear is responsible for some 63% of the annual cobalt release in BWRs and a more modest 14% in PWRs. Some 12% of the cobalt input was attributed to the wear and corrosion of the cobalt-base pins and rollers in BWR control blades. The calculated annual cobalt releases from key reactor systems and the total releases are shown in Figure 2.

The approach used in all these cobalt inventory calculations consisted of (1) identifying each system with a pathway to the coolant, (2) identifying the construction materials used in these systems and the cobalt content of each alloy, (3) calculating the area of each alloy in contact with the coolant, and (4) using available literature information on corrosion release rates and wear rates to estimate the annual cobalt release rate of the various systems. No effort was made to distinguish between cobalt releases from in-core and out-of-core sources; this underestimates the significance of in-core sources that release activated cobalt.

These calculations were hampered by the fact that only limited corrosion release data were available; wear data were even sparser. Accordingly, projects were initiated to obtain corrosion release data from plant construction materials and wear measurements on components following service in commercial nuclear plants.

Measurements of Component Wear

Wear measurements reported by Dufrane and Naughton (5) and by Dufrane et al. (6) have helped identify the components that are significant cobalt sources. The technique used to obtain wear data bears noting.

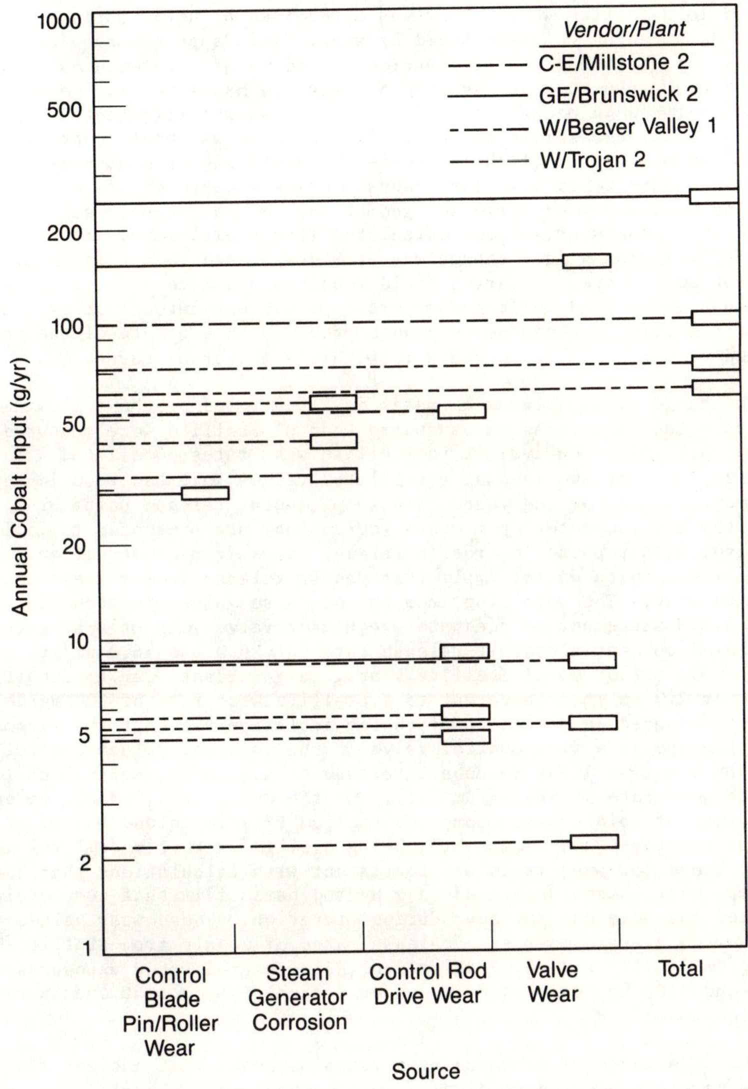

Figure 2 Calculated annual cobalt releases.

The traditional method for determining wear rates is to obtain pre- and posttest specimen weights. Clearly, for measurements of reactor components this procedure is not feasible. Pre-wear measurements are not available, components are welded in place, wear leads to only in a small change in component weight, and radiation fields limit the time that can be spent working near the component of interest. Therefore, an indirect technique was used to determine wear. Wear was determined by using surface profilometry to measure the volume removed by wear. This approach required one to locate a suitable nearby unworn surface to serve as a reference. This approach also assumed that the original surface had a regular geometry and was manufactured in accordance with the original specifications. If the component was inaccessible to the profilometer or was badly contaminated, replicas were prepared of the worn areas. Profilometer measurements then were made on the replicas. The accuracy of this approach was estimated by machining wear scars of different geometries in flat brass specimens. Comparison of the wear volumes calculated from profilometer traces and those derived from weight change measurements agreed to within 5% for four different wear scars. Clearly, field measurements made under nonideal conditions would lead to larger errors. It was estimated that even under adverse conditions, these measurements provided an estimate of the cobalt released by wear that is accurate to within a factor of three.

The range of measured wear rates is illustrated by the following examples. Cobalt release rates due to wear of Stellite were measured at 8.0 and 30.4 gm/yr (equivalent to Stellite wear rates of 112 and 425 $mg/dm^2 \cdot mo$) for two feedwater regulator valves that had been in service at Vermont Yankee for one year. The supplemental release units of mg (Stellite) per decimeter^2 per month ($mg/dm^2 \cdot mo$) are presented to facilitate comparison with reported corrosion release rates from construction materials. These units do not imply that cobalt release from valves occurs by corrosion wear. The worn plug from one of these valves is shown in Figure 3. Measurement of feedwater regulator valve wear at Oyster Creek also showed subtantial cobalt release rates of 20.0 and 56.0 gm/yr (454 and 933 $mg/dm^2 \cdot mo$ of Stellite wear). Significant wear (a cobalt release of 2.0 gm/yr, equivalent to a Stellite wear rate of 200 $mg/dm^2 \cdot mo$) also was measured in a cleanup discharge isolation valve at the Vermont Yankee BWR and in a flow control valve in the chemical volume control system at the Zion 1 PWR (a cobalt release of 1.6 gm/yr, equivalent to a Stellite wear rate of 583 $mg/dm^2 \cdot mo$). On the other hand, cobalt releases due to wear of main coolant pump shafts from Prairie Island and H. B. Robinson reactors were negligible, being 0.04 gm/yr (0.2 $mg/dm^2 \cdot mo$) in both cases. These low wear rates are consistent with calculations that showed the pump shaft journal operated on a hydrodynamic film that completely separated the adjacent surfaces during operation. These wear values compare with average corrosion release rates of cobalt from Stellite 6 of ~1.4 $mg/dm^2 \cdot mo$ (7) and ~0.12 $mg/dm^2 \cdot mo$ (8), measured after exposures of 690 hr and 7100 hr, respectively, under typical PWR coolant chemistry conditions.

The wide range of measured wear rates suggests that nuclear plant components are subjected to diverse wear mechanisms. Clearly, the release of cobalt resulting from the wear of some nuclear plant valves can be substantial. The feature common to all the valves for which high wear rates were measured is their use to control or throttle coolant flow. Under such erosion or cavitation wear conditions, the Stellites may not be the optimum hardfacing alloy.

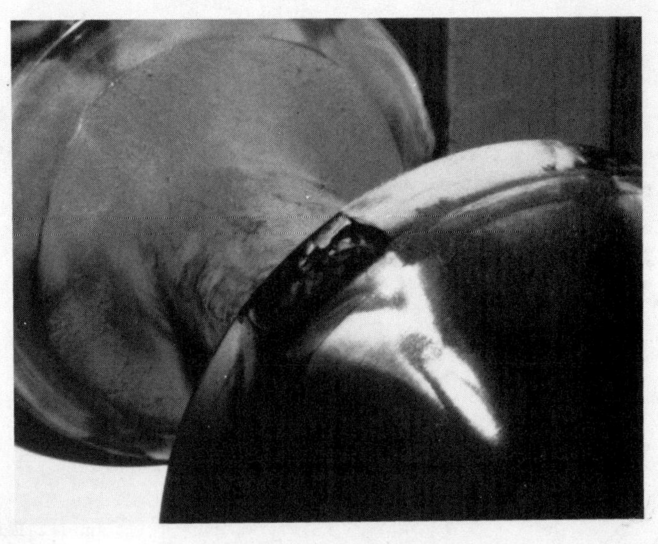

Figure 3 Severely worn plug from a feedwater regulator valve from the Vermont Yankee BWR.

Cobalt also may be introduced into the primary coolant as a result of valve maintenance operations. Heard and Freeman (1) investigated the effect of lapping and grinding of hardfaced valve seats, which are performed as a corrective measure on safety-related valves that fail local leak rate tests. These authors relied on measurements of debris left in a valve after grinding and cleaning, using standard procedures, and on estimates of the amount of valve maintenance required over the course of a year for all valves in a plant, to calculate the annual cobalt input at 10 to 30 gm/yr for a representative PWR and 30 to 90 gm/yr for a representative BWR. These values each represent ~10% of the total cobalt input resulting from wear and corrosion for the two reactor designs. Although these are not negligible quantities, Heard and Freeman identified postmaintenance cleaning procedures that can effectively limit the amount of debris that gets into the coolant. The recent measurements have been used to estimate the primary sources of cobalt released to the primary coolant in PWRs and BWRs, and these results are presented in Tables 1 and 2.

These cobalt source evaluation studies lead one to conclude that an effective cobalt replacement program should address both the construction materials that contain cobalt as an impurity (corrosion release from inconel steam generator tubing being the primary source in PWRs) and the cobalt-base alloys that are used in hardfacing applications. Immediate benefits can be realized if "housecleaning" procedures are given their due during valve maintenance operations.

Table I. PWR Cobalt Sources

Source	Annual Release Rate Per Plant (gm/yr)
Steam Generator Tubing Corrosion	33-55
Valve Maintenance	10-30
Control Element Drive Mechanism Wear	2-5
Check Valve Wear	1
Gate Valve Wear	0.5
Main Coolant Pump Shaft Wear	0.2

Table II. BWR Cobalt Sources

Source	Annual Release Rate Per Plant (gm/yr)
Feedwater Regulator Valve Wear	38-76
Valve Maintenance	30-90
Control Blade Pin/Roller Wear	29
Cleanup Discharge Isolation Valve Wear	4
Main Steam Isolation Valve Wear	1.6
Feedwater Swing Check Valve Wear	0.4-0.7
Steam Flow Control Valve Wear	0.4
Feedwater Pilot Relief Valve Wear	0.06

Qualification of Commercially Available Low-Cobalt Alloys for Nuclear Applications

The introductory remarks noted that components hardfaced with cobalt-base alloys have performed well in reactor environments. Clearly, any alloy proposed as a replacement for the Stellites will be scrutinized carefully, and a significant alloy evaluation and qualification program will be required. Laboratory evaluations of wear behavior will have to be supplemented by prototypical testing of specific components in loop facilities or commercial nuclear reactors in applications not critical to reactor safety. It is instructive to first review applications that have led to the replacement of cobalt-base alloys in reactor components and then discuss studies in which alloy development and evaluations have not proceeded beyond the laboratory stage.

Pins and Rollers for BWR Control Blades

Cruciform-shaped control blades are used to control the power distribution and for reactivity control in a BWR. Each wing of the cruciform contains 21 type-304 stainless steel tubes that are filled with boron carbide. The tubes are held in the cruciform array by a type-304 stainless steel sheath that extends the full length of the tubes.

Four pairs of pins and rollers, representing the "high-tech" analogue of a baker's rolling pin, are located at the top and bottom of each BWR control blade. The upper sets (Figure 4) help the control blades move smoothly between the adjacent Zircaloy fuel channels; the lower sets run against the type-304 stainless steel guide tubes. The upper (handle) pin is 3.2 cm long and 0.32 cm in diameter. The handle roller is 1.0 cm in diameter at the widest point and is 0.87 wide. In domestic designs, rollers have been made of Stellite 3 and the pins of Haynes 25, both of which contain about 50% cobalt. The wear and corrosion resistance of these alloys is outstanding, and no problems have been attributed to deficiencies in pin and roller design or performance. Yet, as noted earlier, Falk (4) estimated that this component is responsible for some 12% of the cobalt released in a BWR. This is a remarkable result, because the surface area of the pins and rollers is only $\sim 10^{-6}$ % of the total surface in contact with the coolant. Consequently, a strong incentive existed to identify cobalt-free alloys for pins and rollers with wear attributes sufficient for this application.

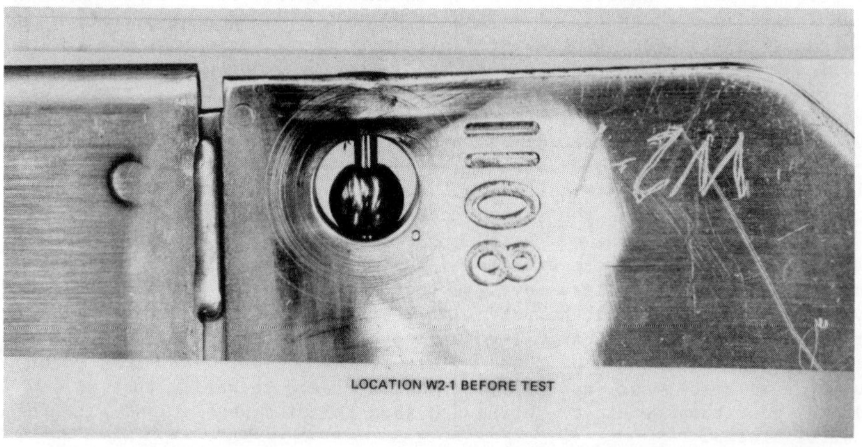

Figure 4 Pin and roller in the handle of a BWR control blade.

The laboratory evaluations performed on candidate replacement alloys are discussed by Aldred (9). The qualification of the low-cobalt alloys relied on prototype and laboratory wear testing of low-cobalt pins and rollers and on other pertinent mechanical property and corrosion tests. Similar data for the reference cobalt-base alloys also was obtained. A brief description of the alloys selected for the prototype tests is given in Table 3 and the test matrix in Table 4. The composition of these alloys is given in Table 5.

In the prototype tests, standard control blades, with the sole exception being the pin and roller combinations identified in Table 3, were subjected to accelerated wear tests under typical BWR coolant temperatures (ranging from 100°F to 400°F), pressures, and water chemistry conditions. Various control blade drive motions reflecting the conditions expected during normal and transient operation of a reactor were imposed.

Surface profilometry of roller tracks on the fuel channels and on the guide tubes showed that the wear from both the cobalt and noncobalt alloys were less than the normal surface roughness of the channels and guide tubes. Energy dispersive analysis (EDS) of scanning electron microscopy specimens were used to examine one handle pin and roller combination from each prototype test to determine if material transfer occurred across the wear surfaces. Standard EDS spectra were obtained for each alloy by scanning the polished ends of Charpy impact specimens. After completion of the prototype tests, the bore and surface of the handle rollers and the pins were examined. With the exception of the Stellite 3 and Haynes 25 cobalt alloys, EDS showed that Zircaloy-4 was transferred from the fuel channel to the roller surface, and material transfer occurred between the pin and the roller bore.

Wear data also were obtained from laboratory tests in which twelve pin and roller specimens were rotated against a Zircaloy-4 or type-304 stainless steel mandrel in an autoclave in which circulating demineralized water at 1220 psi was maintained at 550°F. In these tests, the initial applied load ranged from 22.2N (5 lb) to 800N (180 lb). In both the prototype and laboratory tests wear was evaluated primarily by measuring changes in the radii and in the weight of the pins and rollers. In both types of tests, the total distance travelled by the rollers was 4267 m (14,000 ft). This travel is estimated to be that sustained in reactor over a five-year period. The common specimen dimensions, travel, and wear evaluation methods permitted the unknown load imposed in the prototype tests to be deduced. Clearly, neither of these tests simulated irradiation-induced effects that would be encountered in the reactor core.

Other tests that were used to evaluate the suitability of these alloys for use as pins and rollers were (1) bent-beam tests to evaluate general corrosion resistance and the potential for stress corrosion cracking and crevice corrosion, (2) Charpy V-notch impact and dynamic loading tests to determine the resistance of these alloys to the loading that might occur during a seismic event, and (3) plug welding tests to verify that no adverse conditions would be introduced when the pins were welded into the control blades.

Analyses of the prototype wear test data showed that the loads on the pins and rollers in the prototype tests were low, typically in the range 2.2 to 8.9 N (0.5 to 2.0 lb.); the maximum load was 28.9 N (6.5 lb.). Under these loads, and using acceptance criteria that restricted pin and roller weight changes and dimensional changes, the cobalt-free pin/roller combinations Nitronic-60/CFA (developed by Toshiba Corp.) and PH 13-8 Mo/Inconel X-750 were found to perform as well as, if not better than, the standard high-cobalt Haynes 25/Stellite 3 combination. These alloys do not represent a generic solution to the quest for a Stellite substitute, however, because at loads $\gtrsim$44 N (10 lb.) the wear resistance of these alloy combinations typically was poorer than that of the cobalt-base alloys.

Table III. Alloys Evaluated In Pin/Roller Program

Alloy	Type	Condition
Stellite 3	Cast cobalt-base alloy	Cast
Haynes 25	Wrought cobalt base alloy	30-35% cold-worked; aged 1125°F (6 hr)
Nitronic-60	Austenitic stainless steel	40% cold-worked
PH 13-8 Mo	Martensitic precipitation hardened stainless steel	Aged 1100°F (4 hr)
Inconel X-750	Precipitation hardened austenitic Ni-Cr alloy	Annealed 2000°F (1 hr); aged 1300°F (20 hr)
CFA	Precipitation hardened austenitic Ni-base alloy	Aged 1300°F (3 hr)

Table IV. Prototype Control Blade Test Matrix

Test	Pin Alloy	Roller Alloy	
		Handle	Velocity Limiter
1	Haynes 25	Stellite 3	Stellite 3
2	Nitronic-60	CFA	CFA
3	Nitronic-60	Inconel X-750	Inconel X-750
4	PH 13-8 Mo	Inconel X-750	Inconel X-750

On the other hand, the pin/roller combination Nitronic-60/Inconel X-750 showed poorer wear resistance than the cobalt-base alloys, even at low loads. The measured weight loss of the handle pin/roller alloys as a function of the applied lead is shown in Figure 5. The average weight loss of the various pin and roller combinations is plotted against the acceptance criterion that limits wear in the candidate alloys to less than that observed in the cobalt-base alloys after 15 years (Figure 6).

Other properties of these alloys were found to be more than adequate for the intended application. The impact strength and the dynamic loading response of the cobalt-free alloys were superior to those of the cobalt-base alloys. The general and crevice corrosion behavior of these alloys was comparable, and the welding tests showed that sound weld joints were achieved when standard production welding techniques and parameters were used.

In light of these results, it was concluded that testing of the superior cobalt-free alloys in a commercial power reactor was warranted. Accordingly, four control blades containing Nitronic-60/CFA pins and rollers and four containing PH 13-8 Mo/Inconel X-750 were fabricated. Two control rods with each alloy combination were installed in a control-cell-core BWR in March 1981, and the other four control rods were installed in a conventional-core BWR in April 1981. In the control-cell core, the control rods experience higher irradiation levels but travel less than in the conventional core. After more than two years of exposure, the performance of the cobalt-free alloys is satisfactory.

BWR Feedwater Regulator Valves

The identification of cobalt-free alloys suitable for use as pins and rollers described above is the result of a traditional applied research and development program. If a utility must contend with high occupational radiation exposures, laboratory evaluations may be held to a minimum before a cobalt-free alloy is selected for use in a commercial nuclear plant. Such was the case when Vermont Yankee personnel reviewed alternatives to using Stellite to hardface feedwater regulator valves. As noted earlier, wear measurements showed significant cobalt releases from these valves. The utility then decided to use type-440C stainless steel as a replacement alloy for this valve. The results from sliding wear tests in oxygenated water at 550°F reported by Freund (10) noted that this alloy showed wear low enough that it should be satisfactory in bearing applications. Schumacher (11) and Budinski (12) showed that the galling wear resistance of type-440C stainless steel is higher than that of most stainless steels.

The performance of the type-440C stainless steel has proved to be outstanding. Recent measurements by Dufrane et al. (6) show that after one year of service, the wear of the type-440C stainless steel is about 100 times less than that of the Stellite hardfacing: the maximum wear scar in the Stellite hardfaced valve was $\sim 1.5 \times 10^{-1}$ cm and $\leq 1.2 \times 10^{-3}$ cm in the type-440C stainless steel valve. These results have led the utility to keep the type-440C feedwater regulator valves in service for a second year. The use of Stellite hardfacing necessitated annual replacement.

Table V. Composition of Hardfacing Alloys

WEIGHT PERCENT

Cobalt-base alloys	Ni	Co	Cr	C	Si	Fe	W	Mo	Mn	Other
Stellite 3	<3.0	bal	29.0-33.0	2.0-2.7	1.0	3.0	11-14		1.0	
Haynes 25	9.0-11.0	bal	19.0-21.0	0.05-0.15	1.0	3.0	14-16		1.0-2.0	
Co-156	3	bal	29	1.6	1.2	0.75	4.5	1	1	
Co-12	2.5	bal	30	1.4	1.7	2.5	8	1	0.25	
Coast Metals 6	3	bal	28	1.2	1.5	3	4	1	0.5	
Low-cobalt alloys										
CFA	bal	<0.5	35.0-39.0	<0.10	<1.0			<2.0	<1.0	Al 3.4-4.0
Deloro 40	bal	<1.5	7.5	0.35	3.5	1.5				B 1.7
Haynes 711	bal	12	27	2.7	1	23	3	8	1	
Vertx 4776	bal	0.2	13	0.6	4	4		3		
Cobalt-free alloys										
Nitronic-60	8.0-9.0		16.0-18.0	<0.1	3.5-4.5	bal			7.0-9.0	N 0.08-0.18
PH 13-8 Mo	7.5-8.5		12.25-13.25	<0.05	<0.1	bal		2.0-2.5	<0.1	P <0.10, S <0.008, Al <0.9-1.35, N <0.01
Inconel X-750	bal		14.0-17.0	<0.08	<0.5	5.0-9.0			<1.0	S <0.01, Cu <0.5, Al 0.4-1.0, Ti 2.25-2.75, Nb+Ta 0.7-1.2
Type 440C stainless steel			16.0-18.0	0.95-1.20	<1.0	bal		<0.75	<1.0	P <0.04, S <0.03
Colmonoy 74	bal		10.4	0.48	3.7	4.2	4.0			B 2.1
Colmonoy 84	bal		29	1.1	1.95	2	7.5			B 1.3
RHDIC			28	1.5	1.5	bal	1.5	1.5		
Tribaloy T-700	bal		15.5	0.08	3.4			32.5		
Carbide-strengthened iron-base alloys	0-8		18-28	1.0-2.0	1-4	bal		0-4	4-8	N 0.15-0.35
Intermetallic-strengthened iron-base alloys	0-8		18		1-4	bal		10-15	4-8	N 0.15-0.35

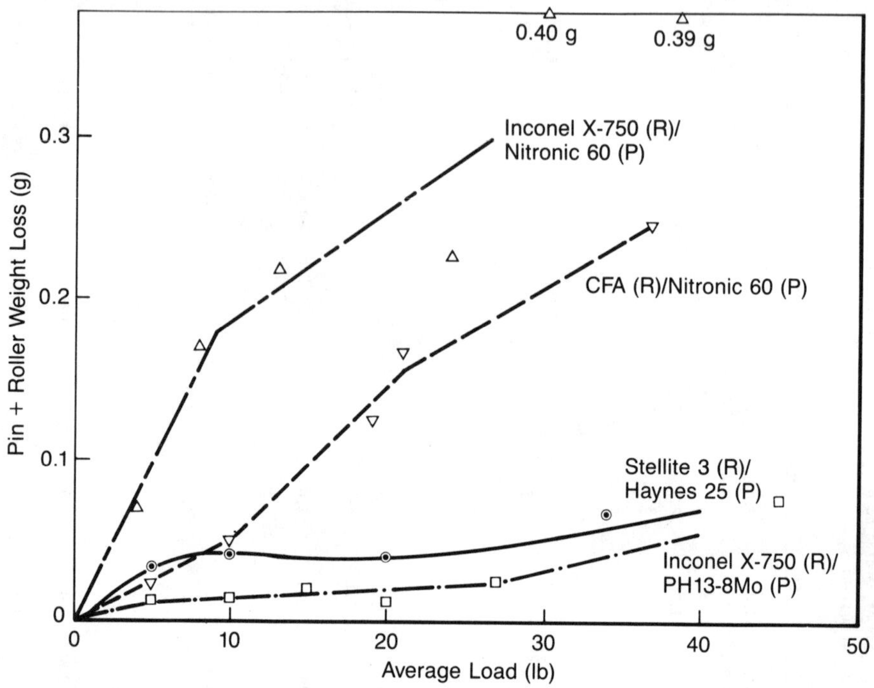

Figure 5 Measured handle pin and roller weight losses as a function of load in laboratory wear tests.

Laboratory Evaluations of Hardfacing Alloys

Landerman et al. (13) recently have completed a study that evaluated low-cobalt and cobalt-free hardfacing alloys for nuclear applications. Wear tests were used to generate sliding wear data on (mainly) nickel-base alloys and on reference cobalt-base alloys. Testing of self-mated couples at low loads initially served to identify the most attractive low-cobalt alloys, which were then tested at higher loads. The low-cobalt alloys most extensively evaluated were Colmonoy 74, Colmonoy 84, Deloro 40, Haynes 711, RHDIC, Tribaloy T-700, and Vertx 4776; their compositions are given in Table 5. The tests were conducted on a rig that provided concurrent information about the wear and friction behavior of the specimens. In each test, two pairs of moving blocks slide across a stationary block. The test conditions are ambient temperature and simulated PWR primary coolant that is borated and lithiated. Test loads are 34.5, 103.4, and 206.8 MPa (5000, 15,000, and 30,000 psi), which were selected to bracket calculated design loads for various nuclear plant components. Each test consists of 1000 cycles, with the displacement being 1.91 cm during each cycle. The velocity of the moving block is 0.11 cm/s. The wear response of each couple was evaluated from weight loss measurements, measurement of the static and dynamic coefficients of friction, and visual examination of the post-test surface condition.

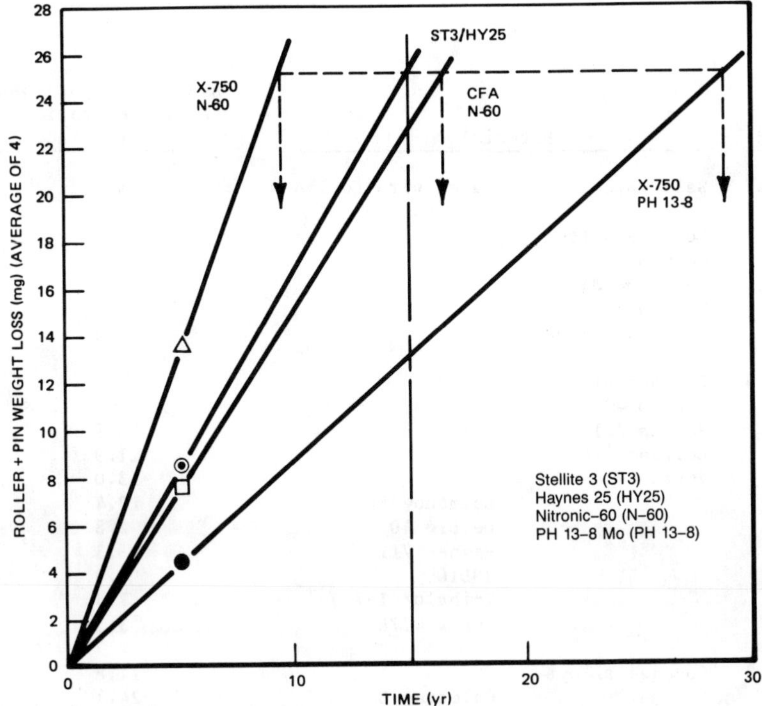

Figure 6 Pin and roller weight losses (measured in prototype wear tests) related to years of service.

The response of the wear couples was evaluated according to the following scheme. The average value of the total weight loss (i.e., the sum of the weight loss of the moving and stationary blocks) measured at each load for the self-mated cobalt-base alloys was taken as the standard. The response of other wear couples was rated "excellent" if the total weight loss is less than twice and "good" if the total weight loss is two to five times the standard value. The most wear-resistant alloys are identified in Table 6. At the lowest load of 34.5 MPa (5,000 psi) self-mated Colmonoy 74, Colmonoy 84, Deloro 40, and Tribaloy T-700 are highly rated. The superior wear resistance of the cobalt-base alloys is most evident at the highest load of 206.8 MPa (30,000 psi). No self-mated couples are rated highly. However, Colmonoy 84 and Vertx 4776 are rated excellent and Deloro 40 and RHDIC are rated good when mated with high-cobalt Co-156.

For those applications where the designer is confident of his ability to calculate accurately the anticipated service loads and is willing to use dissimilar hardfacing alloys, it is clear that alloys with wear resistance comparable to that of the Stellites are available. It is anticipated that the most attractive of these alloys will be evaluated further in a loop facility or in a noncritical application in a commercial nuclear plant.

Table VI. Wear Response of Most Attractive Low-Cobalt Alloys

Load	Material		Average Weight Loss (Moving & Stationary Blocks
(MPa)	Self-mated	Mated with Co-156	(mg)
34.5	Co-base alloys		1.3
	Colmonoy 74		1.8
	Colmonoy 84		1.1
	Deloro 40		0
	Tribaloy T-700		5.5
103.4	Co-base alloys		7.5
	Deloro 40		0
	Haynes 711		5.9
	Colmonoy 74		21.9
	Vertx 4776		13.0
		Colmonoy 84	12.4
		Deloro 40	2.8
		Haynes 711	9.3
		RHDIC	9.6
		Tribaloy T-700	7.3
		Vertx 4776	3.5
206.8	Co-base alloys		11.8
		Colmonoy 84	24.3
		Vertx 4776	33.8
		Deloro 40	55.5
		RHDIC	53.1

Hardfacing Alloy Development

The examples discussed above show that for certain specific applications commercially available cobalt-free alloys can be substituted for the cobalt-base Stellites without compromising component performance. These successes still leave unanswered the question: can a cobalt-free alloy be found whose responses to laboratory wear and corrosion tests suggest that it could replace the cobalt-base Stellite hardfacings? This section discusses forthcoming EPRI activities that will address this issue. It is first instructive to identify the typical laboratory wear tests used to evaluate hardfacing alloys and the performance of the Stellites that make them the premier hardfacing alloys. The response of the Stellites to these tests has been reported by Anthony (14).

One of the key attributes of the Stellites is their high resistance to galling. Galling wear refers to the transfer of material between contacting surfaces and their resultant roughening. Galling occurs most frequently where high surface contact stresses develop or in applications in which lubrication is difficult or impossible. Hardfacing alloys with good resistance to galling wear are required in valves and bearings. Poor galling resistance might lead to seal failure or seizure of the contacting surfaces. Although a standard galling test does not exist, galling resistance typically is measured by rotating a test pin that is subjected to a high load against a fixed test block. Galling resistance is deduced from quantitative analysis of the wear scars as determined by surface profilometry or by determining the threshold stress at which material is transferred across the interface between the test pin and the test block. The Stellites have superior galling resistance. This behavior appears to be related to the low-stacking fault energy of cobalt and the transformation by mechanical deformation of the cobalt alloy matrix from a metastable FCC structure to the HCP phase characteristic of cobalt at low temperature. Studies on pure cobalt and cobalt alloys indicate that resistance to galling wear depends on this transformation, which appears to be unique to cobalt. These factors have been discussed in detail by Bhansali and Miller (15).

Resistance to sliding wear is measured in the laboratory by adhesive wear tests. Here, a rotating ring (typically made of a case-hardened steel) rotates against a stationary block that serves as the test material. The load is varied and the volume lost is measured as a function of the load. The relative sliding velocity and the total sliding distance are fixed values. Hypereutectic Stellite alloys tested against case-hardened type-4620 steel show low wear that is linearly related to loads of up to 1373 N (140 kg). Hypoeutectic alloys exhibit severe adhesive wear at loads >637 N (65 kg). Resistance to adhesive wear apparently is related to the nature of the oxide glazes that form on the wear surfaces and not to any inherent property of cobalt. The adhesive wear of the Stellites has been discussed by Bhansali (16).

Another type of wear that is of concern in nuclear environments is cavitation. Cavitation is associated with turbulent liquid flow, which is caused by bubbles or gas cavities forming and suddenly collapsing. The sudden bubble collapse causes mechanical shock and wear on metallic surfaces. Cavitation-erosion wear resistance typically is evaluated by measuring the thickness or weight loss of a specimen attached to the end of an ultrasonic horn that pulsates in a constant-temperature water bath at fixed frequency and amplitude. The specimen's volume loss is measured periodically. The steady-state cavitation-erosion rates for Stellite 6 and other cobalt-base alloys are among the lowest for any commercial alloys. This response has been ascribed to the high work hardening rate of these alloys, which is a consequence of their low-staking fault energy. Significant quantities of the strain-induced HCP phase have been observed in cobalt-base alloys subjected to cavitation erosion. Heathcock et al. (17) have compared the cavitation erosion behavior of the Stellites and other wear-resistant alloys.

EPRI is planning to support work at AMAX of Michigan, Inc., whose goal is to develop a cobalt-free hardfacing alloy to substitute for the cobalt-base Stellites. The approach will be to characterize the Stellite 6 alloy, several commercially available cobalt-free hardfacing alloys, and a group of experimental iron-base hardfacing alloys. The response of these alloys will be determined by using laboratory galling wear tests, adhesive wear tests, and vibratory cavitation-erosion tests. Microstructural evaluations will be undertaken to identify alloy systems that are most likely to approach properties equivalent to those of Stellite 6. The mechanical properties of the candidate alloys will be evaluated by using tensile tests, hardness tests, and unnotched Charpy impact tests. On the basis of current understanding of the wear properties of cobalt-base alloys, the alloy system that will serve as the primary subject for evaluation is an austenitic iron-base alloy system. The matrix composition of this system is similar to that of the Nitronic-60 stainless steel, an alloy that Schumaker (11) has shown is extremely resistant to galling wear. The project will evaluate a two-phase alloy with a hard volume fraction similar to Stellite 6 and a matrix composition similar to that of the galling-resistant iron-base Nitronic-60 stainless steel. One group of alloys will use carbides as the hard phase, the other intermetallic Laves phases. The range of alloy compositions proposed for study is identified in Table 5. A second iteration will then serve to optimize the wear resistance of the most promising alloys. A few alloys then will be characterized with regard to their corrosion behavior. It is anticipated that uniform corrosion, crevice corrosion, cavitation corrosion, and stress corrosion cracking properties will be evaluated. Plans then call for the field testing of the most attractive of these alloys in nuclear components.

Clearly, the identification and qualification of the best iron-base alloys for nuclear hardfacing alloys will be a time-consuming effort. Also, given the long-standing lead of the Stellites for hardfacing applications, one must admit to the possibility that the alloy development work may not prove completely successful. As good as the Stellites are, is there any prospect of improving their performance? The use of advanced surface processing techniques such as lasers, electron-beam melting, and ion implantation has shown that unique surface structures can be generated. In some cases where attributes that reflect the response of the surface, such as wear and corrosion, have been determined, the treated material responds better than the untreated material. Ayers et al. (18) Dearnaley (19), and Ruff and Ives (20) have reported improvements in the wear of laser-processed, ion-implanted and electron-beam surface melted alloys. The question arises as to whether the wear resistance of the Stellites can be improved by using such approaches. Dillich et al. (21) have shown that titanium implantation increased the abrasive wear resistance of the cobalt-base Stoody 3 hardfacing alloy. EPRI currently is planning work with IIT Research Institute on this topic. Specimens hardfaced with Stellite will be surface glazed (a local surface area is rapidly melted and then solidified at ultrafast rates) and surface alloyed (the candidate alloying agents being TiC and WC) by laser. The wear properties of the resulting specimens will then be evaluated by AMAX of Michigan, Inc., by using the same wear tests that will be used to evaluate the iron-base alloys discussed above.

Conclusions

Occupational radiation exposure continues to be a significant issue facing nuclear utilities. One attractive approach to reducing radiation field buildup is to reduce the cobalt inventory in the plant.

Corrosion of the Inconel 600 steam generator tubing is the primary cobalt source in PWR. Wear of feedwater regulator valves and of control blade pins and rollers are the primary sources in BWRs. Valve maintenance operations are a moderate source of cobalt, but attention to cleaning and post-maintenance procedures can effectively minimize cobalt input from this source.

Commercially available cobalt-free alloys are now being used in BWR control blade pins and rollers and in BWR feedwater regulator valves. Nitronic-60/CFA and PH 13-8 Mo/Inconel X-750 are being used as pins/roller alloys at two plants; type-440C stainless steel is being used in feedwater regulator valves at another plant. Preliminary data indicate that the performance of these cobalt-free alloys has been outstanding. Laboratory wear tests of some commercially available low-cobalt hardfacing alloys at ambient temperature in simulated PWR coolant chemistry showed that the wear resistance of some low-cobalt alloys is comparable to that of the cobalt-base alloys when tested at low loads. At higher loads, some dissimilar wear couples show wear resistance comparable to the self-mated cobalt-base alloys.

There is still a need for a cobalt-free hardfacing alloy that can compete with the Stellites, especially at high applied loads. EPRI is initiating a program to determine if austenitic iron-base alloys can meet these needs. Another program will determine if the wear resistance of the Stellite can be increased by using advanced surface process techniques.

Reducing radiation field buildup and occupational radiation exposure will require continued vigilance and attention. The reduction in radiation exposure at U.S. plants over the past two years suggests that some of the measures discussed in this paper may have contributed to this turnabout.

References

1. D. B. Heard and R. J. Freeman, "Cobalt Contamination Resulting from Valve Maintenance", EPRI NP-3220, Electric Power Research Institute, August 1983.

2. T. R. Young, J. T. LaDuca, T. E. Quattrochi, M. Wiatrowski, "Cobalt Source Identification Program," EPRI NP-2685, Electric Power Research Institute, October 1982.

3. C. A. Bergmann, E. I. Landerman, D. R. Lorentz, D. D. Whyte, "Evaluation of Cobalt Sources in Westinghouse-Designed Three- and Four-Loop Plants," EPRI NP-2681, Electric Power Research Institute, October 1982.

4. C. F. Falk, "BWR Cobalt Source Identification," EPRI NP-2263, Electric Power Research Institute, February 1982.

5. K. F. Dufrane and M. D. Naughton, "Nuclear Component Wear Measurements," Nuclear Technology, 63 (10) (1983) pp. 102-109.

6. K. F. Dufrane, T. R. Young, J. T. LaDuca, and P. L. Mellor, "Wear Measurements of Nuclear Power Plant Components," EPRI NP-3444, Electric Power Research Institute, in press.

7. D. H. Lister, E. McAlpine, and W. H. Hocking, "The Release of Corrosion From Surfaces in the Primary Coolant of Light-Water-Cooled Reactors," pp. 242-255 in Proceedings of the 43rd International Water Conference, Engineers Society of Western Pennsylvania, Pittsburgh, PA, 1982.

8. N. K. Taylor, I. Armson, "Corrosion Product Release From Stellites and Stainless Steel in High Pressure, High Temperature Lithiated Water," pp. 141-151 in Proceedings of the Third International Conference on Water Chemistry of Nuclear Reactor Systems, British Nuclear Energy Society, London, England, 1983.

9. P. Aldred, "BWR Control Rod Cobalt Alloy Replacement," Electric Power Research Institute, EPRI NP-2329, March 1982.

10. G. A. Freund, Materials for Control Rod Drive Mechanisms, pp. 118-121; Rowman and Littlefield, Inc., New York, N.Y., 1963.

11. W. J. Schumaker, "Metals for Nonlubricated Wear," Machine Design, 48 (6) (1976) pp. 57-59.

12. K. G. Budinski, "Incipient Galling of Metals," Wear, 74 (1) (1981) pp. 93-105.

13. E. I. Lauderman, D. J. Boes, P. Bowen, and M. J.Huck, "Evaluation of Low-Cobalt Alloys for Hardfacing Applications in Nuclear Components," EPRI NP-3446, Electric Power Research Institute, in press.

14. K. C. Anthony, "Wear-Resistant Cobalt-Base Alloys," Journal of Metals, 35 (2) (1983) pp. 52-60.

15. K. J. Bhansali and A. E. Miller, "The Role of Stacking Fault Energy on Galling and Wear Behavior," Wear, 75 (2) (1982) pp. 241-252.

16. K. J. Bhansali, "Adhesive Wear of Nickel and Cobalt-Base Alloys," Wear, 60 (1) (1980) pp. 95-110.

17. C. J. Heathcock, A. Ball, and B. E. Protheroe, "Cavitation Erosion of Cobalt-Based Stellite Alloys, Cemented Carbides, and Surface-Treated Low Alloy Steels," Wear, 74 (1) (1981) pp. 11-26.

18. J. D. Ayers, R. J. Schaefer, W. P. Robey, "A Laser Processing Technique for Improving Wear Resistance of Metals," Journal of Metals, 33 (8) (1981) pp. 19-23.

19. G. Dearnaley, "Practical Applications of Ion Implantation," Journal of Metals, 34 (9) (1982) pp. 18-28.

20. A. W. Ruff and L. K. Ives, "Sliding Wear Behavior of Electron Beam Surface-Melted O-2 Tool Steel," Wear, 75 (2) (1982) pp. 285-301.

21. S. A. Dillich, R. N. Bolster, and I. L. Singer, "Effects of Ion Implantation on the Tribological Behavior of a Cobalt-Based Alloy," in Proceedings of the NATO Advanced Study Institute on Surface Engineering, R. Kossowsky and S. C. Singhal. ed.; Elsevier Press, Amsterdam, Netherlands, (1983), in press.

FACTORS AFFECTING IN-CORE DIMENSIONAL STABILITY OF ZIRCALOY-2

CALANDRIA TUBES

V. Fidleris, A. R. Causey

Atomic Energy of Canada Limited
Chalk River Nuclear Laboratories
Chalk River, Ontario CANADA KOJ 1JO

and

R. A. Holt

Physical Metallurgy Research Laboratories
CANMET, Ottawa, Ontario CANADA K1A OG1

Abstract

In CANDU PHW reactors, the heavy water moderator is contained in a cylindrical vessel (calandria) which is penetrated by 380 horizontal fuel channel assemblies. The outer Zircaloy-2 tube of each assembly (the calandria tube) is rolled into the end shields to seal the calandria. The calandria tubes operate at ∿340 K with axial stresses that range from -10 to +40 MPa and experience fast neutron fluxes as large as 3×10^{17} n m^{-2} s^{-1}, E > 1.0 MeV. In this environment tubes elongate and sag due to irradiation-induced creep and growth. Our understanding of these irradiation effects is based on creep, stress relaxation and irradiation growth experiments on calandria tube materials irradiated to neutron fluences of 7×10^{25} n m^{-2}, E > 1.0 MeV. Both creep and growth strains decrease with the proportion of grains that have basal plane normals in the direction of testing. Cold work increases the creep rate but appears to introduce a negative component of growth in the working direction due to neutron induced stress relief that persists up to at least 7×10^{25} n m^{-2}. Thermal stress relief restores the positive growth rate in the working direction. There is little effect of grain size in the range 10 to 30 μm. This information can be used to select fabrication routes that will minimize dimensional changes of tubes during service.

Introduction

In CANDU PHW reactors, the heavy water moderator is contained in a cylindrical vessel (calandria) which is penetrated by 380 horizontal fuel channel assemblies, as shown in Figure 1. The reactor designation is derived from CANanda Deuterium Uranium-Pressurized Heavy Water.

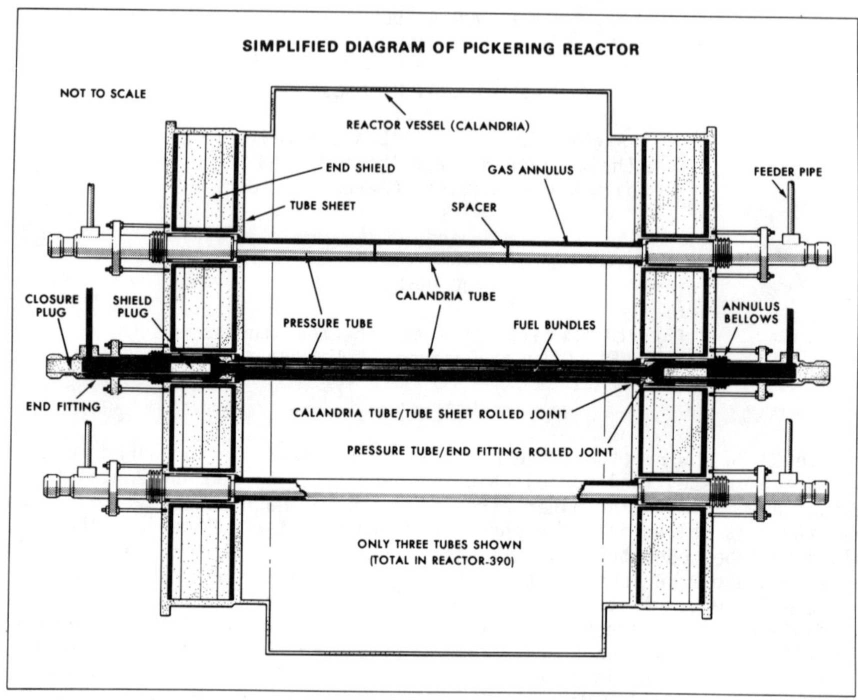

Figure 1 - A schematic diagram of a CANDU PHW reactor vessel.

The outer Zircaloy-2 tube of each assembly (the calandria tube) is rolled into the end shields to seal the calandria. The tubes operate at temperatures and stresses so low that significant changes in dimensions would not occur in the absence of a neutron flux. Irradiation causes the tubes to sag and change in length, diameter and ovality as a result of irradiation growth and irradiation enhanced creep of the material. "Irradiation growth" is the term used to describe the dimensional changes at constant volume caused by a neutron flux in an unstressed material (1). "Irradiation enhanced creep" refers to the fact that at low stresses the creep rate in an irradiated metal can be several orders of magnitude higher than the creep rate observed in the absence of the neutron flux (1).

There are many publications describing the relationships between stress, temperature, fast neutron flux and fluence, microstructure and the in-reactor deformation of zirconium alloys used for fuel cladding and pressure tubes in the temperature range of their operation, i.e. 520 K to 620 K (1,2,3). Much less is known about the in-reactor deformation of zirconium alloys at lower temperatures, corresponding to the operating conditions of Zircaloy-2 calandria tubes (about 340 K and stresses ranging from -10 to +40 MPa). We report here the effects of microstructural properties on the in-reactor creep and growth of different Zircaloy-2 calandria tube materials and how these properties are related to tube fabrication procedures.

Experimental Details

The calandria tubes are not normally accessible for measurement of dimensional changes in operating reactors. Their deformation must, therefore, be determined indirectly from the measurements of end shield displacements, sag of pressure tubes and from creep and growth experiments conducted in research reactors.

Calandria tubes are made by seam welding of brake-formed annealed Zircaloy-2 sheet whose chemical composition may vary within limits specified by the ASTM-B350 code and listed in Table I. Fabrication procedures of the sheet and of the finished tubes have varied slightly from one production batch to the next. Representative samples were therefore selected from different production batches of calandria tubes for in-reactor testing together with a series of Zircaloy-2 strip materials. These latter have controlled microstructures and have been prepared from a common starting material. Their preparation details are given in Table II. The grain size, texture and dislocation density of each material were determined by techniques used previously (4) and are also compiled in Table II.

Table I. Composition of Zircaloy-2 Calandria Tube Materials

Element	Composition in Percent by Weight
Sn	1.2 - 1.7
Fe	0.07 - 0.20
Cr	0.05 - 0.15
Ni	0.03 - 0.08
Hf	<50 ppm
V	<50 ppm
O	<1600 ppm
H	<25 ppm
Zr	Balance

The test specimens were in the form of small flat strips, 38 mm x 6.3 mm x 1.5 mm for growth testing and 76 mm x 3.8 mm x 0.7 mm for stress relaxation testing. These were machined either from longitudinal or transverse sections of the material. In the case of calandria tube materials only longitudinal specimens were tested for growth, because it would have been necessary to flatten transverse specimens, thereby altering their microstructure.

The irradiations were carried out in water-cooled test facilities of the National Research Universal (NRU) reactor at Chalk River Nuclear Laboratories (CRNL) and the Advanced Test Reactor (ATR) at Idaho Falls. The test temperatures were in the range 320 K to 350 K and the fast neutron flux was about 1.6×10^{17} n m^{-2} s^{-1} in NRU and about 3×10^{18} n m^{-2} s^{-1} in ATR.

The growth was determined by measuring the length changes of the growth specimens at three fixed positions using a sensitive electronic comparator. The results, corrected for temperature fluctuations during measurements, were accurate to ± 0.5 μm.

Table II. Fabrication and Microstructural Details of Zircaloy-2 Test Materials

Code	Material	Metallurgical Condition	Average Grain Size (μ m)	Resolved Basal Pole Components F_L	F_T	F_R	Dislocation Density* ($m^{-2} \times 10^{13}$)
A	Calandria Tube	α-annealed, cold-worked 1.5%, stress relieved 1 h at 770 K	16	.10	.18	.72	2.3 - 4.3
B	"	α-annealed, cold-worked 1.5%	19	.11	.22	.67	8.4 - 10.5
F	"	α-annealed, cold-worked 1.5%, plus 1/2 h at 1018 K	16	.11	.17	.71	1.5 - 2.2
K	Experimental Strip	α-annealed	31	.07	.40	.53	0.7 - 2.4
M	"	α-annealed, plus 0.25% c.w.	25	.06	.39	.56	7.6 - 8.0
N	"	α-annealed, plus 1.5% c.w.	28	.06	.39	.56	17.0
P	"	α-annealed, plus 0.25% c.w. + 1 h at 770 K	27	.05	.40	.55	0.7 - 1.7
R	"	α-annealed, plus 1.5% c.w. + 1 h at 770 K	28	.06	.37	.57	4.4 - 5.4
S	"	α-annealed	12	.05	.35	.60	1.2 - 3.3
V	"	β-treated	33	.26	.46	0.28	4.1 - 5.9
Z	Pressure Tube	25% cold-worked, stress relieved 24 h at 670 K	5	.06	.48	.46	25.1

*from X-ray line broadening

The creep information was obtained from bent-beam stress relaxation tests (5). The rate at which elastic strain, $\underline{e}$ is converted to plastic strain ε_c is given by

$$\dot{\varepsilon}_c = K_c \; C_d \; \sigma \; \phi, \tag{1}$$

where ϕ is the fast neutron flux, $\sigma = E_d\underline{e}$ is the stress in a specimen with elastic modulus, E_d, deformed to an elastic strain, $\underline{e}$, K_c is the specific creep rate (a material constant), the units of which depend on the units of flux, and C_d is the creep anisotropy factor in the "d" direction. From Hooke's law and Equation 1 the stress relaxation in a bent beam can be described by

$$\sigma = \sigma_o \exp(-K_c \; C_d \; E_d \; \phi \; t), \tag{2}$$

where σ_0 is the stress at the surface of the beam at t = 0. A plot of $\ln(\sigma/\sigma_0)$ vs. ϕt has a slope of $-K_cC_dE_d$ from which K_cC_d can be determined for a given testing direction. Eighteen calandria tube and pressure tube materials were tested at 320 K $\pm$ 10 K up to fast neutron fluences of 1.7×10^{25} n m^{-2}, E > 1.0 MeV.

Results

Irradiation Growth. The growth of calandria tube materials is shown in Figure 2. Beyond the initial positive strain transient they behaved quite differently, even though all three had similar textures, grain sizes and had been cold-worked the same amount (Table II). Thus the material without a final heat treatment (code B) showed a negative growth rate, the material given a high temperature anneal (code F) had a positive growth rate and the stress relieved material (code A) fell in between the other two.

The growth behavior of the experimental Zircaloy-2 strip materials is summarized in Figures 3 to 6. The annealed materials (Figure 3) exhibited a marked growth anisotropy. This observation supports the findings of other investigators (6,7) in which the direction and magnitude of growth depended on the predominant orientation of grains in the direction of measurement, being zero for a random orientation distribution of grains. Since a single crystal of zirconium shortens during irradiation along the c-axis and expands along the a-axis, in highly textured zirconium alloys such as the experimental Zircaloy-2 strip under investigation, the growth would be expected to follow a (1-3F) type of dependence, where F is the fraction of basal poles in the direction of measurement. To a first approximation the results of Figure 3 appear to follow such a dependence (Table III). The finer grained material exhibited slightly lower growth than that of the coarser grained material.

The effects of cold work on growth are shown in Figures 4 and 5. At low levels of cold work the growth of longitudinal specimens decreases with increasing cold work such that the long term growth rate after cold work is negative. The large scatter of the 0.25% cold worked material reflects the difficulty in introducing such small amounts of cold work uniformly throughout the material. A reversal in the direction of growth with small amounts of cold work is also observed in the transverse direction (Figure 5). These results contrast with those of earlier observations obtained on more heavily cold worked Zircaloys (4,6,7). In these experiments the growth in the longitudinal direction was found to increase approximately linearly with the amount of cold work, i.e. in agreement with the current results on 25% cold worked and stress relieved Zircaloy-2 pressure tube (material Z).

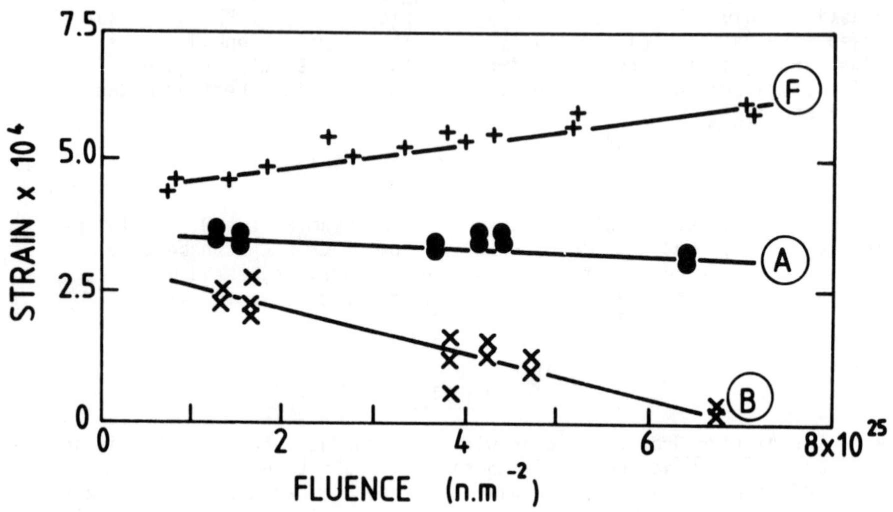

Figure 2 - Irradiation growth of calandria tube specimens at 320 K (longitudinal direction). A = 1.5% cold worked and stress relieved at 770 K, B = 1.5% cold worked with no stress relief, F = 1.5% cold worked and annealed at 1018 K.

Table III. Correlation of Irradiation Growth of Annealed Zircaloy-2 with Texture and Grain Shape

Material	Direction of Testing	Basal Pole Fraction F_d	Grain Shape Anisotropy Factor, A_d	$1-3F_d$	$1-F_d-2A_d$	Measured Growth (%) after $4x10^{25}$ n m^{-2} (E > 1.0 MeV)
Strip K	Long.	0.07	0.306	0.79	0.318	0.110
Strip K	Trans.	0.40	0.326	-0.20	-0.050	-0.025
Strip V	Long.	0.26	0.290	0.22	0.160	0
Strip V	Trans.	0.46	0.275	-0.38	-0.010	-0.030
Strip S	Long.	0.05	0.332	0.85	0.286	0.085

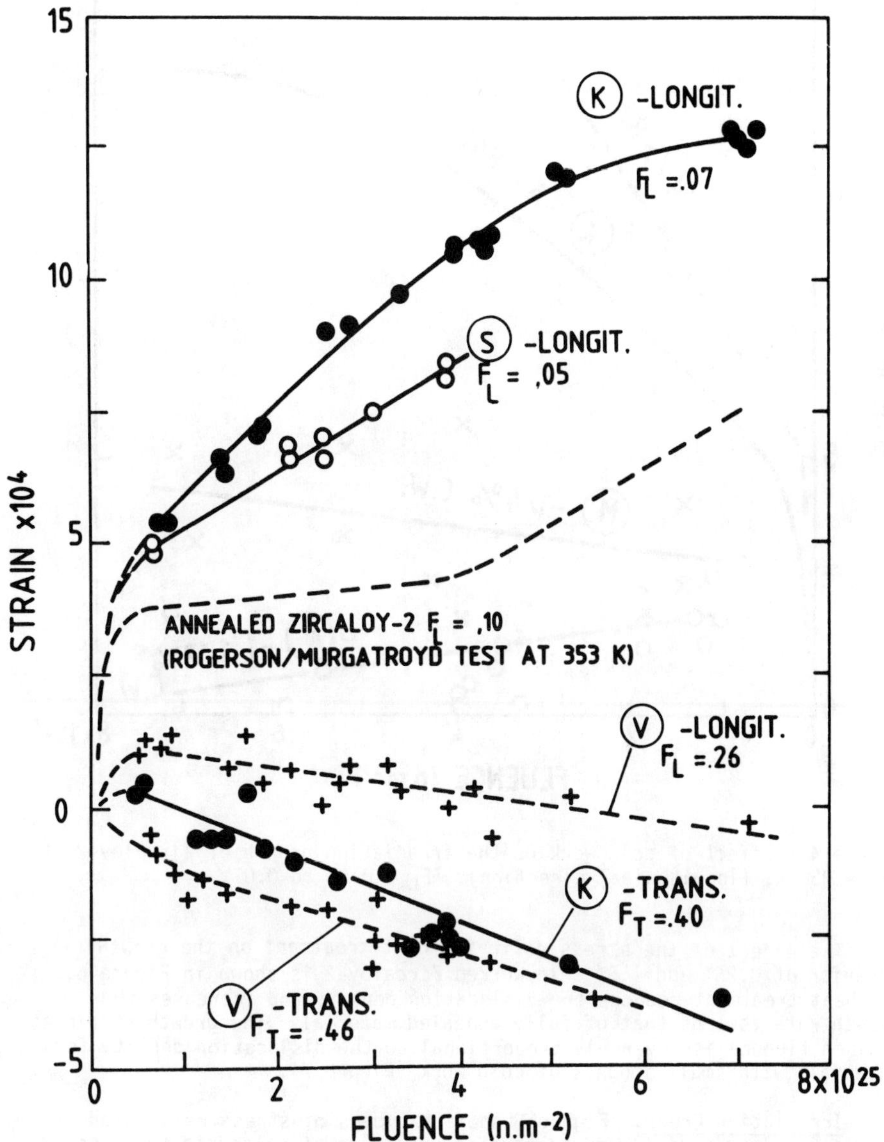

Figure 3 - Irradiation growth of annealed strip materials at 320 K. K = alpha phase annealed, coarse grained, S = alpha phase annealed, fine grained, V = beta phase annealed, coarse grained.

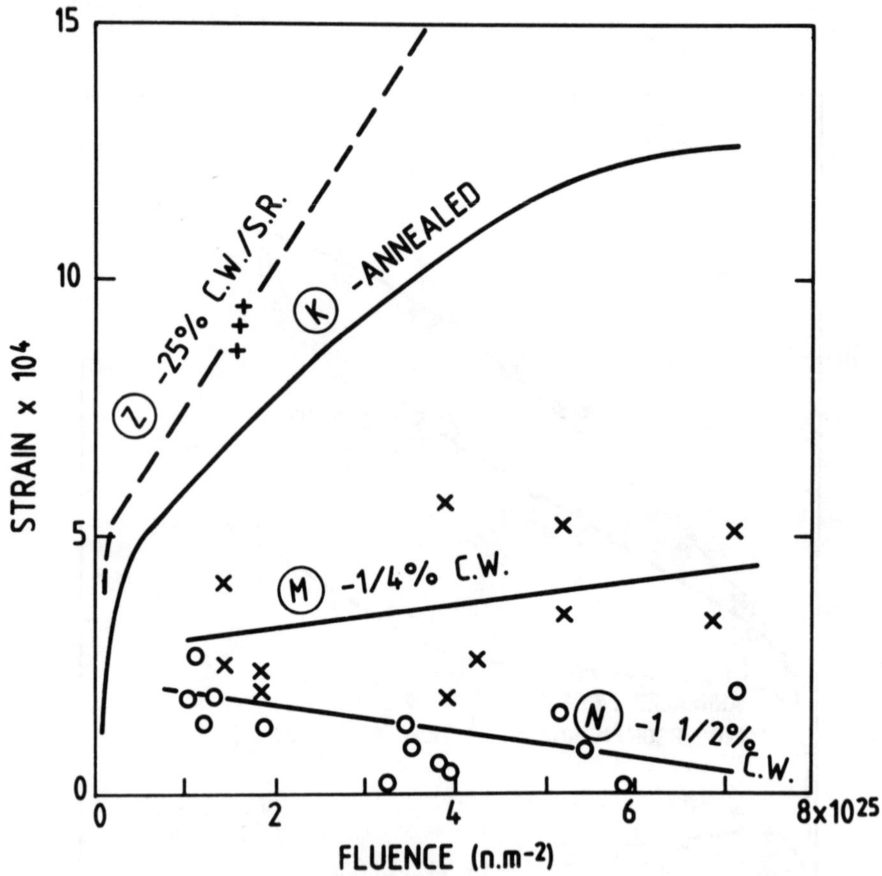

Figure 4 - Effect of cold work on the irradiation growth of Zircaloy-2 at 320 - 350 K, (longitudinal direction). F_L = 0.06 to 0.07.

The effect of the stress relieving heat treatment on the growth behavior of 0.25% and 1.5% cold worked Zircaloy-2 is shown in Figure 6. The heat treatment reduces the dislocation density and increases the growth rate towards that of fully annealed material. The growth strain at a given fluence is inversely proportional to the dislocation density for materials with small amounts of cold work (Figure 7).

Irradiation Creep. Figure 8 shows examples of stress relaxation curves for longitudinal and transverse specimens of calandria tube (code B) and pressure tube (code Z) material. The calandria tube behavior is very nearly identical in both directions, reflecting their similiarity of textures (F_L = 0.10, F_T = 0.17). The average ratio of creep rates from the longitudinal and transverse pressure tube specimens was found to be 1.61 ± 0.28 (one standard deviation).

Creep occurs primarily through glide or climb of <a> type dislocations with secondary contributions from <c+a> types (5). Assuming it to be glide, 64% of <a> type dislocations on {1$\bar{1}$20} prism planes and 36% of <c+a> type dislocations on {10$\bar{1}$1} pyramidal planes are required to account for the observed creep anisotropy. Using these fractions the creep anisotropy factors for longitudinal and transverse specimens of each material were used to calculate the specific creep rate, K_C, shown plotted in Figure 9 as a function of dislocation density. The specific creep rate varies with dislocation density raised to the power of 0.40 $\pm$0.06 (95% confidence limit).

Discussion of Results

The long term in-reactor deformation of zirconium alloys consists of at least three components (2): irradiation creep, irradiation growth and thermal creep. At the operating temperature of calandria tubes (∿340 K) thermal creep is negligibly small. Irradiation creep, the stress-induced component of in-reactor deformation, can be measured in bending or torsion (5), or, assuming creep and growth to be simply additive, by subtracting irradiation growth from the total strain.

The crystal structure of alpha-zirconium is anisotropic. The close packed hexagonal lattice is weakest in directions contained in the basal plane and, during working, develops strong crystallographic texture and mechanical anisotropy. Thus reactor components generally have anisotropic microstructures which cause growth and anisotropic creep during irradiation.

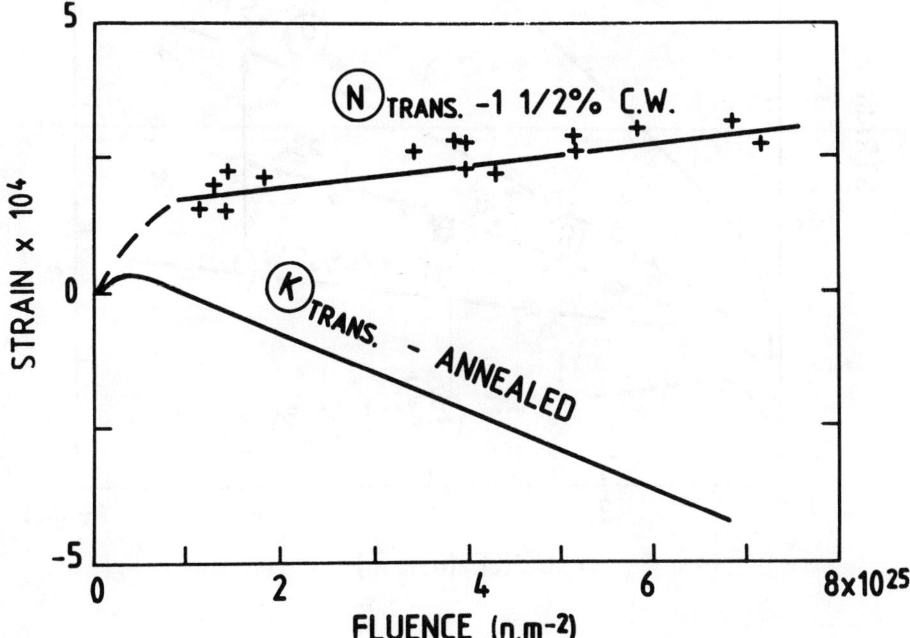

Figure 5 - Effect of 1.5% cold work on irradiation growth of Zircaloy-2 at 320 K (transverse direction). F_T = 0.39 to 0.40.

If creep and growth are additive the deformation rate is given by

$$\dot{\varepsilon}_{total} = \dot{\varepsilon}_{creep} + \dot{\varepsilon}_{growth} =$$

$$= K_c \, C_d \, \bar{\sigma} \, \phi + K_g \, G_d \, \phi, \qquad (3)$$

where $\bar{\sigma}$ is the normalized applied stress and K_c, K_g are specific creep and growth rates (material constants), and C_d, G_d are creep and growth anisotropy factors in the "d" direction respectively. K_c and K_g depend on the metallurgical structure (e.g. dislocation density) and temperature. The anisotropy factors C_d and G_d depend on the anisotropy of the microstructure, e.g. crystallographic texture, orientation of dislocation Burgers vectors and grain boundary anisotropy.

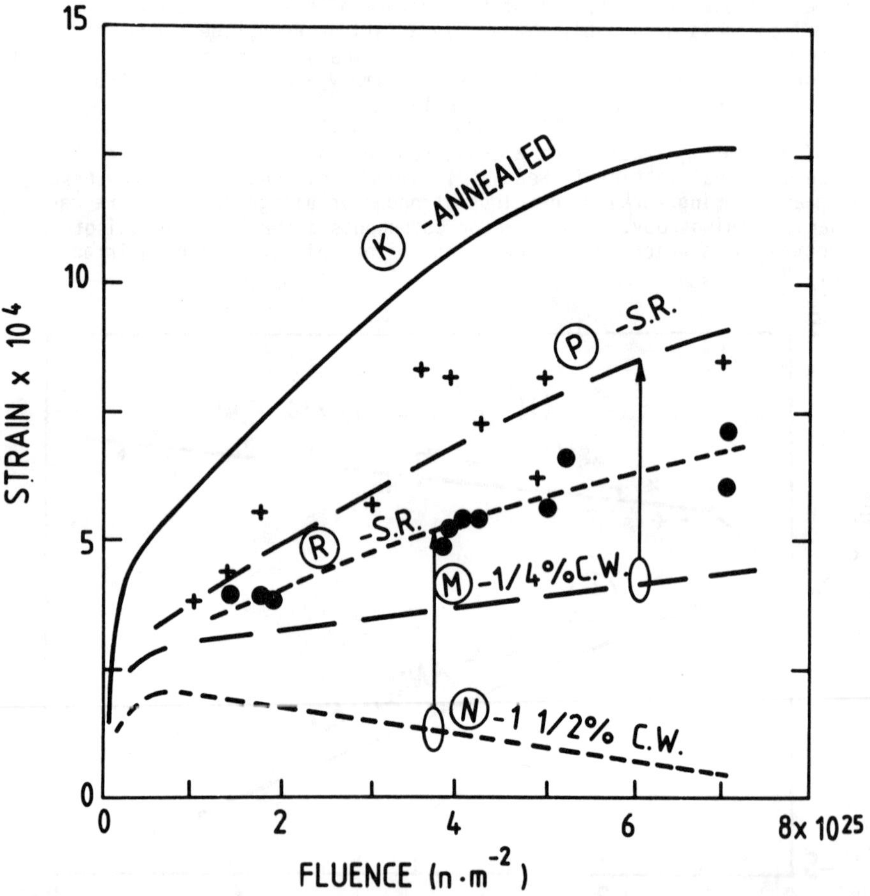

Figure 6 - Effect of stress relieving for one hour at 770 K on irradiation growth of Zircaloy-2 strip material at 320 K (longitudinal direction, with $F_L \sim 0.06$). P = material M plus stress relief, R = material N plus stress relief.

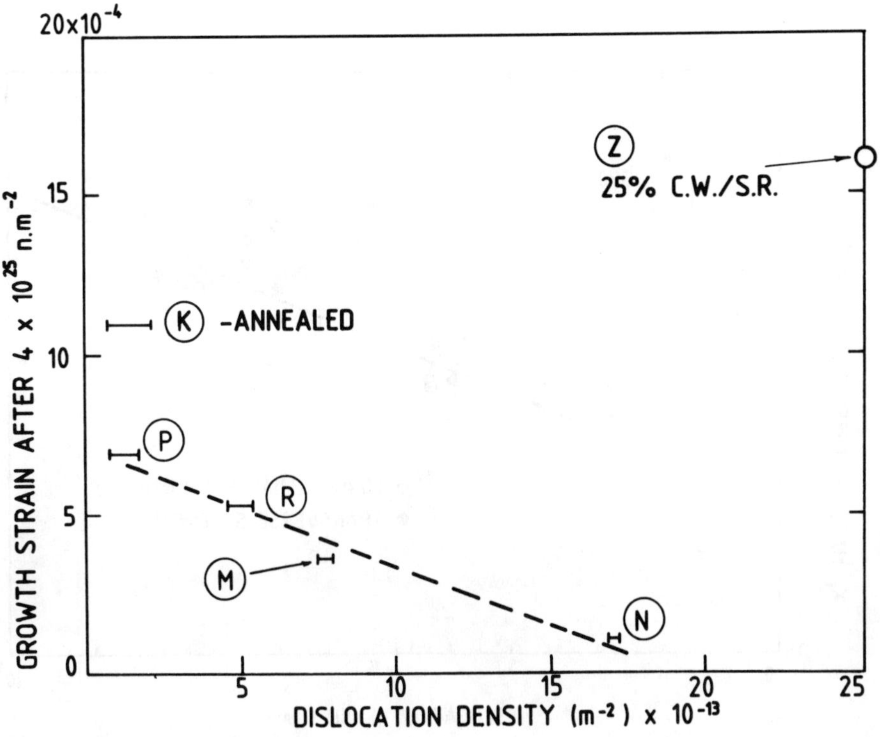

Figure 7 - Apparent inverse dependence of irradiation growth of Zircaloy-2 at 320 K on dislocation density for low levels of cold work, based on results in Figure 6 (longitudinal direction).

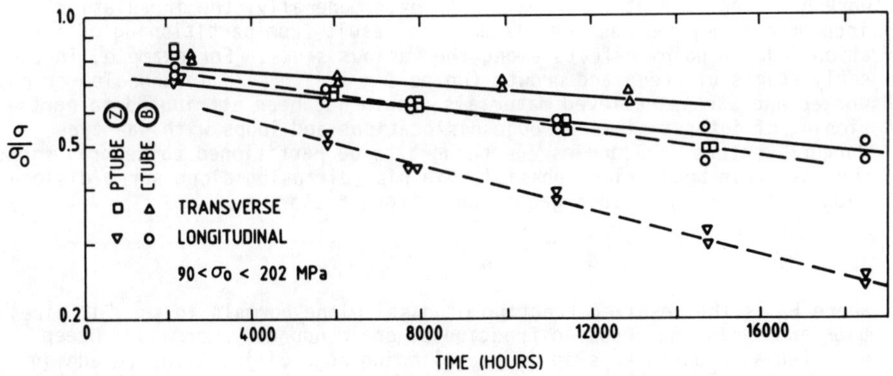

Figure 8 - In-reactor stress relaxation of Zircaloy-2 calandria tube and pressure tube material at 320 K and a neutron flux of 1.6×10^{17} n m^{-2} s^{-1}, E > 1.0 MeV.

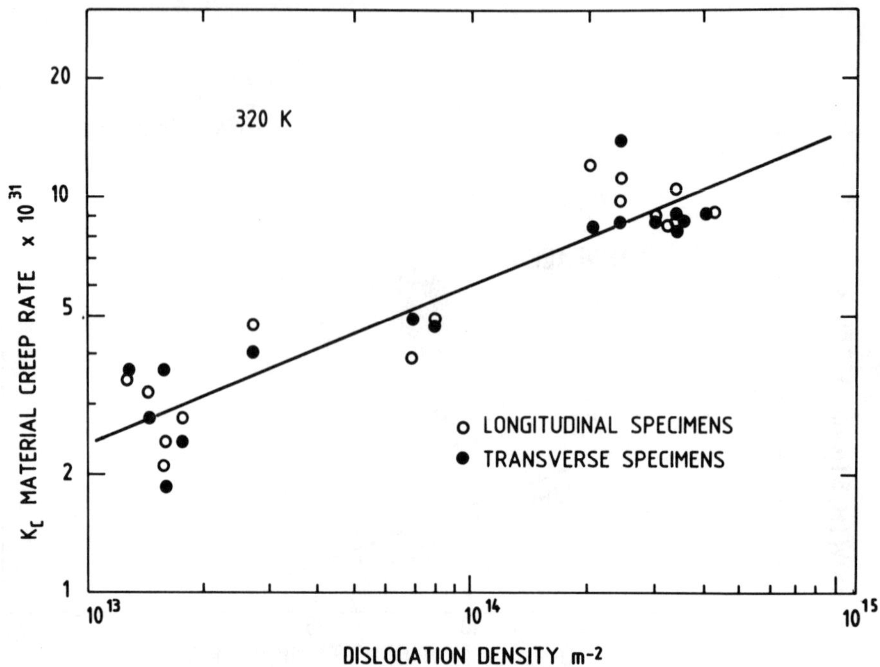

Figure 9 - In-reactor creep of Zircaloy-2 at 320 K, showing variation of material constant K_c with dislocation density.

Material constants and anisotropy factors have been derived for cold worked Zircaloy-2 and Zr-2.5 wt% Nb deformed in-reactor at about 570 K based on various models assumed for creep and growth (2,4,5,8), but little work has been done at lower temperatures. Generally, the irradiation induced deformation has been assumed to result from partitioning of irradiation induced point defects among the various sinks. For example, in the early stages of creep and growth (up to 3 x 10^{25} n m^{-2}) in annealed or cold worked and stress relieved materials growth has been attributed to partitioning of interstitials to edge dislocations and loops with <a> type Burgers vectors. Vacancies are thought to be partitioned to neutral sinks, such as grain boundaries, possibly via pipe diffusion along screw dislocations. This results in a growth anisotropy factor

$$G_d = 1 - F_d - 2\ A_d, \quad (4)$$

where F_d is the resolved fraction of basal plane normals in the "d" direction and A_d is the resolved fraction of grain boundary normals. Creep has been attributed to slip of the climbing edge dislocations (predominantly 1/3 $\langle 11\bar{2}0 \rangle$ $\{10\bar{1}0\}$ but modified by some basal or pyramidal slip) which results in higher creep rates along directions containing a predominance of prism plane normals.

Growth has also been attributed to segregation of interstitials to dislocations with <a> type Burgers vectors and vacancies to dislocations with <c> component Burgers vectors (9,10). This results in a growth anisotropy factor of

$$G_d = 1 - 3\ F_d. \tag{5}$$

Experimentally, the observed anisotropy of irradiation creep and growth have supported these types of models. Therefore the microstructure of zirconium alloy pressure tube materials, which operate at 570 K, have been tailored to achieve good in-reactor dimensional stability.

The irradiation creep behavior observed in stress relaxation tests at 320 K is qualitatively similar to that observed at higher temperatures, with the exception that the dependence of the creep rate on dislocation density is somewhat stronger. Creep rates are weakest along directions containing large proportions of prism plane normals (usually the working direction).

The previously reported (6,7) general characteristics of irradation growth behavior of zirconium alloys have been a strong dependence on the pre-irradiation dislocation density (cold work) and except for some short transients in the initial stages (up to 1×10^{25} n m^{-2}, E > 1.0 MeV) growth is always positive in directions containing low fractions of basal plane normals, in accordance with equations 4 and 5. There is some evidence (11) that at high fluences ($\sim 5 \times 10^{25}$ n m^{-2}, E > 1.0 MeV) irradiation-induced microstructural features, possibly <c> component dislocations increase the irradiation growth rate of annealed materials (Figure 3).

Several novel characteristics of irradiation growth are reported here, i.e. the apparent reversal of growth anisotropy up to 5 - 7 x 10^{25} n m^{-2} by low levels of cold work and its restoration by stress relieving (which was common to all the cold worked materials tested), and the large variation in the magnitude and duration of the transient strain in annealed material. The precise agreement of the growth of the 25% cold worked and stress relieved material with previous tests (7) at lower flux verify the validity of these new results. Clearly some factor affects the long term growth behavior that has previously not been considered.

Recently, neutron diffraction studies on zirconium alloys containing low levels of cold work have shown (12) that large intergranular stresses develop normal to the working direction due to the high degree of plastic anisotropy of the hexagonal close packed structure. These stresses are compressive along the <c> directions and tensile along the <a> directions. Anisotropic irradiation-induced relaxation of these stresses would explain qualitatively some of the newly observed growth characteristics. Because irradiation creep is more rapid along the <a> directions, relaxation of the intergranular stresses would occur predominantly by extension along the <a> directions normal to the working direction with a corresponding contraction along the working direction. The rate of this process would increase with increased dislocation density (cold work) while the magnitude of the negative component of strain would increase with cold work due to an increase in the intergranular stresses. Thermal stress relieving would be expected to produce a similar strain while diminishing the subsequent irradiation-induced effect. This has been verified experimentally. Specimens of the 1.5% cold worked plate, stress relieved 1 h at 770 K, contracted 0.07% in the working direction while expanding 0.01% in the transverse direction. The apparent reduction in growth with increased dislocation density (Figure 7) could be explained by an increasing large negative growth rate due to the relaxation of residual stresses combined with the normal positive transient growth thought to be associated with a volume change (7,13).

Finally, the intergranular stresses due to differential thermal contraction in annealed materials during cooling from the annealing temperature would be in the opposite direction from those induced by cold work. Variations in annealing conditions, particularly the cooling rate after annealing, could explain the variability observed in the annealed Zircaloy-2 materials (6,10). Similar differences in the magnitude of the transient growth strain have been reported by Zee et al. (14) between Zr-0.1 wt% Sn and Zr-1.5% wt% Sn alloys irradiated at 353 K. They attributed the more rapid attenuation of growth in the 1.5% Sn alloy to enhanced recombination of point defects as a result of vacancy trapping by Sn atoms.

One implication of the residual stress relief model is that the negative component of growth rate in the working direction of cold worked materials would eventually diminish with increasing fluence and "normal" growth would resume. Higher fluence data combined with microstructural investigations will help to distinguish the validity of this model from that of other possibilities, e.g. those relating to the distribution of dislocation Burgers vectors.

Design Implications

The results reported here provide a wide range of data as a function of metallurgical structure which can be used to select fabrication routes to optimize the dimensional behavior of Zircaloy-2 for service as calandria tubes in CANDU reactors.

The most serious dimensional concerns are elongation, due to a combination of growth and axial tensile stress imposed by the force of the moderator pressure on the end shields, and the sag due to the weight of pressure tube, coolant and fuel acting through the spacers. Both sag and axial creep can be diminished by modifying the crystallographic texture to reduce the creep anisotropy factor, C_d, for the working direction (e.g. by the β-treatment, Table II) and by reducing the dislocation density. The lowest axial growth strains over fluences of 7×10^{25} n m^{-2} (E > 1.0 MeV) are observed in material containing 1.5% cold work. Increasing cold work might produce a negative growth strain in this fluence range, but at the expense of an increasing creep rate due to increasing dislocation density. Beta-treated material exhibits a small negative growth rate which could help to counteract elongation due to creep. Even if the negative component eventually disappears and growth reverses and accelerates, the low degree of anisotropy of the beta-treated structure should enhance its dimensional stability.

There is still a need to monitor the dimensional behavior of these materials to neutron fluences expected in calandria tubes of a modern CANDU reactor near the end of its design life (i.e. 3×10^{26} n m^{-2}, E > 1.0 MeV). The results reported here demonstrate that only small dimensional changes are to be expected in calandria tubes to fluences of about 1×10^{26} n m^{-2}, E > 1.0 MeV, corresponding to some 25 years of operation in a Pickering CANDU reactor.

Conclusions

During irradiation tests at temperatures near 340 K Zircaloy-2 calandria tube materials exhibited anisotropic creep and growth which is related in a complicated way to material texture and microstructure. The results confirm earlier findings that both creep and growth decrease with the proportion of grains that have basal plane normals in the direction of testing. Cold work increases the creep rate, but appears to introduce a negative component of growth which at low levels of cold work can dominate the growth behavior of the material. Flux-induced relief of residual stresses appears to be responsible for the negative component of growth persisting up to at least 7×10^{25} n m^{-2}, E> 1.0 MeV. Thermal stress relief restores the positive growth rate in the working direction. This information can be used to select fabrication routes that will minimize dimensional changes of calandria tubes during irradiation. Based on the available test data good dimensional stability can be expected in CANDU PHW calandria tubes of current design up to fluences of at least 1×10^{26} n m^{-2}, E > 1.0 MeV.

Acknowledgments

This work has been jointly funded by Atomic Energy of Canada Limited and Ontario Hydro. We wish to thank F. J. Butcher, R. M. Condie, S. A. Donohue, I. R. Emmerton, R. W. Gilbert and J. E. Winegar for technical assistance and M. Leger for technical advice. We are grateful to A. Rogerson for supplying us with the cold worked Zircaloy-2 pressure tube material used in his experiments.

References

1. V. Fidleris, "Summary of Experimental Results on In-reactor Creep and Irradiation Growth of Zirconium Alloys", Atomic Energy Review, 13 (1975) pp. 51-80.

2. R. A. Holt, A. R. Causey and V. Fidleris, "Correlation of Creep and Growth of Pressure Tubes with Operating Variables and Microstructure", Proceedings of British Nuclear Energy Society Conference, London, 1983, Vol. I, pp. 175-178.

3. Proceeding of ASTM Conferences on Zirconium in the Nuclear Industry, ASTM Special Technical Publication, STP 633 (1977), STP 681 (1979), STP 754 (1982).

4. R. A. Holt, "Effect of Microstructure on Irradiation Creep and Growth of Zircaloy Pressure Tubes in Power Reactors", Journal of Nuclear Materials, 82 (1979) pp. 419-429.

5. A. R. Causey, R. A. Holt and S. R. MacEwen, "In-reactor Creep of Zr-2.5 wt% Nb," ASTM Special Technical Publication No. 824 (1984) pp. 269-288.

6. R. B. Adamson, "Irradiation Growth of Zircaloy," ASTM Special Technical Publication No. 633 (1977) pp. 326-343.

7. R. A. Murgatroyd and A. Rogerson, "An Assessment of the Influence of Microstructure and Test Conditions on the Irradiation Growth Phenomenon in Zirconium Alloys," Journal of Nuclear Materials, 90 (1980) pp. 240-248.

8. R. A. Holt and E. F. Ibrahim, "Factors Affecting the Anisotropy of Irradiation Creep and Growth of Zirconium Alloys," Acta Metallurgica, 27 (1979) pp. 1319-1328.

9. R. A. Holt and R. W. Gilbert, "c-Component Dislocations in Neutron Irradiated Zircaloy-2," Journal of Nuclear Materials, 116 (1983) pp. 127-130.

10. R. P. Tucker, V. Fidleris and R. B. Adamson, "High Fluence Irradiation Growth of Zirconium Alloys at 644 to 725 K," ASTM Special Technical Publication No. 824 (1984) pp. 427-451.

11. A. Rogerson and R. A. Murgatroyd, "'Breakaway' Growth in Annealed Zircaloy-2 at 353 K and 553 K," Journal of Nuclear Materials, 113 (1983) pp. 256-259.

12. S. R. MacEwen, J. Faber Jr. and A. P. L. Turner, "The Use of Time-of-Flight Neutron Diffraction to Study Grain Interaction Stresses," Acta Metallurgica, 31 (1983) pp. 657-676.

13. J. E. Harbottle and F. Herbillon, "Shape and Volume Changes During Irradiation Growth of Zircaloy-2," Journal of Nuclear Materials, 90 (1980) pp. 249-255.

14. R. H. Zee, A. Rogerson, G. J. C. Carpenter and J. F. Waters, "Effect of Tin on the Irradiation Growth of Polycrystalline Zirconium," Journal of Nuclear Materials, 120 (1984) pp. 223-229.

CONTROL OF MICROSTRUCTURE IN BRAZED ZONE OF ZIRCALOY-4 NUCLEAR FUEL SHEATHING BY OPTIMIZATION OF Σ(C+P+Si) CONTENTS AND COOLING SCHEDULES

Vo Quach and Derek O. Northwood

Department of Engineering Materials
University of Windsor
Windsor, Ontario
Canada N9B 3P4

Summary

In the production of fuel elements for the CANDU-PHW reactor, induction brazing is used to attach appendages (bearing and split spacer pads) onto the outside wall of the Zircaloy-4 sheathing. The brazing process, 40 to 60 seconds at temperatures in excess of 1000°C, produces 3 heat-affected zones amounting to about 30% of the thickness. These heat affected zones quite often contain large grains and either a basketweave or a parallel plate type of Widmanstatten structure. Small grains and a basketweave structure are preferred. Using simulated brazing treatments, it is demonstrated that by control of the impurity content, Σ(C+P+Si), and cooling rate from the brazing temperature, the desired microstructure can be obtained in the braze heat-affected zone. The formation of the basketweave structure is promoted by higher impurity contents, with the second phase impurity particles acting as nuclei for the basketweave structure in preference to the β-grain boundaries where the parallel plate structure is nucleated.

Introduction

A high corrosion resistance, a high strength-to-weight ratio and a low thermal neutron absorption cross-section, are some of the major features of Zircaloy-4 that make it an ideal fuel cladding material for the CANDU-PHW (CANada Deuterium Uranium-Pressurized Heavy Water) reactor. One of the functions of fuel cladding is to safely contain the radioactive fission products generated in the sintered fuel pellets of high density uranium dioxide (UO_2). In order to re-fuel the reactor without shutting it down, a fuel channel (pressure tube) design is employed and the fuel elements are assembled into the fuel bundles. To ensure a good efficiency of heat flow and to eliminate inter-element fretting and wear damage to the fuel cladding, spacers and bearing pads are attached to the fuel elements. These appendages are brazed onto the thin-walled (0.4 mm) fuel cladding using induction brazing with a Zr-5wt%Be braze alloy in an inert atmosphere (1). During the brazing cycle, 40 to 60 seconds at temperatures over 1000°C, the base material (α-phase, h.c.p.) is transformed into the β-phase (b.c.c.) which subsequently transforms back into the α-phase on cooling. Three heat-affected zones amounting to about 30% of the fuel sheathing are produced during brazing. Microstructurally, the transformed β-phase can have two types of Widmanstatten morphologies, namely, a basketweave or a parallel plate structure which are created by the $\alpha \to \beta \to \alpha$ transformation. The grain size of Zircaloy-4 fuel sheathing in the braze heat affected zone also increases to a size which is dependent to a large extent on the brazing time/temperature conditions (2).

The microstructure of the heat affected zone of Zircaloy-4 fuel cladding is one of the main factors in determining the life of the fuel bundle in reactor. The fuel bundle is subjected to both tensile and compressive stresses from fuel expansion, coolant pressure, power changes, etc. and is exposed to aggressive fission products, e.g., iodine (3). To better withstand these conditions a basketweave structure and fine grain size (3 or more grains through thickness) is recommended.

Second-phase particles are believed to be nucleation sites for the basketweave structure (4,5,6,7). These second-phase particles in Zircaloy-4 have been variously identified as zirconium carbide (4), zirconium phosphide (5), elemental silicon (6) and Zr-Fe-Cr compounds (8). The basketweave structure is thought to form when the abundant second-phase particles are abundant and are randomly distributed on numerous habit planes within the prior β grains. However the exact type of second-phase particles or impurity that can act as a nucleation site is still a subject of debate. Okvist and Kallstron (4) found that elements such as oxygen, tin, iron and chromium did not promote a basketweave structure, but that a basketweave structure was created when the carbon content was fairly high (100-200 wt ppm). Holt (5), however, argued that the presence of oxygen, nitrogen or phosphorus as well as carbon could lead to the nucleation of a basketweave structure if the amount of that element was sufficient and it was distributed randomly. The influence of oxygen has been confirmed by Woo and Tangri (7). Recently, elemental silicon (2,6) has also been identified and correlated with the percentage of the basketweave structure. Phosphorus has also been shown to affect the formation of the 'basketweave-type' structure, with phosphorus contents >45 wt ppm, giving approximately 50% 'basketweave-type' structure (9).

Apart from the impurity content or presence of second phase particles, the cooling rate from the β-phase region can also affect the microstructure. Woo and Tangri (7) found that the microstructure of Zircaloy-4-O alloys was changed with increasing cooling rate in the following sequence: lenticular

$\alpha \rightarrow$ parallel plate $\alpha \rightarrow$ basketweave $\alpha \rightarrow$ martensitic α'. The parallel plate structure was formed at cooling rates of approximately 90°C/s. Control of the cooling rate from the braze temperature is thus another parameter (apart from impurity content) that can be used to produce the desired microstructure in the heat affected zones of the fuel cladding.

The present work is aimed at more closely defining the effects of the impurity content and cooling rate from the β-phase field on the microstructural features of Zircaloy-4 fuel cladding. Commercial Zircaloy-4 fuel cladding containing a wide range of impurity (C+P+Si) contents was cooled at various rates from the β-phase field and the microstructural features quantitatively investigated using optical microscopy. Relationships were derived between the microstructural features and the impurity contents/cooling rates using a Statistical Analysis System.

Experimental Methods

Commercial Zircaloy-4 fuel sheathing tubing with a wide range of impurity contents was used for this investigation. The C, Si, and P contents of the ten batches are given in Table I. The other impurity/alloying element contents were reasonably constant at: Sn 1.50-1.55%; Fe≈0.19%; Cr≈0.05%; O = 1120-1440 ppm and N = 60-120 ppm.

Table I. Impurity Contents of the 10 Batches of Zircaloy-4 Nuclear Fuel Sheathing

Impurity Content (ppm by wt.)	A	B	C	D	E	F	G	H	I	J
Carbon	120	130	150	150	140	170	180	190	245	260
Phosphorus	6	16	6	7	12	14	14	12	34	60
Silicon	25	<25	<25	38	79	86	79	96	43	73
C+P+Si*	151	171	181	195	231	270	273	298	322	393

(Columns A–J: Tube Identification)

* if Si<25 taken as 25 in Σ.

The different batches were sectioned into 20 mm lengths using a low speed diamond saw. The samples were then sealed in evacuated quartz capsules at 5×10^{-5} Torr. The brazing was simulated by heating the samples to 1100°C for 5 minutes then subsequently cooling. Six different cooling rates were used, namely: furnace cooling, air cooling, oil quench, water quench, brine quench and dry ice-acetone quench. The air cooling would most closely approximate present commercial practice. After heat treatment, the samples were sectioned and the longitudinal section mounted in a cold resin, rough polished on the silicon-carbide papers down to 600 grit, and then chemical polished/etched with 40% H_2O:40% H_2SO_4:20% HF solution. The amount of each type of microstructure was measured on a volume fraction basis using a 187 intersecting points-counting grid which was directly superimposed on the bright field illumination screen of a Leitz optical microscope. At least 40 intersecting areas were counted for one sample. Grain size (GS), plate width (PW) and plate length (PL) were measured using a linear scale. Measurements were made in both the horizontal and vertical directions of each grain. The grain sizes measured in these two directions were summed and then divided by the total number of measured grains. 80 to 100 grains were measured for any one sample. The grain size measurement method can be expressed mathematically as follows:

$$GS(\mu m) = \frac{\sum_{i=1}^{N} (x_i + y_i)}{2N} \cdot \left(\frac{1}{M}\right)(f_c) \quad (1)$$

where, x_i = the total grain size in horizontal direction;
y_i = the total grain size in vertical direction;
N = the total number of the measured grains;
M = magnification of the OM lens;
f_c = the conversion factor.

For the PW and the PL at least 40 spacings were measured in 20 different areas.

Results

Type of Microstructure

Depending on the impurity (C+P+Si) content and the cooling rate, the microstructure of the Zircaloy-4 fuel cladding after the simulated brazing treatment is generally one of five types, namely: tending to parallel plate structure (Tp), parallel plate structure (Pa), tending to basketweave structure (Tb), basketweave structure (B), and martensitic structure (M). These microstructures are shown in Figure 1.

The quantitative results detailing the volume fraction of each type of microstructure are correlated with the impurity (C+P+Si) content and cooling rate in Table II.

The Influence of the Impurity Content on the Microstructure

The total impurity content (defined as C+P+Si) in Zircaloy-4 plays a major role in affecting the microstructure of the braze heat affected zone. This not only shows up in the type of microstructure formed but also in the fineness of the structure. Figure 2 is a series of micrographs illustrating the variation of the basketweave structure with the impurity (C+P+Si) content for the same heat treatment condition (heated at 1100°C for 5 minutes; air cooled).

The Influence of the Cooling Rate on the Microstructure

As was the case for impurity content, the cooling rate also influences the microstructure of Zircaloy-4 in the braze heat affected zone. In general, the B (basketweave) structure is refined as the cooling rate is increased. This is illustrated in Figure 3, for a sample containing C+P+Si = 393 ppm cooled at different rates.

Grain Size (GS), Plate Width (PW) and Plate Length (PL)

Values for the grain size (GS), plate width (PW) and plate length (PL) for the 10 different batches of tubing cooled at the 6 different rates are given in Table III.

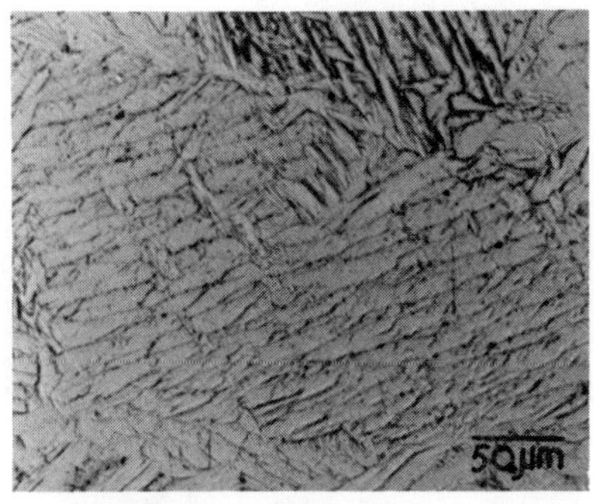

Tp, Tending to Parallel Plate
C+P+Si = 151 ppm; Oil Quench

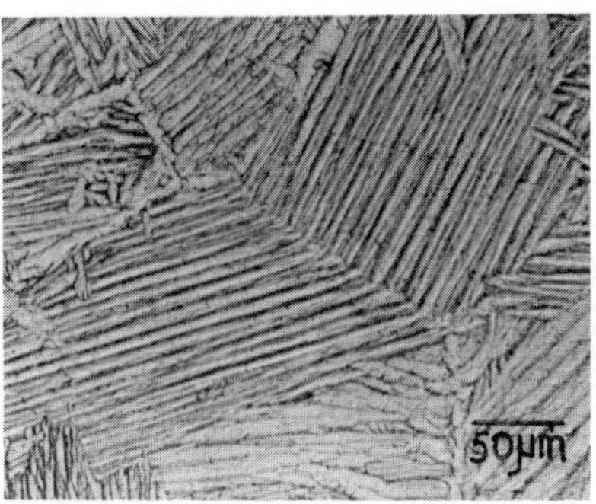

Pa, Parallel Plate
C+P+Si = 151 ppm; Water Quench

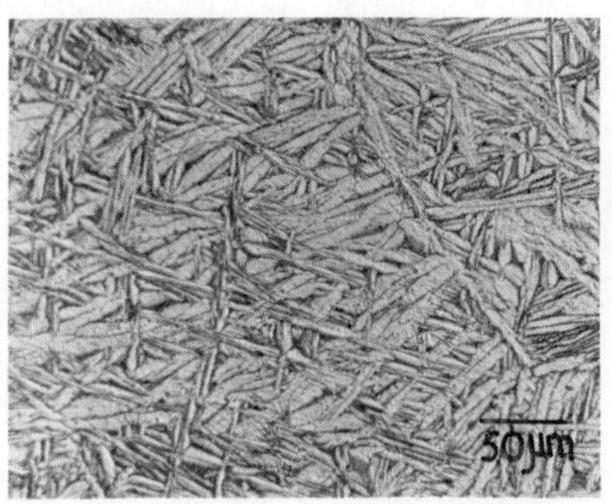

Tb, Tending to Basketweave
C+P+Si = 273 ppm; Oil Quench

B, Basketweave
C+P+Si = 393 ppm; Air Cooled

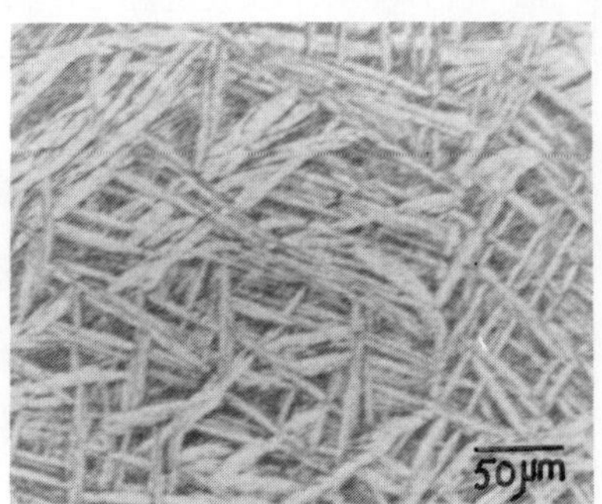

M, Martensite
C+P+Si = 151 ppm; Dry Ice-Acetone Quench

Figure 1 - The microstructure of Zircaloy-4 after a simulated brazing treatment.

Table II. A Summary of the Effect of the Impurity Content and Cooling Rate on the Percentage of the Microstructure in the Simulated Braze Heat-Affected Zone

Cooling Rate	Type of Microstructure (%)	A (151)	B (171)	C (181)	D (195)	E (231)	F (270)	G (273)	H (298)	I (322)	J (393)
		Tube Identification $\Sigma(C+P+Si)$									
Furnace Cooling (0.01°C/s)*	Tp	42	40	30	39	30	14	25	13	7	5
	Pa	15	12	18	12	17	15	10	9	6	6
	Tb	43	48	52	59	51	54	54	41	42	45
	B					2	17	11	37	45	44
Air Cooling (100°C/s)*	Tp	30	37	27	29	33	12	12	13		
	Pa	25	6	11	8	8	3	5			
	Tb	36	48	50	46	42	54	61			
	B	9	9	12	17	17	31	22	87	100	100
Oil Quench (750°C/s)*	Tp	36	31	32	28	30	10	16	11		
	Pa	10	9	8	3	9	5	3			
	Tb	45	50	43	49	58	52	30			
	B	9	10	17	20	3	33	51	89	100	100
Water Quench (5000°C/s)*	Tp	24	22	25	24	24	4	9			
	Pa	20	15	12	10	14	3	6			
	Tb	39	30	41	44	46	59	42			
	B	11	27	19	17	16	34	43	100	100	100
	M	6	6	3	5						
Brine Quench (5700°C/s)*	Tp	24	19	26	22	24	9	11			
	Pa	15	11	10	4	11	2	2			
	Tb	41	48	48	39	36	50	57			
	B	11	14	10	25	25	39	30	100	100	100
	M	9	8	6	10	4					
Dry-Ice Acetone Quench (7000°C/s)*	Tp	24	15	27	14	2					
	Pa	15	12	10	10	3					
	Tb	24	32	38	24						
	B	9	19	11	37	88	100	100	100	100	100
	M	28	22	14	15	7					

* Nominal cooling rates. These would be operator dependent.

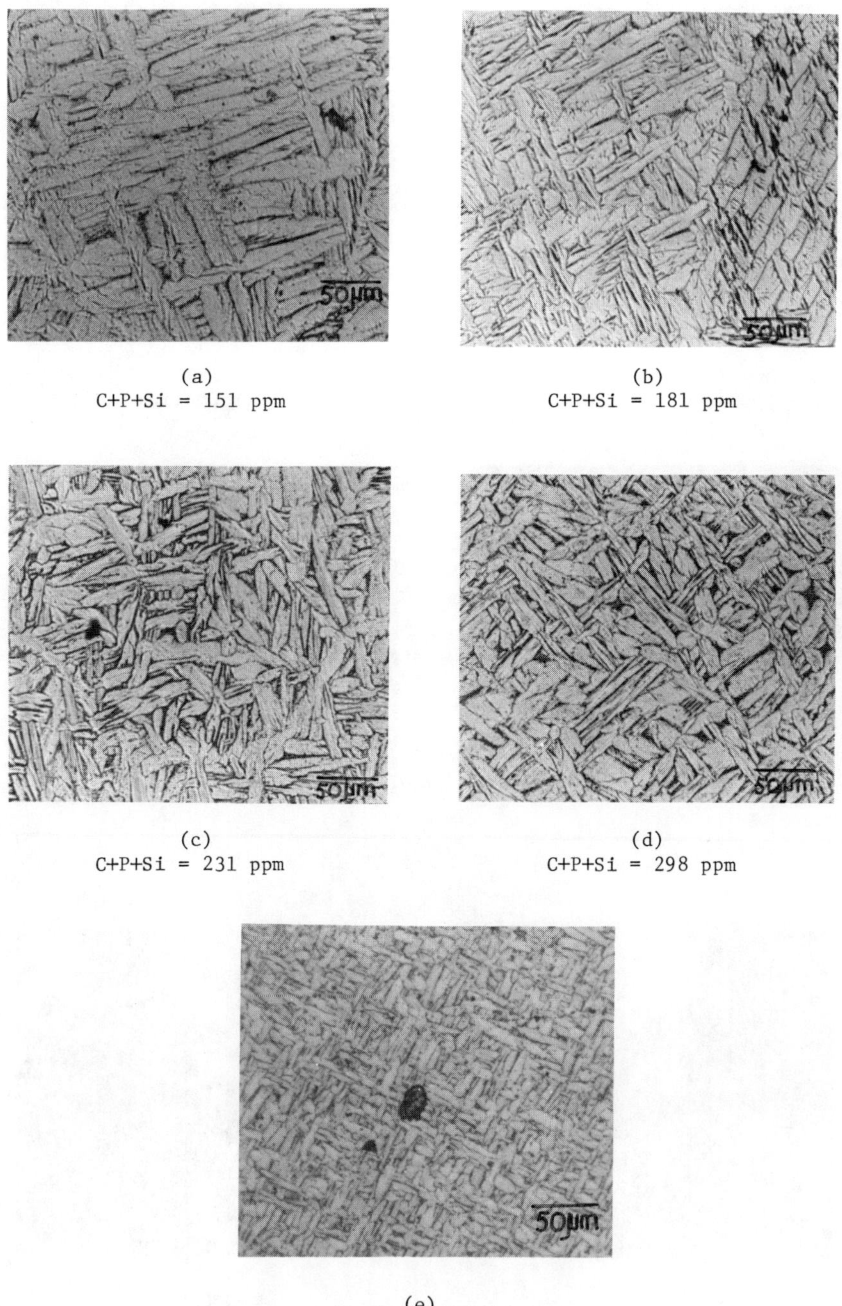

(a) C+P+Si = 151 ppm

(b) C+P+Si = 181 ppm

(c) C+P+Si = 231 ppm

(d) C+P+Si = 298 ppm

(e) C+P+Si = 393 ppm

Figure 2 - The influence of the impurity (C+P+Si) content on the basket-weave microstructure of Zircaloy-4 after air cooling.

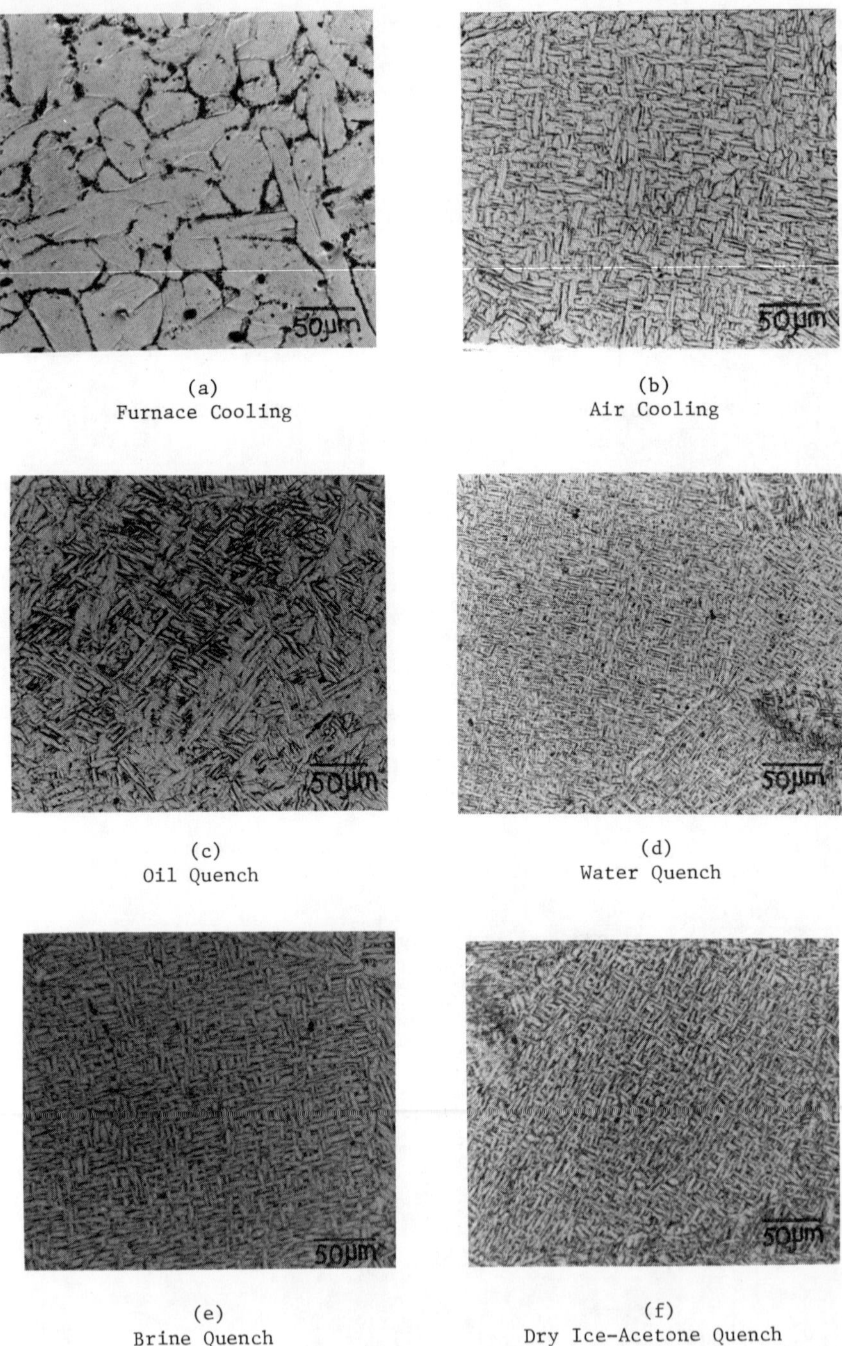

(a) Furnace Cooling

(b) Air Cooling

(c) Oil Quench

(d) Water Quench

(e) Brine Quench

(f) Dry Ice-Acetone Quench

Figure 3 - The influence of the cooling rate on the basketweave microstructure (B) of Zircaloy-4 in the braze heat-affected zone, $\Sigma(C+P+Si)$ = 393 ppm.

Table III. A Summary of the Quantitative Results for the Microstructural Features (GS,PW,PL) of Zircaloy-4 Fuel Cladding Subjected to a Simulated Braze Treatment

Cooling Rate	Structural* Features (μm)	Tube Identification $\Sigma(C+P+Si)$ A (151)	B (171)	C (181)	D (195)	E (231)	F (270)	G (273)	H (298)	I (322)	J (393)
Furnace Cooling ($0.01^{o}C/s$)	GS	475	458	464	441	447	428	435	399	417	393
	PW	53.8	42.3	50.6	44.0	41.9	32.5	36.3			
	PL	278	546	319	443	439	381	263			
Air Cooling ($100^{o}C/s$)	GS	386	367	372	357	361	338	350	325	328	316
	PW	14.8	11.3	14.4	14.1	13.3	9.5	10.9			
	PL	251	191	171	198	158	119	108			
Oil Quench ($750^{o}C/s$)	GS	384	357	363	351	358	334	344	313	322	310
	PW	9.9	7.3	9.2	9.1	8.1	6.3	6.4			
	PL	131	133	126	132	112	105	110			
Water Quench ($5000^{o}C/s$)	GS	381	350	353	348	351	325	335	303	318	293
	PW	8.3	6.5	7.7	6.8	6.7	5.0	5.5			
	PL	191	105	140	147	158	121	103			
Brine Quench ($5700^{o}C/s$)	GS	374	346	347	338	341	317	328	290	301	283
	PW	7.2	5.6	7.0	6.3	6.4	4.6	5.1			
	PL	119	145	191	124	154	100	80			
Dry Ice-Acetone Quench ($7000^{o}C/s$)	GS	362	344	345	332	338	306	326	287	290	278
	PW	6.4	4.5	6.2	6.1	3.1					
	PL	171	151	133	115	31					

* Grain Size Variations are ±10%; Plate Width and Plate Length Variations are ±25%.

The Empirical Equations

Empirical equations were derived using the SAS program and the quantitative data on the type of microstructure, GS, PW, PL given in Tables II and III. These equations are summarised in Table IV. The equations can be divided into two groups, referred to as the general group and the specific group, respectively.

The general group is a set of equations which include all of the data available, i.e., impurity content, cooling rate, type of microstructure, GS, PW, PL, and the six cooling rates. The specific group is a set of equations for the air cooled condition only.

Table IV The Empirical Equation of the Quantitative Data of the Microstructural Features of Zr-4 at the Brazing Affect Zone

Independent Variable*	Dependent Variable+	Empirical Equations	
C,P,Si,CR	B	B(%) =-102.13+0.72C-0.36P+0.26Si+0.004CR	(i)
	B+Tb	B+Tb(%)= - 8.44+0.41C-0.31P+0.24Si+0.001CR	(ii)
	GS	GS(μm) = 474.22-0.44C+0.09P-0.33Si-0.010CR	(iii)
	PW	PW(μm) = 34.29-0.05C-0.35P-0.02Si-0.003CR	(iv)
	PL	PL(μm) = 405.88-1.04C+0.84P-0.50Si-0.020CR	(v)
C,P,Si	B+Tb	B+Tb(%)=- 6.58+0.45C-0.32P+0.14Si	(vi)
	GS	GS(μm) = 432.38-0.39C+0.07P-0.30Si	(vii)
	PW	PW(μm) = 21.87-0.04C-0.37P+0.003Si	(viii)
	PL	PL(μm) = 450.87-1.54C-3.12P-0.34Si	(ix)

* C=Carbon, P=Phosphorus, Si=Silicon, CR=Cooling Rate.
\+ B=Basketweave, Tb=Tending to basketweave, GS=Grain Size, PW=Plate Width, PL=Plate Length.

NOTE: Eqs. (i)-(v) = General Group; Eqs. (vi)-(ix) = Specific Group.

Discussion and Implications

As is evident from the data presented in Tables II and III, the percentage of the 'basketweave-type' microstructure can be increased by increasing either the impurity (C+P+Si) content or the cooling rate or by decreasing the prior β grain size. The α-plates making up the Widmanstatten structure are finer and shorter the higher the impurity content or the faster the cooling rate. No basketweave (B) and 'basketweave-type' (B+Tb) structure would be formed without the presence of a certain level of impurities (Equations (i),(ii),(vi) of Table IV). Table IV indicates that phosphorus is a retarding factor for the formation of either a B- or a (B+Tb)-type structure. This is surprising since in an independent study where the (C+Si+O+N) content was kept low and constant and the phosphorus content varied from 42 ppm to 160 ppm, phosphorus promoted the formation of either a B or B+Tb structure (9). In commercial Zircaloy-4, the phosphorus content is relatively low compared with the carbon and silicon contents. As was illustrated in Figure 2, when the impurity content varied from 151 ppm (C+P+Si) to 393 ppm (C+P+Si), the microstructure changed in the sequence: Tp→Tb→B. When the impurity content is above about 300 ppm and the cooling rate is faster than furnace cooling, then the braze heat affected zone has a basketweave structure. This is the desired microstructure because of mechanical property and corrosion resistance considerations. When the cooling rate is increased from the presently used air cool to water quenching, brine quenching or dry ice-acetone quenching, the formation of a small percentage of the martensitic structure results. However, no martensitic structure was formed even at fast cooling rates when the impurity (C+P+Si) content is >270 ppm. As well as forming more basketweave structure a finer structure is obtained by increasing the cooling rate, e.g., in Figure 3, the α-plates become finer and more interweaved with increased cooling rate. An α-parallel-type plate is generated for the large prior-β grain sizes and the plates are longer and wider with decreasing impurity (C+P+Si) contents or cooling rates. The cooling rate has a stronger effect on the grain size and plate length (Equations (iii), (v), respectively) than on the other microstructural features.

In summary, the desired microstructure of Zircaloy-4 in the brazed heat-affected zone is a basketweave-type structure with a small prior-β grain size. This requirement is met by increasing either the impurity (C+P+Si) content or the cooling rate. Increasing the cooling rate beyond air cooling, is difficult given the present procedures for brazing. A better method to get the desired microstructural features in Zircaloy-4 would be to optimize the impurity content. If the total impurity content is kept slightly above 300 ppm and a cooling rate of air cooling or faster is employed, then a basketweave structure will be formed. Increasing the (C+P+Si) content much above 300 ppm should be avoided since this might lead to a general loss of ductility.

Conclusions

1. More basketweave and/or basketweave-type microstructure is formed when either the impurity (C+P+Si) content or cooling rate from the β-phase region is increased.

2. The grain size (GS), plate width (PW) and plate length (PL) decrease with increasing impurity (C+P+Si) content and cooling rate. A decrease in plate length with increasing Σ(C+P+Si) content also suggests an increasing incidence of basketweave structure.

3. If the impurity (C+P+Si) content is above about 300 ppm, the $\beta \rightarrow \alpha$ transformation gives a 100% basketweave structure.

4. Since it is impractical to increase the cooling rate to get the desired microstructural features, control of the impurity (C+P+Si) content is recommended.

Acknowledgements

One of the authors (D.O. Northwood) is grateful to the Natural Sciences and Engineering Research Council of Canada for the provision of a Research Grant (#A4391).

References

1. D.G. Hardy and N.A. Graham, "The Performance, Strength and Corrosion Resistance of Brazed Joints on CANDU Fuel Bundles," Transaction, American Nuclear Society, 19 (1974) pp. 122-123.

2. W.L. Fong and D.O. Northwood, "The Relationship Between the Impurities

Content and the Brazed Zone Microstructure in Zircaloy-4 Nuclear Fuel Sheathing," The Metallurgical Society of AIME, Paper Selection, A81-3 (1981) 8 pages.

3. R.A. Holt, W. Evans and B.A. Cheadle, "The Role of Zirconium Alloy Metallurgy in the Fabrication of CANDU Fuel," Atomic Energy of Canada Limited Report AECL-5107, June 1975.

4. G. Okvist and K. Kallstrom, "The Effect of Zirconium Carbide on the $\beta \to \alpha$ Transformation Structure in Zircaloy," Journal of Nuclear Materials, 35 (1970) pp. 316-321.

5. R.A. Holt, "The Beta to Alpha Phase Transformation in Zircaloy-4," Journal of Nuclear Materials, 35 (1970) pp. 322-334.

6. W.L. Fong and D.O. Northwood, "Identification of Second-Phase Particles in Zircaloy-4 Nuclear Fuel Sheathing," Metallography 15 (1982) pp. 27-41.

7. O.T. Woo and K. Tangri, "Transformation Characteristics of Rapidly Heated and Quenched Zircaloy-4-Oxygen Alloys," Journal of Nuclear Materials, 79 (1979) pp. 82-94.

8. J.B. VanderSande and A.L. Bement, "An Investigation of Second Phase Particles in Zircaloy-4 Alloys," Journal of Nuclear Materials, 52 (1974) pp. 115-118.

9. V. Quach, M.A.Sc. thesis, University of Windsor, Windsor, Ontario, Canada (1984).

OPTIMIZATION OF MARTENSITIC STAINLESS STEELS FOR NUCLEAR REACTOR APPLICATIONS

D. S. Gelles

Westinghouse Hanford Company
Richland, Washington 99352
USA

Summary

Martensitic stainless steels show great promise for structural components in high radiation damage regions of fission and fusion reactors. However, irradiation causes many microstructural changes in these materials and optimization for improved properties in irradiation environments should be possible. A brief review of the physical metallurgy of martensitic steels and a description of the effects of irradiation on their properties are provided as a basis for improving alloy response. Examples are given showing alloy design approaches for 1) concurrent toughness and creep rupture strength, 2) low activation alloys and 3) prevention of precipitate phases which form during irradiation.

Introduction

Martensitic stainless steels such as the Super 12Cr Steels (1) appear to be increasingly attractive materials both for in-core structural components in Liquid Metal Fast Breeder Reactors and for first wall materials in Fusion Reactors (2). For both applications (3,4) the commercial alloy Sandvik* HT-9, a Super 12Cr Steel, is being qualified as a potential structural material and a similar alloy Modified 9Cr-1Mo steel, now designated as T91, is being studied as an alternative because it exhibits several properties that are even better than those of HT-9. The process of qualifying a material for reactor applications requires years of testing and historically it has been difficult to make significant compositional modifications to a candidate alloy specification because it takes several additional years of testing to qualify the modification. Therefore, few attempts have been made to optimize martensitic steels for nuclear reactor applications. However, the U.S. Department of Energy's materials development program for fusion reactors, known as the Alloy Development for Irradiation Performance (ADIP) Program, has recently reached a crossroad. Martensitic stainless steels have been demonstrated to be strong candidate materials for fusion systems, however commercialization of such systems is not anticipated before the year 2020. The ADIP program has therefore resolved to begin a testing program to optimize the martensitic alloy class for fusion applications. This paper presents one viewpoint on the considerations which are most important in that optimization.

Physical Metallurgy

The physical metallurgy of martensitic stainless steels is described in detail elsewhere (see for example references 1 and 5). This section is intended to briefly review that information. Martensitic stainless steels are a unique alloy class which may be described as ferrous alloys containing approximately 8 to 12%Cr and 0.1-0.2%C (by weight). The chromium levels ensure good corrosion resistance and the carbon additions (along with other austenite formers such as nickel) are added in order to allow complete transformation to martensite at room temperature following an austenitization treatment. Formation of pearlite and bainite during cooling is very slow and therefore air cooling is sufficient. Following the martensite transformation, the material must be tempered in order to regain satisfactory ductility levels. Tempering precipitates carbides such as $M_{23}C_6$ and relaxes the dislocation structure. This behavior is shown by the example of the continuous cooling transformation diagram for a 12CrMoV steel in Figure 1. Note that the austenite revision temperature Ac_1 is 810°C, and the martensite start temperature M_S is 250°C. In general, molybdenum levels of 1% are employed for improved high temperature strength.

The compositional specifications must be set carefully for 12Cr-1Mo-0.2C martensitic stainless steels. Slight increases in chromium or molybdenum or slight decreases in carbon will result in incomplete austenitization because ferrite will remain stable at the austenite formation temperatures. For ferrite content levels of about 10 percent, strengths are reduced and weldability is degraded severely. The sensitivity to chromium content is demonstrated by the phase diagram for steels containing 0.1%C, Figure 2. As shown in Figure 2, Fe-Cr-0.1C alloys with Cr contents greater than 12 percent will always contain delta ferrite. Minor additions of N,

---

*HT-9 is a trademark of Sandvik AB, Sandviken, Sweden.

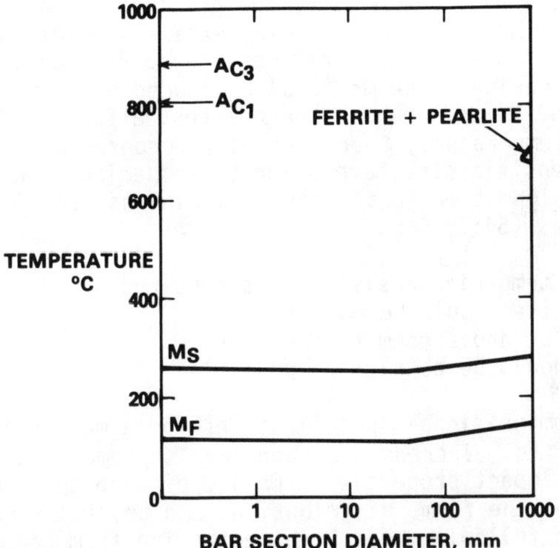

Figure 1 - Continuous cooling transformation diagram for air cooled 12CrMoV steel (after reference 6). Ac_1 and Ac_3 define the austenite reversion temperature and the temperature at which the austenite transformation is complete, respectively. M_S is the martensite start temperature and M_F the temperature at which the martensite transformation is complete.

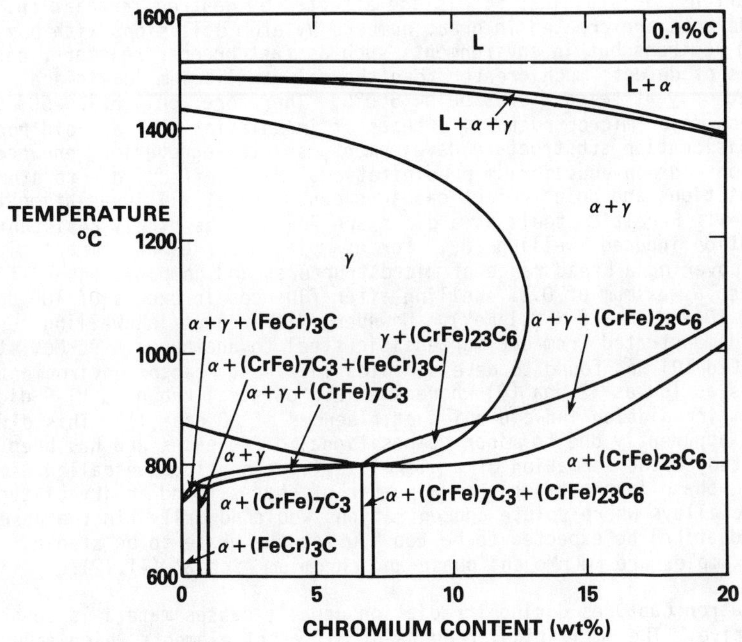

Figure 2 - Fe-Cr-0.1C phase diagram (after reference 7).

C, Ni, V and Al will alter the delta ferrite content greatly (5). Minor element additions will also change Ac_1, the greatest effect being due to V, Ni, Mo, Al, Si and Mn (5). For example, excessive additions of nickel will require reduced tempering temperatures in order to avoid austenite formation during tempering. The M_S is also reduced by minor element additions; C, Mn, Mo, Ni and Cr have the greatest effect. If the M_F is reduced below room temperature, then special procedures will be required to get a fully martensitic structure prior to tempering. The design criteria for 12%Cr high strength structural stainless steels have been summarized as follows (5):

1) a maximum tempering resistance is required
2) delta ferrite should be absent
3) M_F should be above room temperature
4) the Ac_1 should be as high as possible.

Selection of compositional specifications for a martensitic steel involves many tradeoffs. Increased carbon levels promote high temperature strength but reduce impact properties. Choice of carbide forming minor additions can control the form, distribution, and crystal structure of the carbide precipitates following heat treatment. The form can vary from large blocky MC particles in the case of Ti and Nb additions, to fine MC for V additions, fine M_2X precipitates for N additions and blocky $M_{23}C_6$ for W additions. As a consequence of the many choices possible a large number of martensitic stainless steels have been produced commercially (1). However, it is not yet clear which of these choices provides an optimum composition for irradiation environments.

Effects of Irradiation on Properties

Martensitic stainless steels are affected by neutron irradiation. Point defects are created in great numbers by atom collisions with bombarding neutrons but in environments such as fast breeder reactors, concentrations of defects much greater than thermal equilibrium levels are retained only at temperatures below 500°C. Therefore, only below 500°C do we expect major microstructural effects of irradiation such as void formation, dislocation substructure development, solute segregation, enhanced diffusion and non-equilibrium precipitation. Minor effects due to atomic transmutations and point defect cascades can occur at all irradiation temperatures. Ferritic steels as a class are found to be highly resistant to irradiation induced swelling (8). For example, a series of commercial alloys covering a broad range of microstructures and compositions (2-17%Cr) developed a maximum of 0.6% swelling after fluences in excess of 100 dpa (2.07×10^{23} n/cm^2, E > 0.1 MeV). However, differences in swelling response can be demonstrated from one martensitic steel to another. A 9CrMoV steel designated 191 was found to develop voids in a fast reactor environment at fluences as low as 12 dpa (9) whereas a 12CrMoW steel known as HT-9 did not form irradiation induced voids at fluences of 60 dpa (1). This difference is apparently due to minor compositional differences and has been attributed to the formation of a nickel silicide precipitate called G-phase (10). G-phase is one example of a number of phases found in irradiated ferritic alloys where solute concentrations would normally (in the absence of irradiation) be expected to be too low for the phase to be stable. Other examples are FeCrMo chi phase and chromium rich α' (11,12).

Neutron captures during irradiation usually causes materials to become radioactive. The activation products of different elements which make up an alloy decay with various types of emitted radiation and different half-lives. Therefore, suitable choice of alloy composition can minimize the time during which the alloy retains a radioactivity above a minimum level. Suggested guidelines for low activation composition limits have been established (13,14). A low activation martensitic stainless steel having composition within these guidelines should be possible.

Neutron irradiation produces changes in properties. Martensitic steels show hardening below ∿450°C and softening above ∿450°C following irradiation to ∿30 dpa in a fast reactor (11,15). Hardening is concurrent with substantial upward shifts in the ductile to brittle transition temperature (DBTT). Shifts of ∿120°C have been measured for HT-9 following irradiation at 400°C (16,17). However, the shift in DBTT is apparently very sensitive to composition because measurements on a 9Cr-1Mo martensitic steel irradiated under identical conditions showed a shift of only 55°C (17). (The DBTT prior to irradiation was -25°C for 9Cr-1Mo and 5°C for HT-9). Creep response is also affected. At temperatures below 500°C, the creep rate of HT-9 in a fast reactor is found to be athermal (18,19). As out-of-reactor creep is thermally activated, it is apparent that irradiation enhances creep at low temperatures. However, differences in irradiation creep rate of ferritic steels due to composition variations appear to be small (19).

Optimization for Enhanced Toughness and Rupture Strength

Simultaneous optimization of toughness and rupture strength is generally difficult. The precipitate structure which promotes improvements in rupture strength provides nucleation sites for brittle failure at low temperatures and degrades toughness. Therefore, optimization for both properties requires that the strengthening phase for high temperature behavior be distributed as finely as possible (and be stabilized for long service times). The Modified 9Cr-1Mo martensitic stainless steel, T91, provides an excellent example where this approach has been used (20). Table I contains the chemistry specifications for T91 and the standard 9Cr-1Mo steel (20). Comparison of these two specifications shows that T91 has significantly lower C and Si and controlled levels of V, Nb and N not specified in the standard alloy. Refinement of the precipitate structure is achieved by reducing the carbon content, (adding N as a substitute) and adding the carbide formers niobium and vanadium, to promote formation of MC carbide. The V and Nb additions retard the coarsening kinetics of the $(CrMo)_{23}C_6$ phase. Niobium also controls grain growth during austenitizing because Nb carbide dissolves slowly at austenitizing temperatures. A further insight can be found on this composition specification from publications by Fujita and coworkers. They found that for creep rupture strength the optimum (V + Nb)/C ratio in 10Cr-2Mo-V-Nb alloy is 0.7 where concentrations are in atom percent (21). The T91 specification for this ratio is 0.62. They also found that for optimum long term creep strength in a 10Cr-2Mo alloy, the ratio of V/Nb should be ∿2.0 at 600°C and ∿1.5 at 650°C where concentrations are in weight percent (22). However, their optimized alloy uses a V/Nb ratio of 4.9 (23). The T91 specification gives 2.75. Therefore, it can be concluded that the T91 specification comes close to optimizing the levels of V and Nb for creep rupture strength.

Concurrently, the alloy T91 has also been optimized for toughness. The carbon content has been kept low and the carbide formers which do not readily go into solid solution during austenitization are eliminated (except for Nb). This minimizes the amount of carbide present and keeps its distribution uniformly fine. Fine prior austenite grain size is maintained and the alloy contains no delta ferrite. This avoids heavy carbide concentrations at grain boundaries. All of the above restrictions tend to minimize the propensity for crack nucleation at carbide particles

Table I. Chemical Analysis of Modified 9Cr-1Mo Steel and Standard 9Cr-1Mo Steel (from Reference 20)

Element	Content Range, weight %	
	Modified 9Cr-1Mo (Grade 91)	Standard 9Cr-1Mo (Grade 9)
Carbon	0.80-0.12	0.15 max
Manganese	0.30-0.60	0.30-0.60
Phosphorus	0.020 max	0.030 max
Sulfur	0.010 max	0.030 max
Silicon	0.20-0.50	1.00 max
Chromium	8.00-9.50	8.00-10.00
Molybdenum	0.85-1.05	0.90-1.10
Nickel	0.40 max	
Vanadium	0.18-0.25	
Niobium	0.06-0.10	
Nitrogen	0.030-0.070	
Aluminum	0.04 max	

during low temperature impact testing. At the same time other factors which degrade toughness are controlled. Hardening is nominal and the potentially detrimental segregant phosphorus is limited. There may be further improvements imparted by the nickel content restriction but the reason for the improvement is not yet understood.

It can be concluded that the basic approach needed to optimize both toughness and rupture strength involves reducing the volume fraction and coarseness of the carbide strengthening precipitate phases, removing elements which promote temper embrittlement and optimizing creep rupture properties within those guidelines. It is apparent that T91 is a successful example of this approach for designing an alloy with improved properties (20).

Optimization for Low Activation

The composition limits required to provide a structural material which has sufficiently low radioactivity 80 years after irradiation in the design study of the proposed Starfire Fusion Reactor first wall are given in Table II (13). Of the elements listed in Table II, only the limits on Nb, Mo, N and perhaps Ni will impose restrictions on martensitic stainless steel alloy design. A limit on niobium below the 3 ppm range will be very difficult to achieve on a commercial scale; other limits should be feasible. If niobium is not available, strength optimization based on $(V,Nb)_4C_3$ stability such as that used in T91 would not be possible. If molybdenum is not used (it may be possible to use molybdenum if the isotopes ^{92}Mo and ^{94}Mo are reduced) an alternative solid solution hardening agent will be needed. Possibilities include W and V but there is no agreement on the exact levels needed for equivalent strength. For example, Japanese technology recommends a one for one substitution of W for Mo in weight percent but U.S. technology recommends a two for one substitution. Limits on N and Ni additions are well above standard levels in martensitic stainless steels but alloy design options for high N or Ni contents would not be allowed. Therefore, a low activation martensitic stainless steel should be viable and experiments have been initiated to test the irradiation resistance of such alloys.

Table II. Estimated Composition Limits (in weight percent) Needed to Meet Class C Disposal Requirements Following 10MW-y/m^2 Exposure at the Starfire First Wall (from Reference 13)

Element	Limit, %
Niobium	<0.00029
Molybdenum	<0.0030
Copper	<0.12
Nitrogen	<0.33
Nickel	<0.91
Lead	<3.1
Aluminum	<3.6

Optimization for Service in Irradiation Environments

It appears possible to optimize martensitic stainless steels specifically for service in radiation environments. This claim extends beyond restrictions needed for low activation. As noted, irradiation promotes the formation of several precipitate phases which ordinarily would not be present in martensitic stainless steels. Those identified so far are G-phase (a nickel silicide), chi (an FeCrMo intermetallic phase) and α' (a Cr rich BCC phase). Ni and Si are also readily absorbed into carbide phases during irradiation. If these phases are detrimental, optimization can be accomplished by reducing key elements from which they are formed. Suppression of G-phase should be possible by removal of Ni or Si but Ni appears to be the better choice. Suppression of chi phase must be accomplished by reduction of Mo; as Cr and Fe are the principal ingredients of martensitic stainless steels. Because chi phase can also form from FeCrW (24) a direct substitution of W for Mo will not prevent chi phase formation. It has been proposed that Ni and Si replace Mo to some extent (25). The consequences of V additions on chi phase have not yet been determined. A simple approach for removal of α' apart from reduction of Cr is not available. However, it is not clear that this is necessary. Effects on properties due to α' are only found to be serious for chromium levels above 12 percent.

A composition specification for an optimum martensitic stainless steel for irradiation service cannot be given at this time. It will take several years to complete the irradiation testing program designed to demonstrate irradiation resistance improvements.

Conclusions

Several approaches offer promise for optimizing martensitic stainless steels for nuclear reactor applications. Optimization of creep rupture strength concurrent with improved toughness can be achieved by careful choice of minor element carbide formers (V and Nb). Attainment of a low activation martensitic steel should be possible by careful elimination of Nb impurities, by replacement of Mo with ^{92}Mo and ^{94}Mo-free Mo or V or W, and by elimination or control of N, Ni and Cu. Irradiation produced precipitates, such as G-phase, chi and α', can be minimized by reductions in the Ni, Si, Mo and Cr levels. However, the demonstration of successful optimization will require several years of testing.

References

1. J. Z. Briggs and T. D. Parker, The Super 12%Cr Steels; Climax Molybdenum Company, New York, NY, 1965.

2. D. S. Gelles, "12Cr-1Mo Steel for Fission and Fusion Applications," pp. 197-209 in Ferritic Steels for High-Temperature Applications, Ashok Khare, ed.; ASM, Metals Park, OH, 1983.

3. J. L. Straalsund and D. S. Gelles, "Assessment of the Performance of the Martensitic Alloy HT-9 for Liquid Metal Fast Breeder Reactor Applications," HEDL-SA-2771, paper presented at the AIME Topical Conference on Ferritic Alloys for Use in Nuclear Energy Technologies, June 19-23, 1983, Snowbird, UT.

4. Alloy Development for Irradiation Performance progress reports DOE/ER-0045/1 through 11, US Department of Energy, issues beginning December 1979 to present.

5. F. B. Pickering, Physical Metallurgy and the Design of Steels; Applied Science Publishers Ltd, London, England 1978, pp. 163-182.

6. J. Orr, F. R. Beckitt and G. D. Fawkes, "The Physical Metallurgy of Chromium-Molybdenum Steels for Fast Reactor Boilers," pp. 91-109, in Ferritic Steels for Fast Reactor Steam Generators, S. F. Pugh and E. A. Little, eds.; British Nuclear Energy Society, London, England, 1978.

7. K. Bungardt, E. Kunze and E. Horn, "Untersuchungen über den Aufbau des Systems Eisen-Chrom-Kohlenstaff," Arch Eisenhüttenwes, 29 (1958) pp. 193-203.

8. D. S. Gelles, "Swelling in Several Commercial Alloys Irradiated to Very High Neutron Fluence," Journal of Nuclear Materials, 122 & 123 (1984) pp. 207-213.

9. D. S. Gelles and L. E. Thomas, "Microstructural Examination of HT-9 and 9Cr-1Mo Contained in the AD-2 Experiment," in Alloy Development for Irradiation Performance Semiannual Progress Report for Period Ending March 1982, DOE/ER-0045/8, 1982, pp. 343-361.

10. D. S. Gelles and L. E. Thomas, "Effects of Neutron Irradiation on Microstructure in Experimental and Commercial Ferritic Alloys," pp. 559-568 in Proceedings of Topical Conference on Ferritic Alloys for Use in Nuclear Energy Technologies, J. W. Davis and D. J. Michel, eds; AIME, New York, N. Y., 1984.

11. E. A. Little and L. P. Stoter, "Microstructural Stability of Fast Reactor Irradiated 10-12%Cr Ferritic-Martensitic Stainless Steels," pp. 207-233, in Effects of Radiation on Materials: Eleventh Conference ASTM STP 782, H. R. Brager and J. S. Perrin, eds.; Philadelphia, PA, 1982.

12. D. S. Gelles, "Microstructural Examination of Several Commercial Ferritic Alloys Irradiated to High Fluence," Journal of Nuclear Materials, 103&104 (1981) pp. 975-980.

13. F. M. Mann, "Reduced Activation Calculations for the Starfire First Wall," Fusion Technology, 6 (1984) pp. 273-287.

14. G. J. Butterworth and O. N. Jarvis, "Comparison of Transmutation and Activation Effects in Five Ferritic Alloys and AISI 316 Stainless Steel in a Fusion Neutron Spectum," Journal of Nuclear Materials, 122 & 123 (1984) pp. 982-988.

15. T. Lauritzen and S. Vaidyanathan, "Effects of Irradiation on the Mechanical Properties of Ferritic Alloys HT-9 and 2-1/4Cr-1Mo," ibid reference 10, pp. 623-630.

16. F. A. Smidt, Jr, J. R. Hawthorne and V. Provenzano, "Fracture Resistance of HT-9 After Irradiation at Elevated Temperature," pp. 269-284, in Effects of Radiation on Materials: Tenth Conference, ASTM STP 725, D. Kramer, H. R. Brager and J. S. Perrin, eds.; Philadelphia, PA, 1981.

17. W. L. Hu and D. S. Gelles, "Miniature Charpy Impact Test Results for the Irradiated Ferritic Alloys HT-9 and Modified 9Cr-1Mo," ibid reference 10, pp. 631-645.

18. B. A. Chin, "An Analysis of the Creep Properties of a 12Cr-1Mo-W-V Steel," ibid, pp. 593-599.

19. R. J. Puigh and G. L. Wire, "In-Reactor Creep Behavior of Selected Ferritic Alloys," ibid, pp. 601-606.

20. V. K. Sikka, "Development of Modified 9Cr-1Mo Steel for Elevated-Temperature Service," ibid, pp. 317-327.

21. T. Fujita, K. Asakura and T. Sato, "Development and Properties of new 10Cr-2Mo-V-Nb Heat Resisting Steel," Transactions of the Iron and Steel Institute Japan, 19 (1979) pp. 605-613.

22. T. Fujita and N. Takahashi, "The Effects of V and Nb on the Long Period Creep Rupture Strength of 12%Cr Heat Resisting Steel Containing Mo and B," Transactions of the Iron and Steel Institute Japan, 18 (1978) pp. 269-278.

23. I. M. Park, T. Fujita and K. Asakura, "Microstructure and Creep Rupture Properties of a Low Si-12Cr-Mo-V-Nb Steel," Transactions of the Iron and Steel Institute Japan, 20 (1980) pp. 99-107.

24. H. J. Goldschmidt, Interstitial Alloys, Plenum Press, p. 128, New York, NY, 1967.

25. K. W. Andrews, written discussion to "Equilibrium Structures in Fe-Cr-Mo Alloys," Transactions of the American Society for Metals, 46 (1954) pp. 808-810.

THE DEVELOPMENT OF AUSTENITIC STEELS FOR FAST INDUCED-RADIOACTIVITY DECAY FOR FUSION REACTOR APPLICATIONS*

R. L. Klueh and E. E. Bloom
Metals and Ceramics Division, Oak Ridge National Laboratory
P. O. Box X, Oak Ridge, Tennessee 37831

Summary

Austenitic stainless steels (primarily type 316 and variations on that composition) are leading candidates for the structural components for future fusion reactors. However, irradiation of such steels in a fusion environment produces long-lived radioactive isotopes. These isotopes lead to difficult radioactive waste disposal problems once the structure is removed from service. Such problems could be reduced by developing steels that contain only those elements that produce radioactive isotopes that decay to low levels in a reasonable time (tens of years instead of hundreds or thousands of years). This paper discusses the development of such austenitic steels by making elemental substitutions in the steels now under consideration. Nickel is the most important element that would have to be replaced (molybdenum would have to be replaced; the nitrogen concentration would have to be limited, and the niobium maintained to extremely low levels). Manganese appears to be an appropriate substitution for nickel. Because manganese is not as effective an austenite-stabilizing element as nickel, a simple one-to-one substitution is not possible. Therefore, the first step in the alloy-development program should be the determination of a stable austenitic Fe-Cr-Mn-C composition, after which alloying additions can be made to improve strength and irradiation resistance.

---

*Research sponsored by the Office of Fusion Energy, U.S. Department of Energy under Contract No. DE-AC05-84OR21400 with Martin Marietta Energy Systems, Inc.

Introduction

The most serious safety and environmental concerns for fusion reactors involve induced radioactivity in the first-wall and blanket structures. Public safety could be jeopardized by the accidental release of this induced radioactivity, and the biological hazard to plant personnel would eliminate the possibility of contact maintenance and repair. Another problem is that of special waste storage of the highly radioactive blanket and first-wall structures after service. All these problems should be alleviated by the use of a low-activation structural material. However, as pointed out in a recent report by a U.S. Department of Energy (DOE) panel set up to study this subject, the technology for commercially producing and fabricating material that would meet both the low-activation criteria and engineering properties requirements is not available and is unlikely to be available soon (high-purity silicon carbide was the only material suggested) (1). Although other solutions must be sought for the safety and maintenance problems, it should be possible to simplify the disposal of radioactive reactor components after service.

Radioactive Waste and Storage Guidelines

Guidelines (10 CFR Part 61) for the classification and disposal of low-level nuclear wastes have been issued by the U.S. Nuclear Regulatory Commission (2); these guidelines are summarized in Table I. The DOE Panel on Low Activation Materials set as a goal the development of reactor materials that fall within Class C, but with the additional hope of meeting the more stringent Class B criteria (1). It should be noted that the 10 CFR Part 61 guidelines were developed primarily to treat fission reactor waste and will undoubtedly be extensively revised by the time the first components

Table I. Nuclear Waste Classification and Storage Under Proposed 10 CFR Part 61 Rules

Waste class	Definition	Disposal
Class A, segregated	Decays to acceptable levels during site occupancy	
Class B, stable	Decays within 100 years to levels not dangerous to public health and oafety	Covered to reduce surface radiation to a few percent of background
Class C, intruder	Decays to acceptably safe levels in more than 100 years but less than 500 years	At least 5 m below the surface, with natural or engineered barriers
Waste not meeting Class C intruder waste definition	Decays in more than 500 years	Does not qualify for near-surface disposal; proposed methods will be considered on a case-by-case basis

are discharged from an operating fusion reactor and are ready for disposal. Nevertheless, these guidelines offer a standard to which alloys can be developed for eventual inexpensive near-surface disposal.

The 10 CFR Part 61 guidelines were examined by the DOE panel (1) and more recently by Wiffen and Santoro (3). On the basis of the guidelines, initial concentration limits were calculated for various common alloying elements for the three waste classes for first-wall and blanket structures ten years after the shutdown of a reactor given a 9 MW•years/m^2 exposure. The limits are given in Table II. The limiting values are those for a given element when no other restricted elements are present in the alloy. If other restricted elements are present, the "sum of fractions" rule applies, and the concentration of each restricted element must be divided by its allowable concentration and the fractions thus obtained for all restricted elements added. The sum of these fractions must be less than unity.

Based on the possibility that a material resulting in Class A waste is quite remote (e.g., impurity-free silicon carbide), vanadium alloys (alloyed with chromium and titanium) offer the possibility for Class B waste (residual impurities would appear to make Class A vanadium alloys impossible). It appears that the steels discussed in this paper will have to be handled as Class C waste (1,3).

Table II shows that, for the induced radioactivity of a steel to decay rapidly enough to qualify for Class C, certain common steel alloying elements must be restricted. In particular, niobium must be eliminated (it is the restriction on niobium that will probably make it impossible ever to meet the Class B criteria for any steel). Concentrations of Ni, Mo, and N

Table II. Initial Concentration Level Restrictions from 10 CFR Part 61 Waste Disposal Rules

Element	Initial concentration limit* (at. ppm)		
	Class A	Class B	Class C
N	†	365	3,650
O	250,000	†	10^6
Co	30	10^6	10^6
Cu	12	240	2,400
Fe	350	35,000	10^6
Ni	100	2,000	20,000
Mo	365	†	3,650
Mn	5,000	10^6	10^6
Nb	0.1	†	1
Al, C, Mg, Si, Ti, V	10^6	10^6	10^6

Source: R. W. Conn et al., *Panel Report on Low Activation Materials for Fusion Applications*, UCLA Report PPG-728, University of California at Los Angeles, June 1983.

*Limits apply to first-wall region ten years after shutdown, 9 MW•years/m^2 exposure.

†Storage in this classification is not defined.

must be severely restricted in the steels being considered for first-wall and blanket structure applications. Copper, generally present as an impurity, will also have to be controlled.

In this paper, the development of austenitic steels with "fast" induced-radioactivity decay (FIRD) characteristics are discussed. *Fast* is a relative term, which for this discussion is taken to mean steels with radioactivity decay rapid enough to qualify at least for the Class C waste disposal designation. The term *low-activation material* is often used to describe alloys that would allow hands-on maintenance as well as those that minimize waste disposal problems. Materials that would allow hands-on maintenance would not only have to have low activation, but would also have to decay to extremely low levels very rapidly (e.g., pure SiC).

The DOE panel called attention to the difficulty of defining low activation materials by defining four types of fusion reactors and materials (1). It defined standard activation, very low activation, low activation, and reference activation reactors and materials.

A standard-activation reactor would be constructed of a standard-activation material such as type 316 stainless steel, which does not meet the criteria for shallow land burial of the radioactive waste after reactor shutdown. For such reactors, all maintenance must be performed remotely.

A very-low-activation reactor would be constructed from a very-low-activation material that would be disposable according to the Class A criteria of 10 CFR Part 61 and would allow limited hands-on maintenance behind the blanket within 2 d of shutdown.

The low-activation reactor would be constructed of materials that met the Class B criteria; hands-on maintenance would be possible outside the blanket or outside the shield within 2 d of shutdown.

Finally, a reference-activation reactor would be constructed of a reference-activation material that met Class C criteria; such a material would allow limited hands-on maintenance outside the shield within 2 d of shutdown. The FIRD austenitic steels to be discussed here fall into the class of reference-activation materials.

As stated above, the limits on the initial concentrations of Ni, Mo, Cu, and N (Table II) established by the 10 CFR Part 61 guidelines are mutually exclusive (3), and if any one of the elements is present at the concentration limit, the other elements must be absent. The 1 at. ppm limit for niobium will be the most difficult to meet. Because of this restriction on the niobium, the alloys discussed below will be those for which Mo, Ni, Cu, and N will be kept to a minimum.

The austenitic stainless steels such as type 316 are the leading candidate alloys for fusion reactor structures. The effect of irradiation on such steels has been extensively studied in both the fusion reactor and the breeder reactor development programs. Furthermore, an extensive background of experience is available to draw on for fabricating large and complicated stainless steel structures for elevated-temperature service. Because of this background of experience with such steels, these materials are logical choices for modification to develop FIRD alloys. In the following we discuss possibilities for such a modification and for developing FIRD austenitic stainless steels similar to type 316 stainless steel and the austenitic prime candidate alloy (PCA) developed in the fusion program. The modification will involve the substitution of elements for those that are restricted by 10 CFR Part 61. The initial objective will be the establishment of an austenite-stable base composition that can then be further alloyed for the required elevated-temperature strength, irradiation-damage resistance, and compatibility with potential coolants and breeding materials.

FIRD Austenitic Stainless Steels

Type 316 stainless steel and modifications of that basic alloy composition (e.g., by the adjustment of chromium and nickel contents and the addition of titanium) are the primary austenitic stainless steels under consideration for fusion reactor structural applications (4). In the alloy development approach to be followed here, we will describe a program to develop an austenitic steel with properties similar to those of the PCA, which was developed as a variation of type 316 stainless steel (4). Of the elements contained in PCA and type 316 stainless steel, nickel and molybdenum are not acceptable in a FIRD alloy.

The replacement of nickel is of most importance, because nickel stabilizes the austenite. Common steel alloying elements that can be used in FIRD steels include Mn, Ti, Cr, Si, W, V, Ta, Co, and C. Manganese, like nickel, is an austenite-forming element and has often been used as a replacement for nickel (5–9). However, the austenite-forming tendency of manganese is considerably less than that of nickel, and the development of an Fe-Cr-Mn stainless steel does not follow by simply replacing nickel with manganese.

Other common austenite-forming alloying elements include C, N, Cu, and Co. The carbon concentration will be maintained at 0.05 to 0.1% to minimize welding and corrosion problems. To meet the 10 CFR Part 61 waste storage criteria for Class C waste, the nitrogen and copper concentrations will have to be kept to a minimum (Table II).

Cobalt presents an interesting possibility. As an austenite stabilizer, it is as strong as nickel (10). Its radioactive decay characteristics are such that it is not forbidden in Class B or Class C waste by 10 CFR Part 61 (2). However, immediately after irradiation, steels containing cobalt emit high-energy gamma rays that make the steels difficult to handle. Thus, even though cobalt could increase the possibility of developing a FIRD stainless steel, it would make irradiation studies extremely difficult or impossible. For that reason, cobalt will be considered an unsuitable alloy addition for this discussion.

Alloy Composition Selection

The idea of using less expensive manganese to replace nickel in austenitic stainless steels has appeal; it has been investigated quite extensively, mainly in the 1930s and 1940s (5–9). Although manganese is used to replace some nickel in the 200-series stainless steels, the lower austenite-forming capability of manganese is offset by the addition of up to 0.25% N. The only Cr-Mn stainless steels that appear to have been used commercially were used in Germany (5). These were an 8 Cr-18 Mn and an 18 Cr-8 Mn alloy, both containing 0.1% C. The first alloy was entirely austenite (because of the low chromium concentration), but the second contained about 40% ferrite. The low chromium concentration of the first implies relatively low oxidation and corrosion resistance, and the high ferrite content of the second means it will be prone to sigma phase formation; both are reasons for making an alloy unsuitable for elevated-temperature service. Another reason why the development of manganese-stabilized (nickel-free and low-nitrogen) stainless steels has never been actively pursued is that the increased manganese leads to decreased corrosion resistance to sulfuric, hydrochloric, and nitric acids (11,12).

More recent studies have been published on the effect of manganese on mechanical properties and as an austenite stabilizer (9,11–15). Manganese lowers the stacking-fault energy and therefore increases the rate of work hardening (12). This could lead to problems in fabrication. It also lowers the thermal conductivity relative to nickel (11), a disadvantage for fusion reactor applications. The reduced oxidation resistance of Cr-Mn steels at elevated temperatures should not prove limiting for fusion reactors if the maximum temperature is limited to about 600°C.

Recently, Desforges and Dancoisne investigated a large number of manganese-containing ferrous alloys (15). Because of the low austenite-stabilizing capacity of manganese, most of the alloys proved to be magnetic (i.e., contained some ferrite or martensite), including an Fe–14% Cr–15% Mn steel. Of the 18 manganese-containing alloys melted and cast, only three were nonmagnetic. One of these contained 5% Ni, another 2% Cu. Because copper is generally considered to be a weaker austenite stabilizer than is manganese, the copper could presumably be replaced by manganese and retain the austenite microstructure. However, this Fe–11% Cr–15% Mn–2% Cu alloy suffered from hot shortness when worked. The only other nonmagnetic alloy was an Fe–8% Cr–2% Al–15% Mn steel, which had reasonably good oxidation resistance to 600°C. It had excellent tensile properties relative to the other steels when tested in the cold-rolled condition. After a solution anneal at 900 or 1100°C, it was one of the weakest as measured by the 0.2% offset yield stress; the ultimate tensile strength was similar to that for most of the other steels.

Although the early work on the ternary phase diagrams for Fe-Cr-Mn would indicate that a steel containing 15% Cr and 15% Mn would be entirely austenite, but would be near the alpha-gamma phase boundary, little guidance is available when other elements such as carbon are added (5–9). A common method used to represent the phases present in Cr-Ni stainless steels is a Schaeffler diagram (Fig. 1), in which the phase fields expected at room temperature are shown in terms of nickel and chromium equivalents (10). The nickel and chromium equivalents have been determined empirically for the most common alloying elements, including manganese (10). The equations that apply to Fig. 1 are given by

$$\text{Cr equiv} = (\text{Cr}) + 2(\text{Si}) + 1.5(\text{Mo}) + 5(\text{V}) + 5.5(\text{Al}) + 1.75(\text{Nb}) + 1.5(\text{Ti}) + 0.75(\text{W}) \quad (1)$$

$$\text{Ni equiv} = (\text{Ni}) + (\text{Co}) + 0.5(\text{Mn}) + 0.3(\text{Cu}) + 2.5(\text{N}) + 30(\text{C}) \quad (2)$$

where ( ) is the concentration expressed in weight percent. According to Eq. (2), manganese is only half as effective as is nickel in stabilizing austenite.

The chromium and nickel equivalents were calculated for several alloy compositions (Table III), and the position of these steels is shown in Fig. 1. Of interest is the fact that the 15-15 steel (15% Cr–15% Mn–0.1% C) and the 12-15 steel (12% Cr–15% Mn–0.1% C) are in the austenite-plus-martensite field. The phase diagram indicates that such alloys should be

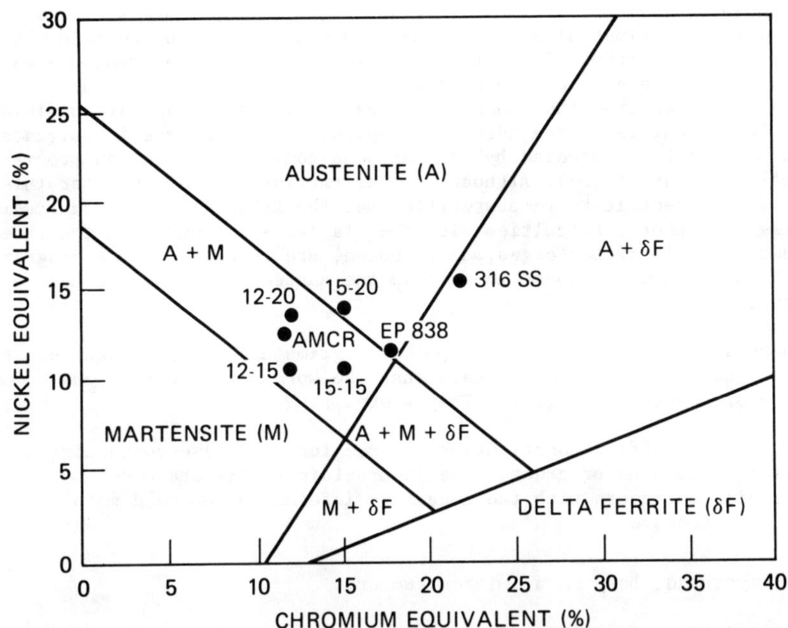

Figure 1 - Schaeffler diagram with points for various steels superimposed. For the double numbers the first number is the weight percent chromium, and the second is the weight percent manganese (e.g., 12-15 is 12% Cr and 15% Mn); all are assumed to have 0.1% C. Alloy AMCR is 10% Cr, 17.5% Mn, and 0.11% C; EP-838 is 11.6% Cr, 13.5% Mn, 4.2% Ni, 0.9% Mo, and 0.02% C; and 316 stainless steel is 17.3% Cr, 1.7% Mn, 12.5% Ni, 2.1% Mo, and 0.06% C.

Table III. Chromium and Nickel Equivalents for Several Commercial and Experimental Steels

Alloy	Composition* (wt %)							Equivalent (wt %)	
	Cr	Mn	Ni	Mo	Si	Al	C	Ni	Cr
316 SS	17.3	1.7	12.4	2.1	0.7		0.06	15.0	21.9
15-15	15	15					0.10	18.0	15.0
12-15	12	15					0.10	18.0	12.0
15-20	15	20					0.10	23.0	15.0
EP-838	11.6	13.5	4.2	0.9	0.4	0.7	0.02	11.6	17.6
AMCR	10.1	17.5	0.10		0.55		0.11	12.2	11.2

*Balance iron.

entirely austenite (7–9). A 12-20 steel falls into the austenite-plus martensite region, and a 15-20 steel lies just outside this field in the austenite field. In this instance, the diagram indicates that an increase in chromium from 12 to 15% (with 20% Mn) leads to an austenitic steel.

Also shown in Table III and Fig. 1 are data for type 316 stainless steel and two experimental steels. The EP-838 is a Russian steel in which much of the nickel from a Cr-Ni-type stainless steel such as type 316 was replaced by manganese. To help maintain austenite, the chromium and molybdenum were reduced from the amounts of these elements in type 316 stainless steel. This steel is in the austenite region. The Cr-Mn steel designated ACMR is a steel being studied by the European communities' fusion program as a possible FIRD steel (16). Although no information is available for this material, we expect it to be austenitic, but the Schaeffler diagram predicts otherwise. Further difficulties with the diagram are found when the three nonmagnetic steels of Desforges and Dancoisne are referred to the diagram (15). None of them was found to fall in the austenite region (or near a boundary).

These inconsistencies between phase diagrams and experimental results with the Schaeffler diagram indicate that the words of Desforges and Dancoisne are appropriate (15). They wrote:

> A Schaeffler constitution diagram for manganese-containing steels would be of great value in providing designers and welding engineers with the necessary information on weld metal microstructure.

And, we might add, base metal microstructure.

Various studies indicate that the 0.5 multiplier for manganese in Eq. (2) may be an overestimate of manganese's austenite-stabilizing effect (17). One study indicated that the effect of manganese may be a constant with a value of 0.35 (17). However, such a low value does not seem reasonable in light of the above discussion. Thus, it appears that much still needs to be learned about the effect of manganese in austenitic steels. The Schaeffler diagram, as now constituted, would appear to be of only limited help in developing manganese-stabilized austenitic steels.

Molybdenum is present in type 316 stainless steel to increase the corrosion resistance against chlorides and pitting and to improve the high-temperature mechanical properties. It is not known whether tungsten, which is analogous to molybdenum in ferritic steels, would achieve the same effect as does molybdenum in these steels. However, tungsten is a ferrite former and in a Cr-Mn steel would probably have to replace chromium if a duplex structure were to be avoided.

Little information is available on the strength of the Fe-Cr-Mn steels. We have recently determined the strength of the Russian steel designated EP-838 (18) (the composition is given in Table III). The tensile properties of this steel in the 20%-cold-worked condition at room temperature (Fig. 2) and 300°C are similar to those of 20%-cold-worked type 316 stainless steel. The effect of manganese on the work-hardening capacity was evident, especially in the room-temperature tests. After irradiation to a displacement-damage level of 5 to 8 dpa, the properties of the two steels were also similar (Fig. 2). These two steels also have similar tensile properties in the solution-annealed condition before and after irradiation (19). From these results it appears that manganese could be used to replace a large portion of the nickel and at the same time to reduce the chromium and molybdenum concentrations but not affect the strength. No information is available on the effect of the reduced amounts of chromium and molybdenum on corrosion resistance.

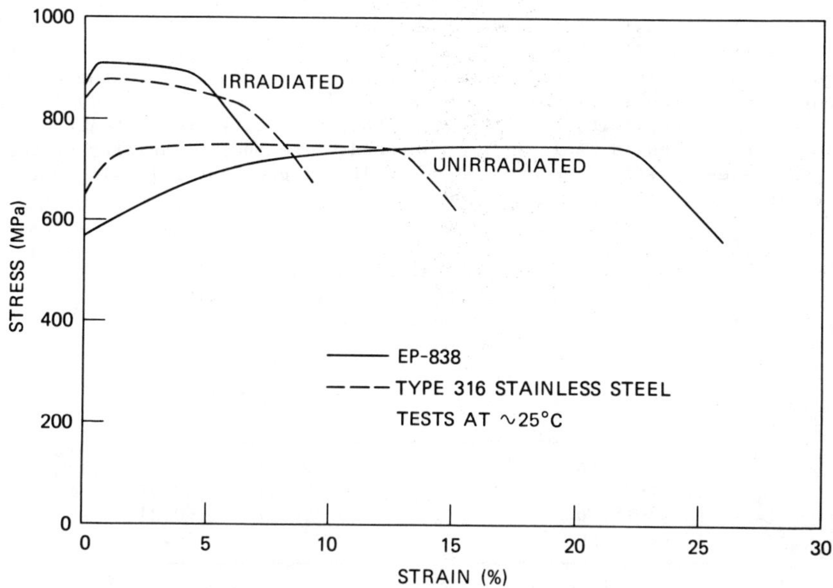

Figure 2 - Engineering stress-strain curves at room temperature and for unirradiated and irradiated 20%-cold-worked EP-838 and type 316 stainless steels. Irradiation was in HFIR at about 50°C to about 5 to 8 dpa.

The major concerns with austenitic steels for fusion reactor applications involve void swelling (20), helium embrittlement (21), and compatibility with coolants. The void and dislocation structure that develops during irradiation in the temperature range 0.35 T_m to 0.6 T_m, where T_m is the absolute melting point of the steel, can lead to large volume increases, which could not be tolerated in a structural component. The large amounts of transmutation helium that will be generated in a fusion reactor first-wall structure are known to affect void nucleation and growth (20). In addition, at temperatures greater than 0.5 T_m, small amounts of helium and other irradiation-induced segregation at grain boundaries lead to embrittlement (21).

Methods being tested for the suppression of void swelling are the use of a high dislocation density (cold work) and/or a high density of fine titanium carbide precipitates (MC precipitates) in the microstructures (4,22). The dislocations and precipitates act as helium collection sites, giving rise to a high density of fine cavities, which subsequently act as preferential sinks for the vacancies and interstitials produced during irradiation. The MC on grain boundaries also imparts resistance to embrittlement at elevated temperatures. The application of these principles has led to an austenitic PCA for fusion reactors (22); this alloy with 14% Cr, 16% Ni, 2% Mo, and 0.25% Ti is being irradiated and tested. In addition to helping control the helium and minimize the swelling, the MC also strengthens by precipitate hardening. The similar use of titanium in a manganese-stabilized austenitic steel would appear to be appropriate for providing swelling resistance and additional strength. Note, however, that titanium is a ferrite stabilizer and that the formation of TiC would eliminate carbon from solution and thus negate its role as an austenite stabilizer.

The above considerations indicate that the development of a FIRD austenitic steel will probably be more complicated than that of a ferritic steel (23). The first step in a developmental program must be the determination of a stable austenitic composition for a Cr-Mn steel. To obtain an indication of the limits of austenitic phase stability, small heats with the compositions given in Table IV are proposed. The phase stability of the proposed steels will be determined by optical microscopy following solution-annealing and thermal aging treatments.

Table IV. Possible Austenitic Stainless Steels for Fast Induced-Radioactivity Decay Alloy Development Program

Alloy	Chemical composition* (wt %)			
	Cr	Mn	C	N
15 Cr-15 Mn	15	15	0.05–0.1	<0.01
15 Cr-20 Mn	15	20	0.05–0.1	<0.01
10 Cr-15 Mn	10	15	0.05–0.1	<0.01
10 Cr-20 Mn	10	20	0.05–0.1	<0.01

*Balance iron.

After a stable composition is determined, it will be necessary to adjust the composition for swelling resistance, strength, and compatibility with possible breeding and cooling media. Experience with type 316 stainless steel and improvements made when the PCA was developed suggests the addition of Ti, Si, and W (as a substitute for Mo). The development of PCA was aided by the knowledge previously obtained from studies on type 316 stainless steel on processes such as solute segregation during cavity growth and the effect of different precipitates on nucleation and growth of cavities (e.g., the difference between MC and $M_{23}C_6$) (4). The substitution of manganese for nickel could well alter these processes, thus making detailed studies of the new alloy base system necessary. Note also that Ti, W, and Si are ferrite stabilizers. Even if a phase-stable quaternary alloy is obtained from the initial studies, that stability may be jeopardized by the addition of these elements, and further compositional adjustments and stability studies will be required.

Alloy Development Strategy

Once an austenite-stable alloy composition is identified, the strength of the alloys must be determined in the cold-worked and solution-annealed conditions. This austenite-stable composition will serve as the base for developing a steel with properties similar to type 316 stainless steel and PCA. Because the life limitation on the austenitic stainless steels is their propensity to swell during irradiation, samples of these alloys should be irradiated to high fluences. Swelling measurements can be used to compare the behavior of the new alloy with type 316 stainless steel and PCA. Transmission electron microscopy on the irradiated steels should be done in an effort to understand the effect of irradiation on phase stability and precipitation processes and the effect these have on swelling.

Information on fabricability will be obtained during the melting, casting, and rolling operations necessary to obtain test material. Simple weldability tests can be conducted on the steels that evolve from the first series of alloys. Compatibility tests with potential cooling and breeding media (e.g., Li, Pb-Li, and solid lithium ceramics) may prove extremely important for this class of alloys. A large reduction of chromium and the elimination of molybdenum may well have a large effect on the compatibility of this class of steel.

Other Austenitic Alloys

Several Japanese steel companies have developed high-manganese, low-chromium austenitic steels. However, most of these either contain nickel or nitrogen above the limits given by 10 CFR Part 61 (2) and thus do not qualify as FIRD alloys. An exception is a steel by Nippon Kokan, designated NM-1, which contains 20% Mn, 2% Cr, 0.5% C, and 0.5% Si. Of course, this steel is not "stainless" and, as such, does not fit the criteria established above in the discussion on types of alloys to be developed. Nevertheless, this alloy might be used to define another class (low-Cr, high-C, Cr-Mn steel) of FIRD steels to be considered, should those discussed above prove to be inadequate. Indeed, in the alloy development approach suggested above, this class of steels is one of the limits being approached.

Summary

If a future fusion reactor could be constructed from a material that develops little or no induced radioactivity during service, a much safer reactor would result, one that would allow hands-on maintenance and minimize radioactive waste disposal problems. No structural materials are currently available to construct such a reactor. However, it appears to be possible to develop alloys that have relative fast induced-radioactivity decay, which would simplify the radioactive waste disposal problem. An approach to develop such austenitic steels has been presented.

The austenitic stainless steels (primarily type 316) are leading candidate structural materials for fusion reactors. These steels do not qualify as FIRD alloys primarily because of the nickel contents. An alloy-development approach for austenitic stainless steels has been outlined in which the substitution of manganese for nickel is proposed. The difficulty of such an approach arises because manganese is not as strong an austenite stabilizer as is nickel.

Acknowledgments

We gratefully acknowledge several people who aided in this work. Detailed discussions with F. W. Wiffen were invaluable. The manuscript was reviewed by V. K. Sikka and F. W. Wiffen. Frances Scarboro prepared the manuscript.

References

1. R. W. Conn et al., Panel Report on Low Activation Materials for Fusion Applications, UCLA Report PPG-728; University of California at Los Angeles, June 1983.

2. Nuclear Regulatory Commission, "Licensing Requirements for Land Disposal of Radioactive Waste," 10 CFR Part 61, Federal Registry, 47(248) 57446–82 (Dec. 27, 1982).

3. F. W. Wiffen and R. T. Santoro, "Control of Activation Levels to Simplify Waste Management of Fusion Reactor Ferritic Steel Components," pp. 195–200 in Ferritic Alloys for Use in Nuclear Technologies, J. W. Davis and D. J. Michel, eds.; TMS-AIME, Warrendale, PA, 1984.

4. P. J. Maziasz et al., "Progress in Alloy Development in the Fusion Materials Program," Journal of Nuclear Materials, 108&109 (1982) pp. 296–98.

5. J.H.G. Monypenny, Stainless Iron and Steel, vol. 2, pp. 261–99, Chapman & Hall, London, 1954.

6. S. Gunzberg, N. A. Aleksandrova, and L. S. Geldermann, "Eigenschaften von Nichtrostenden Chrom-Mangan und Chrom-Nickel-Mangan Stahlen," Archiv Eisenhüttenwesen, 8 (1934) pp. 121–27.

7. M. Schmidt and H. Legat, "Hitzebeständige Chrom-Mangan Stahle," Archiv Eisenhüttenwesen, 10 (1936) pp. 297–303.

8. F. Bruhl, "Gefüge und Eigenschaften von Chrom-Mangan Stahlen mit Gehalten bis 1% C, 15% Mn, und 30% Cr," Archiv Eisenhüttenwesen, 10 (1936) pp. 243–55.

9. C. O. Burgess and W. D. Forgeng, "Constitution of Iron-Chromium-Manganese Alloys," AIME Transactions, 131 (1938) pp. 277–302.

10. H. Schneider, "Investment Casting of High-Hot-Strength 12-Percent Chrome Steel," Foundry Trade Journal, 108 (1960) pp. 562–63.

11. C. D. Desforges, W. E. Duckworth, and T. F. J. N. Ryan, Manganese in Ferrous Metallurgy, pp. 67–70; The Manganese Centre, Paris, 1976.

12. R. A. Lula and W. G. Renshaw, "Corrosion Resistance and Mechanical Properties of Cr-Ni-Mn Stainless Steels," Metal Progress, 69 (1956) pp. 73–77.

13. S. J. Carlisle, J. W. Christian, and W. Hume-Rothery, "The Equilibrium Diagram of the System Cr-Mn," Journal Institute of Metals, 76 (1949) pp. 195–208.

14. F. Nair and M. Semchyshen, "Corrosion Resistance of Molybdenum Modified Cr-Ni-Mn Austenitic Stainless Steels," Corrosion (Houston), 19 (1963) pp. 210–16.

15. C. D. Desforges and P. L. Dancoisne, "The High Temperature Properties of Manganese Containing Ferrous Alloys," pp. 522–37 in Environmental Degradation of High Temperature Materials, vol. 2; Institution of Metallurgists, Northway House, Whetstone, London, 1981.

16. Private communication, M. L. Grossbeck, Oak Ridge National Laboratory, July 1983.

17. E. R. Szumachowski and D. J. Kotecki, "Manganese Effect on Stainless Steel Weld Metal Ferrite," Welding Journal (Miami), 63 (1984) pp. 156S–161S.

18. R. L. Klueh and M. L. Grossbeck, "A Comparison of the Irradiated Tensile Properties of a High-Manganese Austenitic Steel and Type 316 Stainless Steel," Journal of Nuclear Materials, 122&123 (1984) pp. 294–298.

19. S. N. Votinov et al., "Effects of Neutron Irradiation on Mechanical Properties of EP-838 and 316 Steels," Fizika i Khimya Obrabutki Materialov, January 1981(1) pp. 50–52; also ORNL-TR-4812, 1982.

20. P. J. Maziasz and M. L. Grossbeck, "Swelling, Microstructural Development, and Helium Effects in Type 316 Stainless Steel Irradiated in HFIR and EBR-II," Journal of Nuclear Materials, 103&104 (1981) pp. 987–92.

21. E. E. Bloom, "Irradiation Strengthening and Embrittlement," pp. 295–329 in Radiation Damage in Metals; N. L. Peterson and S. D. Harkness, eds.; American Society for Metals; Metals Park, Ohio, 1976.

22. P. J. Maziasz and T. K. Roche, "Preirradiation Microstructural Development Designed to Minimize Properties Degradation During Irradiation in Austenitic Steels," Journal of Nuclear Materials, 103&104 (1981) pp. 797–802.

23. R. L. Klueh and E. E. Bloom, "The Development of Ferritic Steels for Fast Induced-Radioactivity Decay for Fusion Reactor Applications," Nuclear Engineering/Fusion, to be published.

DEPENDENCE OF NEUTRON-INDUCED SWELLING ON COMPOSITION IN IRON-BASED AUSTENITIC ALLOYS

F. A. Garner and H. R. Brager

Hanford Engineering Development Laboratory
Box 1970, Richland, WA 99352
USA

Summary

The neutron-induced swelling of simple Fe-Cr-Ni, Fe-Mn and Fe-Cr-Mn alloys without solute additions has been studied using fast reactor irradiation in the range 400-650°C. It appears that in this temperature range these alloys all eventually swell at ∿1%/dpa after the completion of a transient regime. Although the swelling rate of the post-transient regime is remarkably insensitive to composition and irradiation temperature, the duration of the transient regime is sensitive to these variables. Alloys containing manganese instead of nickel show a much lesser sensitivity to composition and temperature, however. Comparison of these results with those of charged particle simulation studies shows that ion irradiation studies can be somewhat misleading when applied to alloy development and optimization. There does not appear to be any potential for development of an iron-based austenitic alloy optimized for swelling resistance as a consequence of a reduced post-transient swelling rate. Any optimized alloy must be based on attempts to extend the transient regime of swelling.

Introduction

The application of austenitic steels in both fission and fusion neutron environments will be limited in part by their demonstrated potential for large levels of irradiation creep and void-induced swelling. Therefore a necessary step in the optimization of structural alloys for nuclear service must be the reduction of the magnitude of these radiation-induced strains. This paper addresses only the swelling phenomenon, since a reduction in irradiation creep usually accompanies a reduction in swelling.(1)

Swelling is known to be sensitive to environmental variables such as temperature, stress and atomic displacement rate. It is also sensitive to differences in microstructural and microchemical starting conditions arising from preirradiation thermal-mechanical history. It is now known that these variables exert their influence primarily on the duration of the transient regime that precedes the onset of steady-state swelling.(1,2) In this paper it will be shown that the composition dependence of swelling in simple iron-based austenitic alloys also resides in the transient regime.

The compositional sensitivity of Fe-Cr-Ni austenitic alloys is well known and can be divided into two categories, that arising from the major solvent atoms (Fe, Ni, Cr, Mn) and that arising from lesser solutes (C, Si, P, Ti, Mo, etc.) The focus of this paper is primarily on the influence of the solvent components in the Fe-Cr-Ni and Fe-Cr-Mn systems.

While few data have been published on swelling of the Fe-Cr-Mn system, a strong dependence of swelling on both nickel and chromium content of Fe-Cr-Ni alloys has been observed in both solute-free ternary alloys and solute-modified commercial alloys, using ion,(2-7) electron(8) and neutron irradiation.(9-10) None of these studies, however, showed unambiguously whether the influence of compositional variations lies in the transient or post-transient regime of swelling. An extensive irradiation program conducted in both EBR-II (Experimental Breeder Reactor II, Idaho Falls, Idaho) and FFTF (Fast Flux Test Facility, Richland, Washington) has now resolved this issue.

Experimental Details

Fourteen Fe-Cr-Ni ternary alloys in the annealed condition were irradiated in the AA-VII experiment in EBR-II in the form of small disks used for transmission electron microscopy. The dimensions of these disks were 3 mm in diameter and 0.25 mm thick. The compositions of these alloys are given in Table I and shown schematically in Figure 1. They were irradiated in static liquid sodium at eight temperatures between 400 and 650°C to exposures as large as 2.2×10^{23} n/cm^2 (E > 0.1 MeV) or approximately 110 dpa. The displacement rates in this experiment range from 0.6 to 1.1×10^{-6} dpa/sec. The temperatures drifted somewhat during irradiation and varied ±5°C at 400°C and ±15°C at 650°C.

Three binary Fe-Mn and six ternary Fe-Cr-Mn alloys are also currently being irradiated under static sodium in FFTF at roughly twice the displacement rate typical of EBR-II and three irradiation temperatures: 420, 520 and 600°C. These temperatures are actively controlled to within ±5°C throughout the irradiation. To date only specimens irradiated to ≤14 dpa have been examined. The compositions of these alloys are shown in Figure 2 and Table II. All alloys were in the annealed condition.

TABLE I. Compositions of the Simple Fe-Cr-Ni Alloys

Alloy Designation	Weight % Fe	Ni	Cr	C	O	N
E18	Bal	12.1	15.1	.005	.016	.0024
E90	Bal	15.7	15.6	.013	---	---
E39	Bal	20.3	7.5	.004	.016	.0014
E26	Bal	20.1	11.8	.002	.016	.0019
E19	Bal	19.4	14.9	.003	.018	.0015
E27	Bal	24.7	10.2	.001	.012	.0017
E20	Bal	24.4	14.9	.003	.017	.0017
E21	Bal	29.6	15.3	.004	.017	.0020
E37	Bal	35.5	7.5	.002	.016	.0013
E22	Bal	34.5	15.1	.003	.017	.0021
E38	Bal	35.2	20.0	.004	.017	.0010
E25	Bal	35.1	21.7	.004	.015	.0020
E23	Bal	45.3	15.0	.002	.014	.0017
E24	Bal	75.1	14.6	.001	.0077	.0011

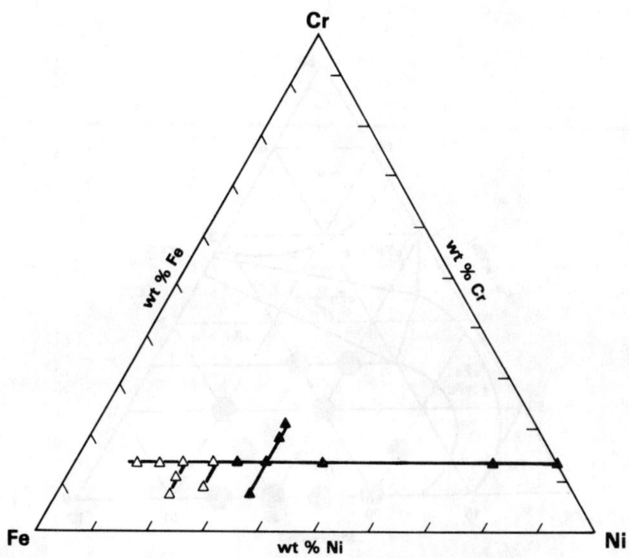

Figure 1 - Fe-Cr-Ni ternary diagram showing matrix of alloys used in this experiment. The open symbols are the alloys whose swelling levels are shown in Figure 3. The binary alloy 85Ni-15Cr is also shown in this figure. Its swelling behavior is atypical of that of iron-based alloys and is reported elsewhere.(11)

TABLE II. Compositions of Simple Fe-Mn and Fe-Cr-Mn Alloys*

Alloy Designation	Weight % Fe	Cr	Mn
R66	80	5	15
R67	70	15	15
R68	80	0	20
R69+	70	10	20
R70	65	15	20
R71	75	0	25
R72	70	0	30
R73	65	5	30
R74	60	10	30
R75	65	0	35

*Minor solute composition has not yet been determined but is known to be small.

+This alloy was found to be brittle and was not irradiated.

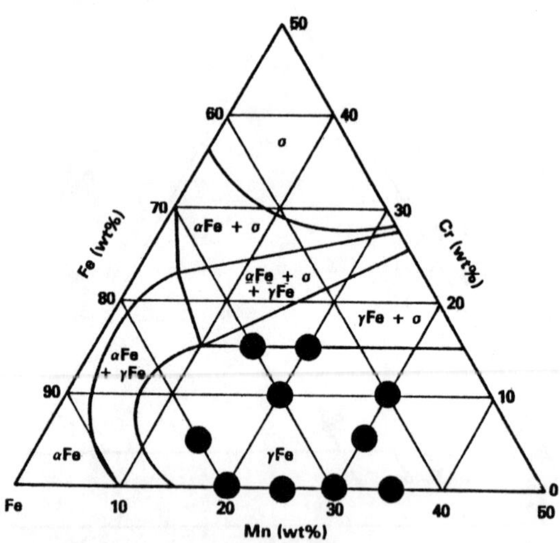

Figure 2 - Fe-Cr-Mn ternary diagram showing matrix of alloys used in this experiment.

In many cases two identical specimens were irradiated side by side. The radiation-induced density changes were measured using an automated immersion density technique. In the majority of cases the swelling of each member of the pair was identical within the system measurement error ($\pm$0.16%). Whenever the swelling varied more than 0.2% between members of the pair, the two values were reported separately. Otherwise they were simply averaged. Some of the alloys densified to levels of ∿1% in the Fe-Cr-Ni system and ∿2% in the Fe-Cr-Mn system. This behavior was unexpected since no recovery or precipitation was anticipated in these simple annealed and solute-free alloys. This densification phenomenon is thought to be quite important and is covered separately in another paper.(12)

Results of Fe-Cr-Ni Irradiations

The data for these irradiations continue to be generated and currently are available for over four hundred combinations of alloy composition, temperature and fluence. Detailed listings of these data are complied in various quarterly progress reports.(13-15) Not all of the 400-plus data will be shown in this report, but the behavior shown below is quite typical of that observed in the entire data field.

Figure 1 shows that the alloy series can be broken down into two subsets, one at 24.4% nickel and below, and another at 29.6% nickel and above. The first subset yields an unanticipated conclusion. As shown in Figure 3, over the range of 12.1 to 24.4% nickel and chromium levels of 7.5 to 15.6%, there is surprisingly little sensitivity of swelling to both composition and temperature between 400 and 510°C. The results of all earlier studies forecast a strong dependence of swelling on both these variables.(2-10) The anticipated dependence on nickel content does become apparent in this study at higher temperatures, however. Figure 4 shows the increasing impact of temperature above 510°C on the duration of the transient regime of swelling in the first subset of alloys. Increasing the nickel content leads to a progressive increase in the duration of the transient regime of swelling. The post-transient swelling rate at all conditions is ∿1%/dpa or ∿5%/10^{22} n cm^{-2} (E > 0.1 MeV). There is roughly a factor of two difference between the highest and lowest displacement rates in this experiment but no consistent trends of transient duration with displacement rate could be found in these data.

Figure 5 shows for one alloy, Fe-24.4Ni-14.9Cr, the rather abrupt transition between temperature-independent and temperature-dependent swelling behavior. At higher nickel levels the influence of temperature on extending the transient regime is observed much sooner in temperature, as demonstrated in Figure 6a.

Increasing the chromium level or decreasing the nickel level tends to counteract the influence of temperature in increasing the transient duration, as shown in Figures 6b and 6c. Figures 7 and 8 confirm that increases in the chromium level at a given temperature and nickel level tend to shorten the duration of the transient regime of swelling. Note that the influence of nickel on swelling, however, is opposite that of chromium, as shown in Figures 4 and 9.

Figure 9 also shows that there appears to be a maximum somewhere near 40% nickel in the duration of the transient regime. This is easier to see when the higher temperature swelling data are plotted vs. nickel content in Figure 10. The minimum in swelling at a given set of irradiation conditions was also observed in various ion irradiation studies published earlier(2-5).

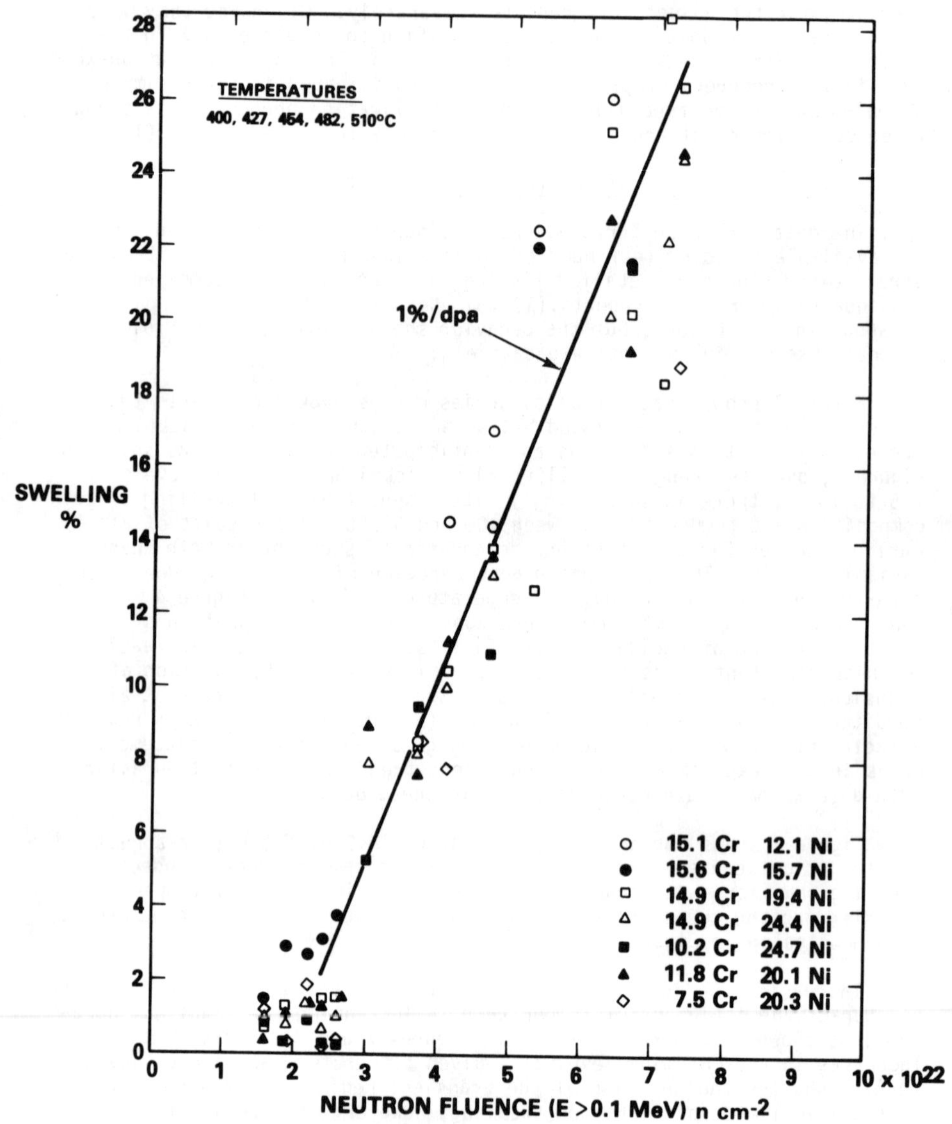

Figure 3 - Swelling of seven Fe-Cr-Ni alloys in EBR-II at five temperatures in the range 400-510°C, showing a remarkable insensitivity of swelling to both composition and irradiation temperature.

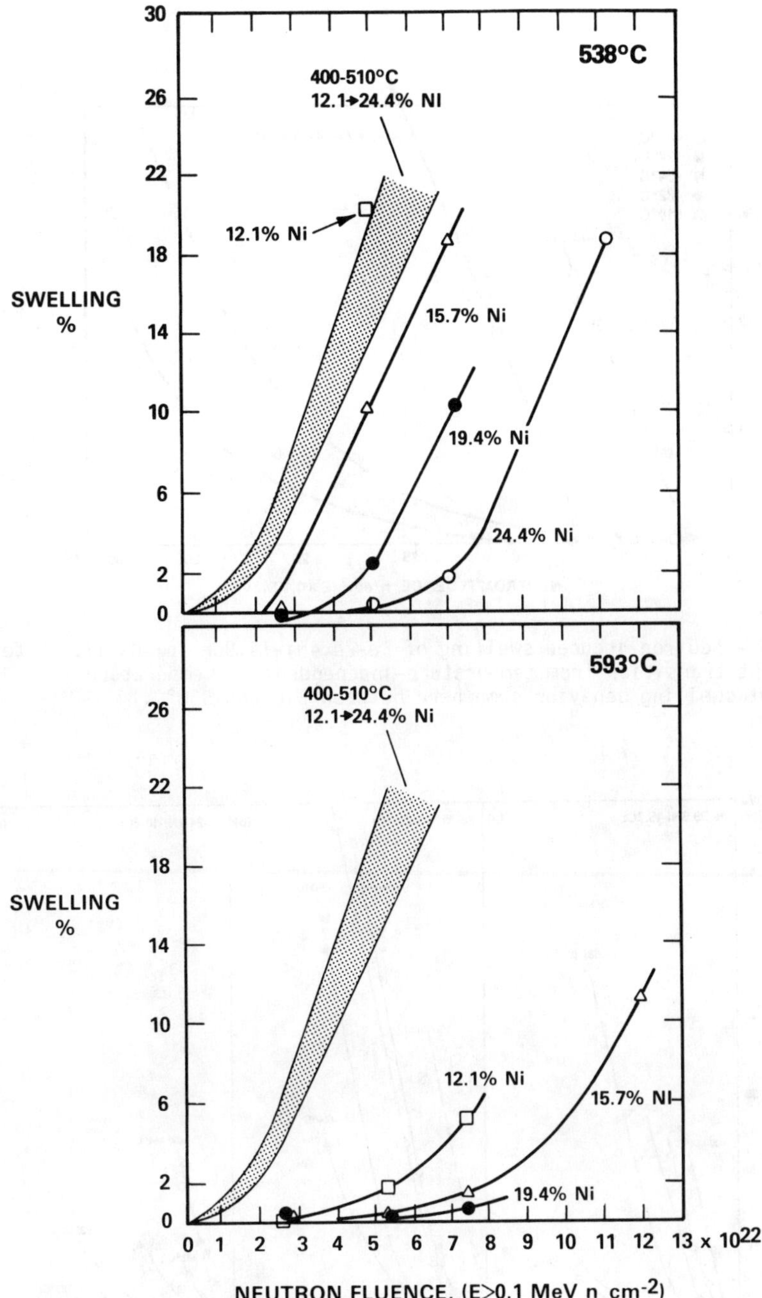

Figure 4 - Effect of temperature and nickel content on the swelling of Fe-15Cr-XNi alloys at two temperatures above 510°C. Some recently acquired data at higher fluence are not included in this figure. The tendency of the duration of the transient regime to further increase with temperature is also preserved at 650°C but is not shown here.

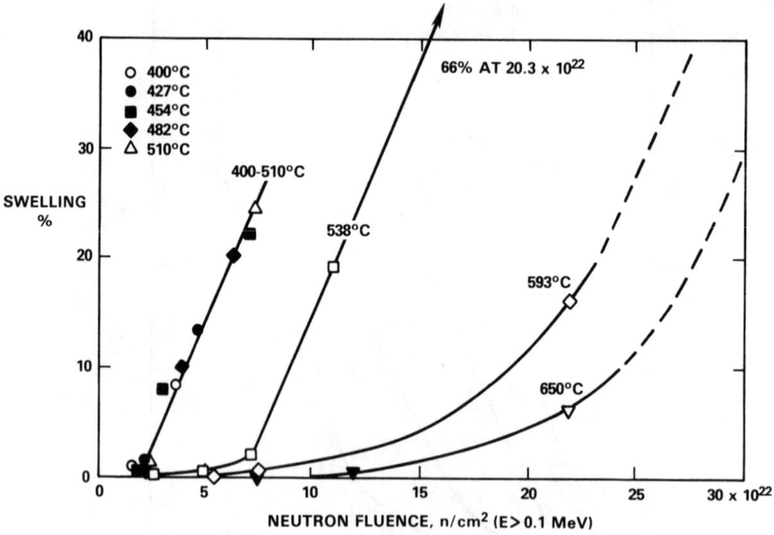

Figure 5 - Neutron-induced swelling of Fe-24.4Ni-14.9Cr in EBR-II. Note the abrupt transition from temperature-independent to temperature-dependent swelling behavior somewhere between 510 and 538°C.

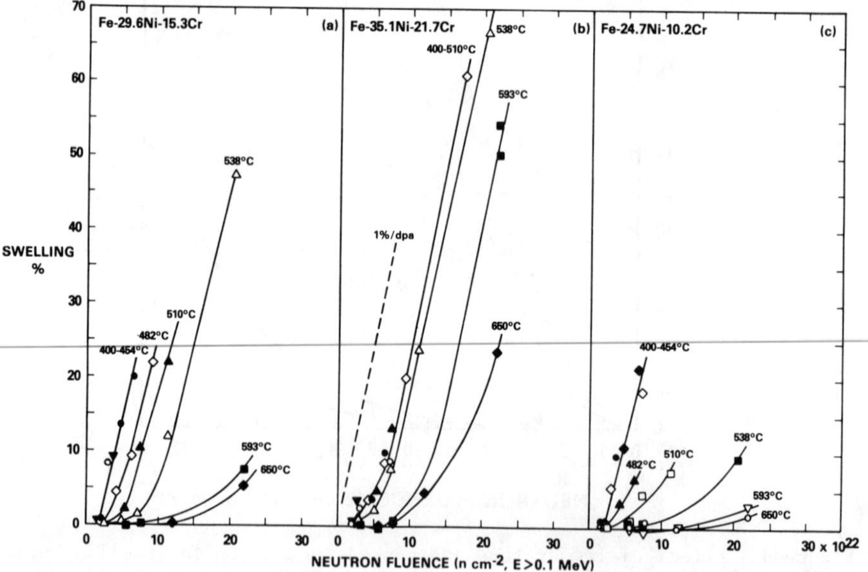

Figure 6 - Temperature dependence of swelling in three Fe-Ni-Cr ternary alloys with relatively high nickel levels and different chromium contents.

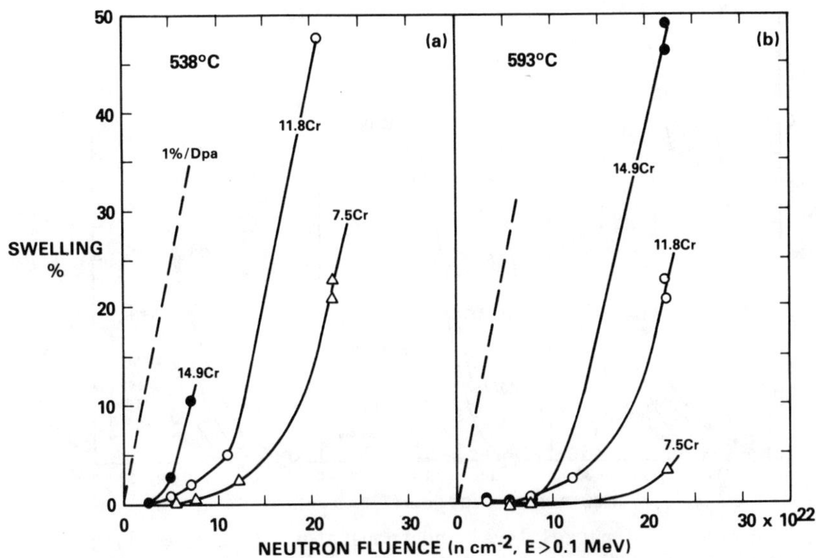

Figure 7. Influence of chromium level and temperature on the swelling of Fe-20Ni-XCr alloys in EBR-II.

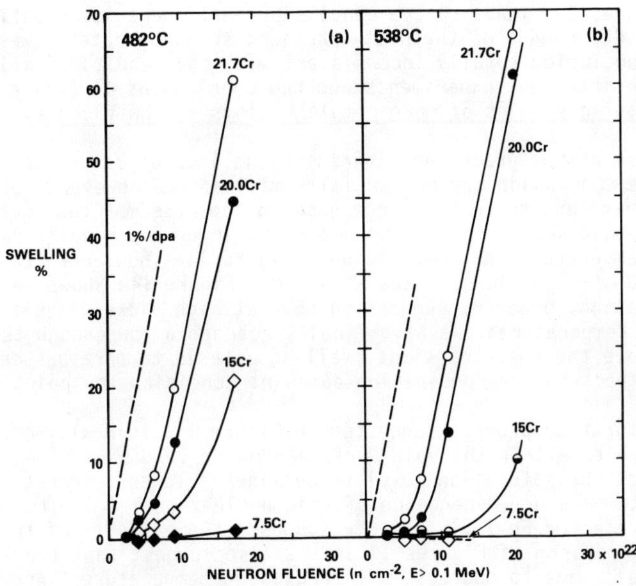

Figure 8 - Influence of chromium level and temperature on the swelling of Fe-35Ni-XCr alloys in EBR-II.

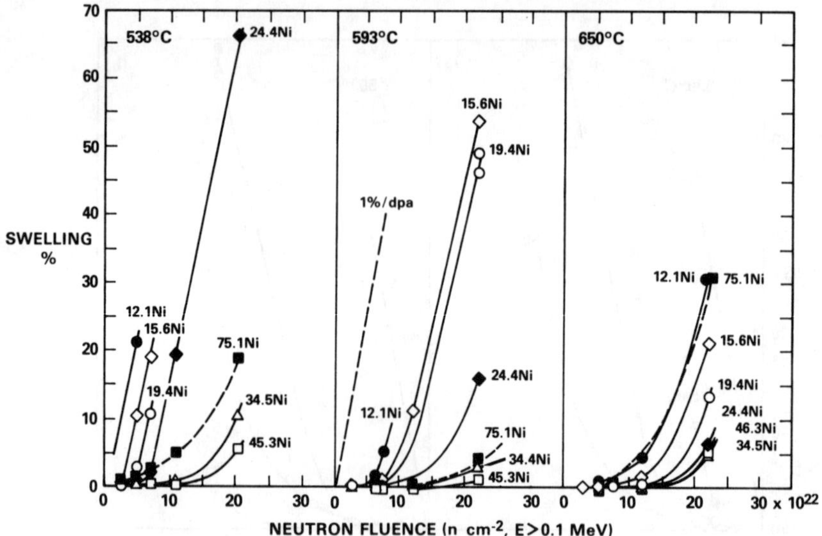

Figure 9 - Influence of nickel level and temperature on the swelling of Fe-15Ni-XCr alloys in EBR-II. Note the reversal in transient duration above ∿45% nickel.

It is important to note that the swelling rate of all alloys shown thus far tends to increase continually with fluence and eventually approaches ∿5%/10^{22} n/cm^2 (E > 0.1 MeV) or ∿1%/dpa. Swelling levels of 50-70% are often shown in these figures, without hint of saturation. Comparison of Figures 3-9 leads to the conclusion that there is essentially no temperature dependence of the post-transient swelling rate between 400 and 650°C, a conclusion totally inconsistent with the results of all published ion bombardment experiments and the conclusions of earlier neutron irradiation studies of more complex alloys at low fluence.

One feature of the preceding figures is the lack of saturation of swelling. This conclusion may be partially misleading, however. At lower temperatures there are currently large gaps in the specimen test matrix but it appears that saturation may be occurring in a manner quite dependent on both temperature and composition. Figure 11a shows one reasonably complete subset of high fluence data at 427°C. Figure 11b shows an alternate interpretation, however, suggesting that at high nickel levels and relatively low temperatures, we have finally reached a low enough temperature regime where the post-transient swelling rate is temperature-dependent as one would expect in a recombination-dominant condition for point defects.

If the saturation process suggested in Figure 11a is really occurring and is sensitive to either the void surface area or void sink strength, one would expect the saturation level to decline at lower temperatures due to the strong temperature dependence of void nucleation. A limited set of data in Figure 11a for Fe-75.1Ni-14.6Cr suggests the validity of this assumption. Comparison of Figures 11 and 12 also suggest that the saturation level tends to decrease with declining temperature, increasing nickel and decreasing chromium. More data at intermediate and higher fluence are required to confirm that saturation is indeed occurring, or to confirm whether the swelling rate is falling below ∿1%/dpa at these lower irradiation temperatures.

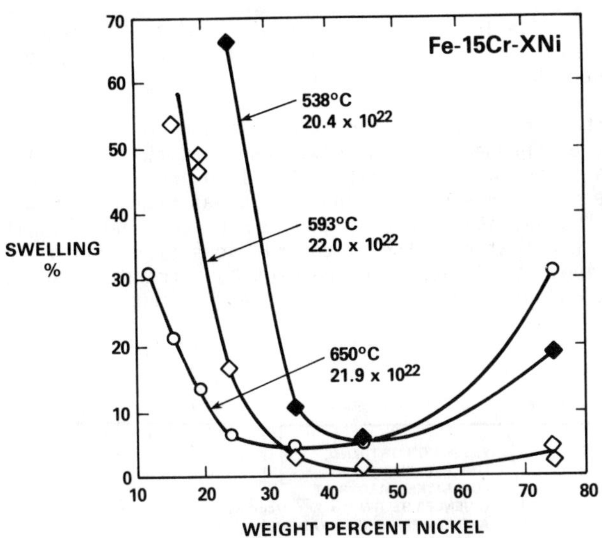

Figure 10 - Dependence of swelling of Fe-15Cr-XNi alloys on temperature, neutron fluence and nickel content, showing minimum swelling at intermediate nickel levels.

Results of Fe-Mn and Fe-Cr-Mn Irradiations

While this data field is still being enlarged, we can draw some conclusions about the swelling of these alloys. Figures 13 and 14 show that the sensitivity of swelling to solvent composition is not as large in the Fe-Cr-Mn system as it is in the Fe-Cr-Ni system at roughly comparable irradiation conditions. The upper half of Figure 15 also suggests that the swelling at 14 dpa for both 520 and 600°C are essentially the same. Note that the lower half of Figure 15 shows that at 9 dpa and 420°C the swelling is small and often negative. The substantial densification (∿2.2%) of Fe-35Mn implies that elemental redistribution, either by spinodal decomposition or phase separation, is occurring.

If we draw on our previous experience with Fe-Cr-Ni alloys it is not unreasonable to plot the Fe-Cr-Mn data and ignore the irradiation temperature, particularly in the lower temperature range. As shown in Figure 16, such a procedure leads to the conclusion that this system also probably swells at ∿1%/dpa, but with a temperature dependence that is smaller than that of the Fe-Cr-Ni system since Figure 15 shows that there is also no temperature dependence between 520 and 600°C. It also appears that the slight dependence of swelling on composition shown in Figure 13 is largely the result of a composition-dependent densification process that is competing with the swelling process to produce the final density change.

Discussion

It is unfortunate that these experiments were conducted in reactors whose inlet temperatures are in the vicinity of 400°C. The swelling vs. temperature profile of these alloys at a given exposure level is obviously more of a plateau than a sharply peaked curve, particularly at low nickel levels typical of the 300 series stainless steels currently used in fast breeder reactors. The low temperature shoulder of the plateau therefore lies somewhere below 400°C, a temperature regime that cannot be examined using either the EBR-II or FFTF reactors.

It is important to note, however, that between 400 and 650°C there is remarkably little dependence of the post-transient swelling rate on temperature. This is in sharp contrast to the long prevailing conception of the temperature dependence of swelling, which was characterized by a steeply-peaked "steady-state" swelling rate with temperature. It has recently been shown that rate theory descriptions of swelling can easily produce a relatively temperature-independent post-transient swelling rate when the latest values of vacancy migration and formation energies are used (2).

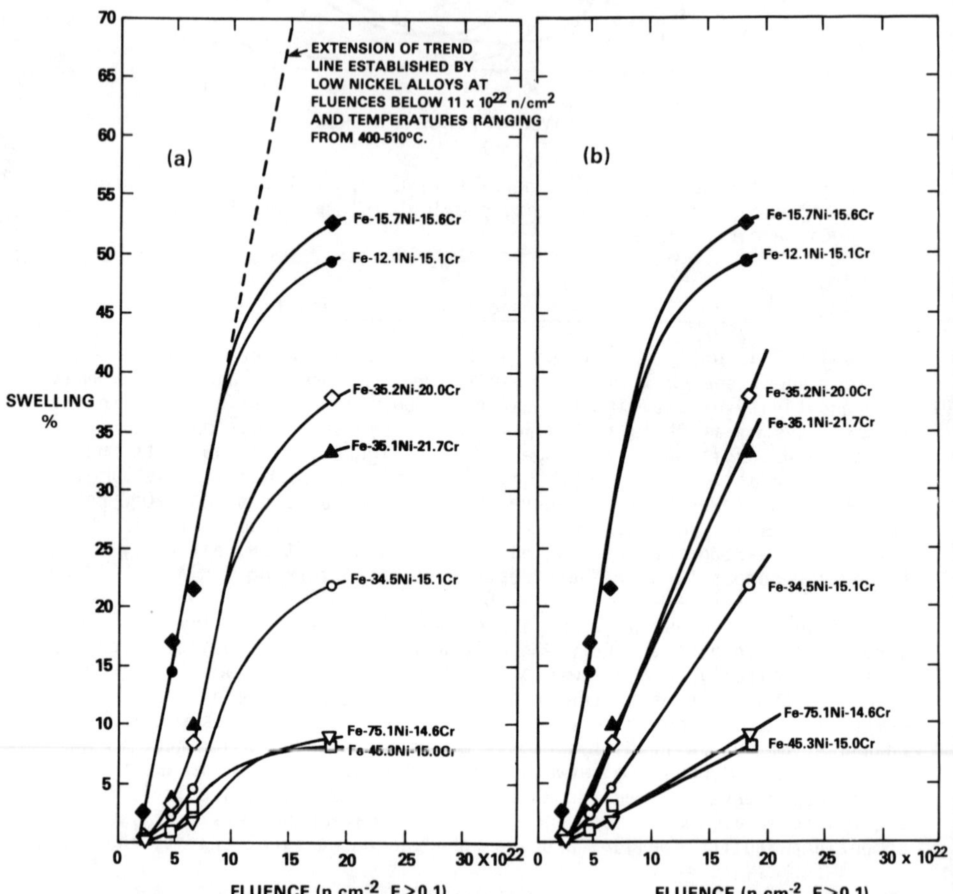

Figure 11-(a) Apparent saturation of swelling occurring in Fe-Cr-Ni alloys irradiated at 427°C in EBR-II. (b) An alternate interpretation suggesting declining swelling rates with increases in nickel and decreases in chromium content.

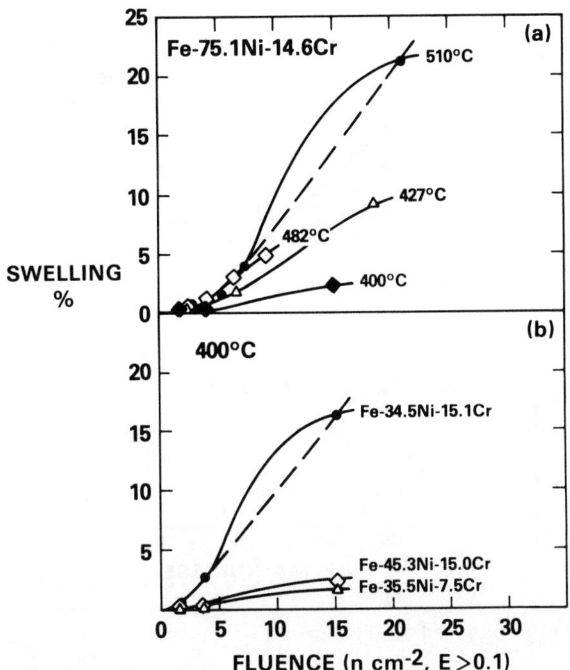

Figure 12-(a) Swelling behavior observed in Fe-75.1Ni-14.6Cr irradiated in EBR-II. (b) Swelling of various Fe-Ni-Cr alloys at 400°C. The dotted lines suggest an alternate interpretation.

The major reason that strong temperature dependencies were expected prior to the results of this study is that all ion irradiations clearly show such a behavior. Figures 17 and 18 show the results of Johnston and coworkers which exhibit a strong influence of nickel and chromium content at all temperatures studied and also a steeply-peaked dependence of swelling on temperature (3-4). A series of ion vs. neutron correlation experiments have now shown that the temperature dependence of both the transient and post-transient regimes of swelling is distorted by the action of the injected interstitial represented by the bombarding ion (16). Figure 19 shows a comparison of the post-transient swelling rates measured in neutron and ion irradiation experiments. It has been suggested that the injected interstitial effect on void nucleation is so large that even coinjected helium atoms effectively function as injected interstitials and further delay void nucleation in dual-ion bombardment experiments conducted at high helium/dpa ratios (17,18). Plumton and coworkers have convincingly demonstrated the strong distortion effect of the injected interstitial on void nucleation at relatively low irradiation temperatures (19,20). More recently Rauh and Bullough have also shown that one would expect a lower interstitial bias and therefore a lower steady-state swelling rate for experiments conducted at $>10^{-3}$ dpa/sec.(21) It therefore appears that a combination of factors, acting synergistically, reduce and distort the temperature dependence of the steady-state swelling rate in ion bombardment experiments, as shown in Figure 19.

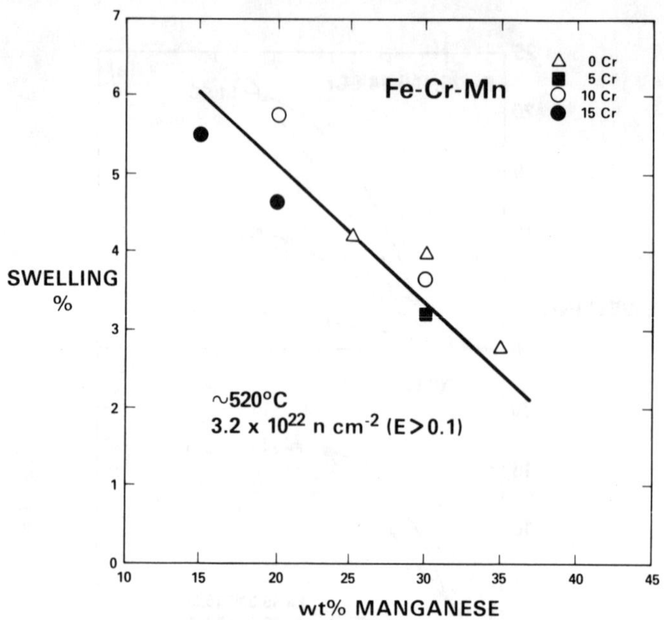

Figure 13 - Swelling observed in Fe-Mn and Fe-Cr-Mn alloys irradiated in FFTF to 3.2×10^{22} n/cm^2 (E > 0.1 MeV) or ~14 dpa at 520°C.

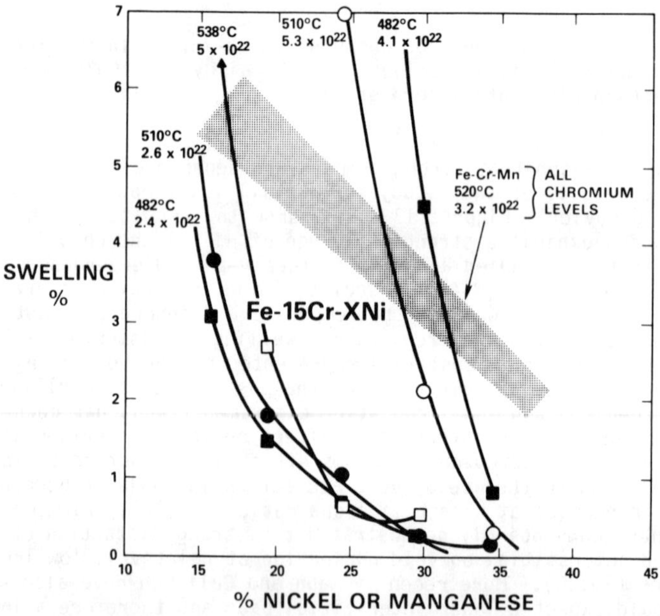

Figure 14 - Comparison of swelling of Fe-Cr-Ni Alloys irradiated in EBR-II with that of Fe-Mn and Fe-Cr-Mn alloys irradiated in FFTF under comparable irradiation conditions.

In general, therefore, early ion bombardment experiments gave a correct but partially misleading impression of the compositional dependence of swelling in simple ternary alloys. The compositional trends shown in the ion data in Figure 20 are slightly distorted (16) but are fully consistent with the trends of the high temperature neutron data of this report, but in no way forecast the composition-independent behavior observed at lower temperatures. This is probably a direct consequence of the low temperature distortion arising from the synergism of the injected interstitial and rate-dependent bias concepts. The ion data also incorrectly predicted that the post-transient swelling rate was dependent on composition. The reasons for these misleading conclusions are discussed in more detail in reference 16.

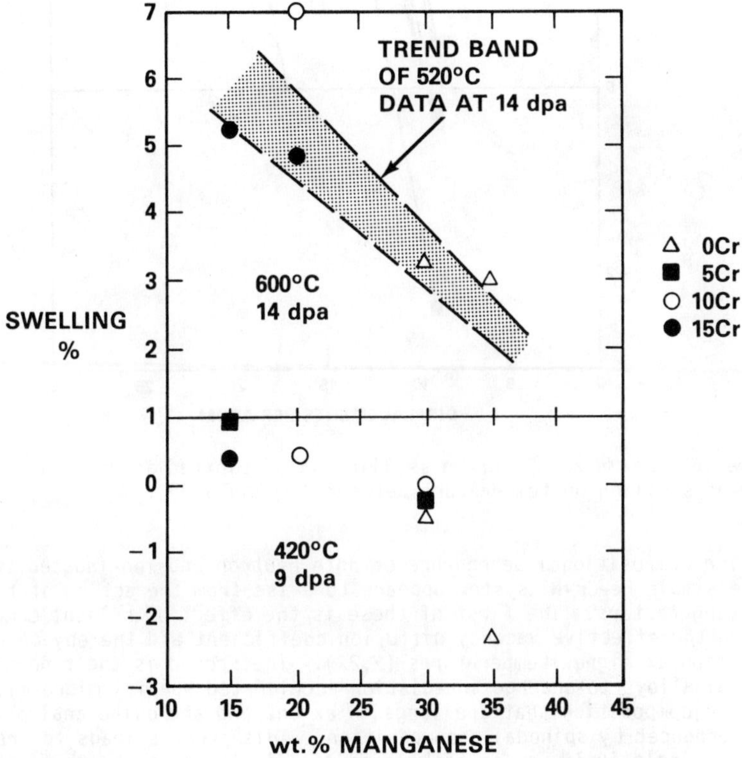

Figure 15 - Swelling of Fe-Mn and Fe-Cr-Mn alloys in FFTF at 600°C and 14 dpa as well as at 420°C and 9 dpa. Note that the swelling at 14 dpa of these alloys appears to be relatively insensitive to temperature in the range 520-600°C.

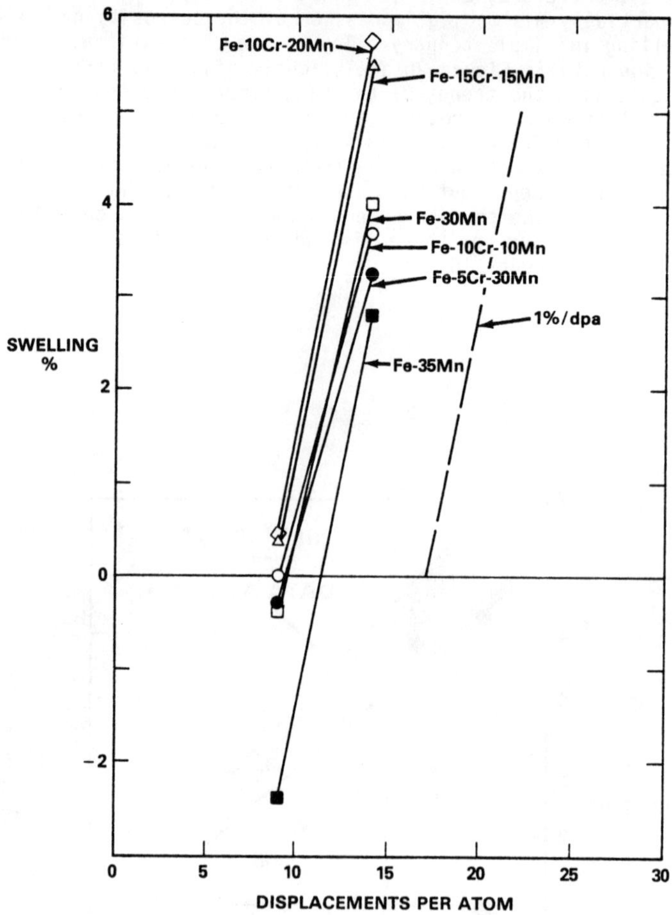

Figure 16 - Fe-Mn and Fe-Cr-Mn swelling data, plotted assuming independence of swelling on temperature between 420 and 520°C.

The compositional dependence of both neutron and ion-induced swelling in the simple Fe-Cr-Ni system appears to arise from the action of two competing factors. The first of these is the effect of solvent composition on the effective vacancy diffusion coefficient and thereby on void nucleation at higher temperatures (2,22). The second is the tendency of Fe-Cr-Ni alloys to undergo irradiation-accelerated spatial microscillations in composition that are large in extent and are quite analogous to those produced by spinodal decomposition. This process leads to areas which are relatively low in nickel and rich in chromium, both of which favor void nucleation. Since the tendency to decompose increases above 30% nickel this appears to account for the minimum in swelling observed in Figure 10. It also accounts for the radiation-induced densification mentioned earlier. Both of these factors are discussed in more detail elsewhere (12,23).

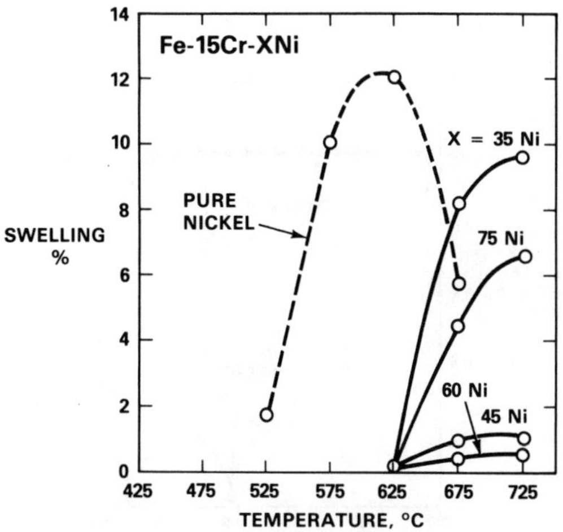

Figure 17 - Composition and temperature dependence of swelling of pure nickel and Fe-15Cr-XNi ternary alloys after irradiation with 5 MeV Ni^{+} ions to 140 dpa (3,4). Note the strong dependence on temperature at all compositions.

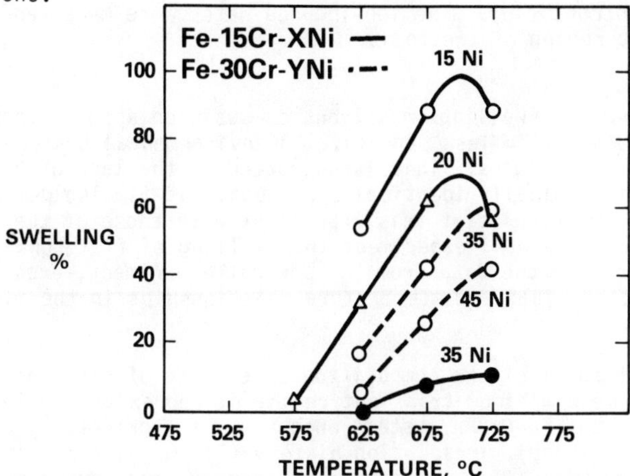

Figure 18 - Temperature-dependent swelling of Fe-15Cr-XNi and Fe-30Cr-YNi alloys after irradiation with 5 MeV Ni^{+} ions to 140 dpa (3,4).

Both Fe-35Ni (12) and Fe-35Mn (24) are known to exhibit the anomalous properties associated with the Invar phenomenon. Radiation destroys the anomalous properties in the Fe-Ni and Fe-Cr-Ni systems and thereby results in a decrease of the lattice parameter and an increase in density . It appears, therefore, that the Fe-Mn and Fe-Mn-Cr systems also may be undergoing spinodal decomposition. Since the densification is at least twice as large in the Fe-Mn system, perhaps the driving force is larger. This would in turn lead to an earlier onset of the radiation-induced spinodal process and the destruction of the resistance to void nucleation. One would therefore expect a lesser parameteric sensitivity of swelling in the Fe-Cr-Mn system compared to that of the Fe-Cr-Ni system.

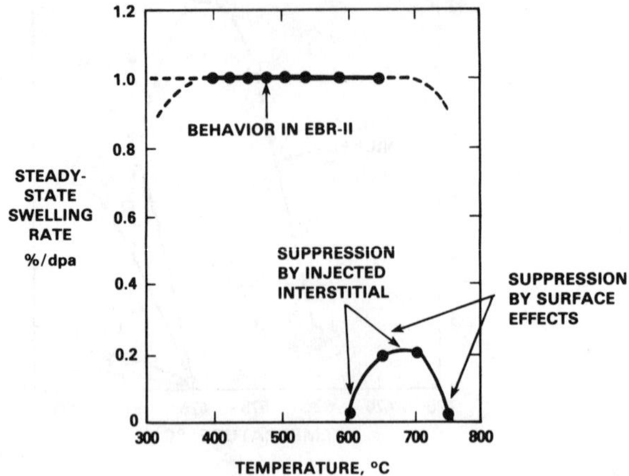

Figure 19 - Comparison of post-transient or "steady-state" swelling rates observed in a comparative irradiation of Fe-24.4Ni-14.9Cr by both Ni^+ ions and fast neutrons (16). The ion-induced rates were measured at the peak displacement region of the foil.

The dependence of swelling on solvent composition appears to be much more reproducible and much less sensitive to environmental history effects than that induced by solutes. This is supported by the lack of variability of swelling in nominally identical specimens. It is also demonstrated by comparison of the results of this experiment with those of the AD-1 experiment(25). In the AD-1 experiment the swelling of flat tensile specimens was colinear with the data from the present experiment, even though the relative fluence, flux and temperature relationships in the two data sets were different.

Another important facet of the composition dependence of swelling is that there appears to be a minimum transient regime of approximately 10 dpa in both the Fe-Cr-Ni and Fe-Cr-Mn systems which is not shortened by composition changes (this study), irradiation history (1), or differences in environmental variables (26). As shown in Figure 21 this minimum does not occur in pure nickel which initially swells at ∿1%/dpa with essentially no incubation period and then saturates at relatively low (≤15%) swelling levels (1). The reason for this minimum incubation period in the iron-based austenitic system is not yet understood.

In various charged particle experiments saturation of swelling has been observed but often at very high levels which are not relevant to engineering design needs. For instance, in the absence of helium, swelling reached ∿260% before saturating in a study of proton-irradiated AISI 316 (27). In neutron experiments swelling levels as large as 72-82% have been reported in controlled materials experiments (1,28) and levels of 40% have been reached in cold-worked AISI 316 fuel pin cladding (29). Both of these studies exhibited relatively temperature-independent post-transient rates of ∿1%/dpa. Reference 1 also contains neutron-induced swelling data on a wide variety of solute-modified commercial alloys which supports the contention that a swelling rate of ∿1%/dpa is the eventual fate of

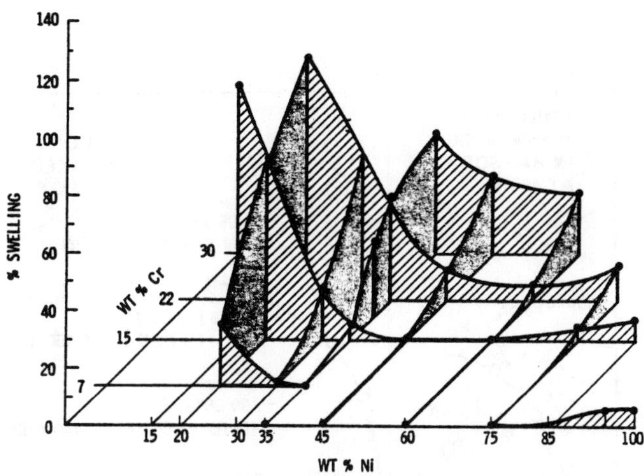

Figure 20 - Swelling of Fe-Ni-Cr alloys after irradiation at 675°C with 5 MeV Ni^+ ions to 116 dpa (3,4).

all austenitic alloys. It therefore does not appear that saturation will occur at design-relevant temperatures in the iron-based austenitic system.

It is interesting to note that reference 1 also shows that the general trends of swelling with respect to the influence of nickel, chromium and temperature are preserved in more complex commercial alloys. Solutes in general only extend the transient regime beyond that dictated by the solvent composition. Exceptions to this rule involve large changes in matrix composition as a result of solute-induced precipitation.

There is currently an initiative in the fusion materials community to develop alloys which will exhibit a reduced level of long-term radioactivity and permit eventual methods of waste disposal that are less restrictive than will be required for most current commercial structural alloys (30,31). The Fe-Cr-Mn system explored in this study represents an attempt to replace nickel which has a persistent radionuclide problem (32). It appears from the results of this study that the substitution of manganese for nickel does not change the basic response of swelling in the simple iron-based austenitic alloy system. A similar conclusion was reached for the more complex manganese-bearing commercial alloy AISI 216 (33).

Finally, there is a compelling simplicity and even beauty in the major conclusions of this study. The post-transient swelling rate is apparently a function of the f.c.c. crystal structure and is not influenced by the large number of variables previously thought to be involved. The duration of the transient regime beyond the ∿10 dpa minimum appears to be describable by simple arguments related to the effect of displacement rate, temperature and composition on vacancy diffusivity and does not require more complex arguments such as the action of interstitial binding, the presence of divacancies, etc. The failure to previously observe this simple behavior arose largely from the neutron-atypical aspects of the ion simulation procedure, the strong gradients of neutron flux in typical fast reactor experiments and the heavy concentration of effort on complex

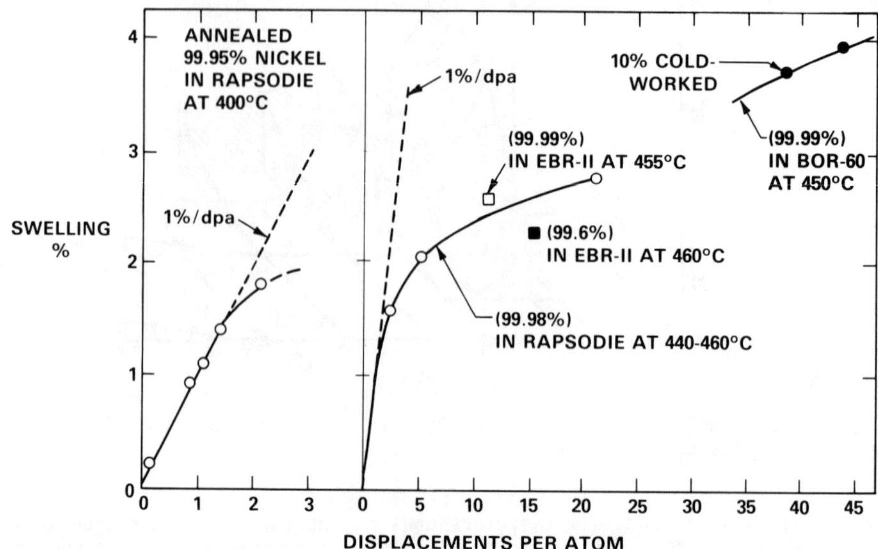

Figure 21 - Compilation by Garner (11) of neutron-induced swelling data observed in relatively pure nickel in fast reactors at 400°C (34) and at 440-460°C (35-38). All data are for annealed specimens except for the one designated as a cold-worked specimen.

solute-bearing alloys whose precipitation behavior often dominates and distorts the influence of the underlying solvent behavior.

Conclusions

The post-transient swelling rate of neutron-irradiated iron-based austenitic alloys is ~1%/dpa over a remarkably wide range of temperature. The compositional dependence of the swelling of these alloys resides primarily in the duration of the transient regime. There does not appear to be any potential in the austenitic system for development of an alloy optimized for swelling resistance as a consequence of a reduced post-transient swelling rate. Any optimized alloy must be based on attempts to extend the transient regime of swelling.

Acknowledgments

This work was initiated in the U.S. Breeder Reactor Program and completed under the auspices of the U.S. Magnetic Fusion Energy Program, both funded by the U.S. Department of Energy.

References

1. F. A. Garner, "Recent Insights on the Swelling and Creep of Irradiated Austenitic Alloys," Journal of Nuclear Materials, 122 & 123 (1984) pp. 459-471.

2. F. A. Garner and W. G. Wolfer, "Factors Which Determine the Swelling Behavior of Austenitic Alloys," Journal of Nuclear Materials, 122 & 123 (1984) pp. 201-206.

3. W. G. Johnston, T. Lauritzen, J. H. Rosolowski, and A. M. Turkalo, "The Effect of Metallurgical Variables on Void Swelling," pp. 227-266 in Radiation Damage in Metals, N. L. Peterson and S. D. Harkness, eds.; ASM, Cleveland, OH, 1976.

4. W. G. Johnston, J. H. Rosolowski, A. M. Turkalo, and T. Lauritzen, "An Experimental Survey of Swelling in Commercial Fe-Cr-Ni Alloys Bombarded with 5 MeV Ni^{+}Ions," Journal Nuclear Materials, 54 (1974) pp. 24-40.

5. D. R. Harries, "Void Swelling in Austenitic Steels and Nickel Base Alloys: Effects of Alloy Constitution and Structure," pp. 287-298 in Proceedings of Consultant Symposium on the Physics of Irradiation Produced Voids, AERE-R7934, Harwell, U. K., September 1974.

6. I. V. Gorynin and A. M. Parshin, "Special Features of Structural Transformations and Radiation Swelling of Constructional Alloys and Steels," Atomnaya Energiya, 50 (5) (1981) pp. 319-321.

7. D. Gulden and K. Ehrlich, "Void Swelling and Microstructural Development in Experimental Alloys," pp. 13-16 in Dimensional Stability and Mechanical Behavior of Irradiated Metals and Alloys, Vol. 1, British Nuclear Energy Society, London, 1983.

8. A. Hishinuma, Y. Katano, and K. Shiraishi, "Swelling and Nickel Segregation Around Voids in Electron-Irradiated Fe-Cr-Ni Alloys," Journal of Nuclear Materials, 103 & 104 (1981) pp. 1063-1067.

9. J. S. Watkin, "Dependence of Void Swelling on the Electron Vacancy Concentration," pp. 270-283 in Irradiation Effects on the Microstruture and Properties of Metals, ASTM STP 611, F. R. Shober, ed; American Society for Testing and Materials, Philadelphia, PA. 1976.

10. J. F. Bates and R. W. Powell, "Irradiation-Induced Swelling in Commercial Alloys," Journal of Nuclear Materials, 102 (1981) pp. 200-213.

11. F. A. Garner, "Swelling in Neutron-Irradiated 85Ni-15Cr at 400-650°C," Damage Analysis and Fundamental Studies Quarterly Progress Report, DOE/ER-0046/19, November 1984, pp. 66-73.

12. H. R. Brager and F. A. Garner, "The Origins of the Large Resistance to Void Swelling Observed in Fe-35.5Ni-7.5Cr," these proceedings.

13. F. A. Garner "Dependence of Swelling on Nickel and Chromium Content in Fe-Ni-Cr Ternary Alloys," Damage Analysis and Fundamental Studies Quarterly Progress Report, DOE/ER-0046/14, August 1984, pp. 133-151.

14. H. R. Brager and F. A. Garner, "Swelling of High Nickel Fe-Ni-Cr Alloys in EBR-II," ibid., pp. 152-159.

15. F. A. Garner and H. R. Brager, "Swelling of Fe-Ni-Cr Ternary Alloys at High Exposure," Damage Analysis and Fundamental Studies Quarterly Progress Report, DOE/ER-0046/16, February 1984, pp. 38-45.

16. F. A. Garner, "Impact of the Injected Interstitial on the Correlation of Charged Particle and Neutron-Induced Radiation Damage," Journal of Nuclear Materials, 117 (1983) pp. 177-197.

17. A. Kumar and F. A. Garner, "Dual Ion Irradiation: Impact of the Conflicting Roles of Helium on Void Nucleation," to be published in the Proceeding of ASTM 12th International Symposium on Effects of Irradiation on Materials, ASTM STP 870, F. A. Garner and J. S. Perrin, eds.; Williamsburg, VA., June 1984.

18. A. Kumar, "Influence of Helium-Generated Interstitials in Dual Ion Irradiation," to be published in Proceedings of First International Conference on Fusion Reactor Materials, Tokyo, Japan, December 3-6, 1985.

19. D: L. Plumton and W. G. Wolfer, "Suppression of Void Nucleation by Injected Interstitials During Heavy Ion Bombardment," Journal of Nuclear Materials, 120 (1984) pp. 245-253.

20. D. L. Plumton, H. Attaya and W. G. Wolfer, "Conditions for the Suppression of Void Formation During Ion Bombardment," Journal of Nuclear Materials, 122 & 123 (1984) pp. 650-653.

21. H. Rauh and R. Bullough, "The Effect of Climb Motion on the Dislocation Bias During Irradiation," UKAEA-Harwell Letter Report TP-1085, 1985.

22. B. Esmailzadeh and A. Kumar, "Influence of Composition on Steady-State Void Nucleation in Irradiated Metals," to be published in reference 17.

23. H. R. Brager and F. A. Garner "Microsegregation Observed in FE-35.5Ni-7.5Cr Irradiated in EBR-II," to be published in reference 17.

24. O. G. Sokolov and A. I. Mel'ker, "Invar Properties of Iron-Manganese Alloys," Soviet Physics-Doklady, 9 (1) (1965) pp. 1019-1021.

25. H. R. Brager and F. A. Garner, "Dependence of Swelling In Fe-Ni-Cr Alloys on Chromium and Nickel Content," Damage Analysis and Fundamental Studies Quarterly Progress Report, DOE/ER-0046/11, November 1982, pp. 221-231.

26. D. L. Porter and F. A. Garner, "Swelling of AISI 304L In Response to Simultaneous Variations in Stress and Displacement Rate," to be published in reference 17.

27. A. Kumar and F. A. Garner, "Saturation of Proton-Induced Swelling in AISI 316," Journal of Nuclear Materials, 117 (1983) pp. 234-238.

28. F. A. Garner, "Overview of the Swelling Behavior of 316 Stainless Steel," these proceedings.

29. B. J. Makenas, "Swelling of 316 Stainless Steel Cold Worked Fuel Pin Claddings and Ducts," to be published in reference 17.

30. H. R. Brager, F. A. Garner and D. S. Gelles and M. L. Hamilton, "Development of Reduced Activation Alloys for Fusion Service," to be published in reference 18.

31. R. L. Kleuh and E. E. Bloom, "The Development of Austenitic Steels For Fast Induced-Radioactivity Decay For Fusion Reactor Applications," these proceedings.

32. F. M. Mann, "Reduced Activation Calculations for the Starfire First Wall," Fusion Technology, 6 (1984) pp. 273-287.

33. D. S. Gelles and F. A. Garner "The Swelling Behavior of Manganese-Bearing AISI 216 Steel," to be published in reference 18.

34. Y. Adda, "Report on the CEA Program on Investigations of Radiation-Induced Cavities in Metals: Presentation of Some Results," pp. 31-83 in Proceedings of International Conference on Radiation-Induced Voids in Metals, CONF-710601, USAEC, Albany, N.Y., 1971.

35. G. Silvestre, A. Silvent, C. Regnard and G. Sainfort, "Alliages De Nickel-Fer et de Nickel-Silicium ne Gonflant das sous Irradiation aux Neutrons Rapides," Journal of Nuclear Materials, 57 (1975) pp. 125-135.

36. N. P. Agapova and none coauthors, "Swelling of Steels and Alloys Irradiated in the BOR-60 Reactor to a Fluence of $1.1 \cdot 10^{23}$ Neutrons/cm^2*," Atomnaya Energiya, 45 (6) (1978) pp. 433-439.

37. K. B. Roarty, J. A. Sprague, R. A. Johnson and F. A. Smidt, "Effect of Ni^+ -Ion Bombardment on Nickel and Binary Nickel Alloys," Journal of Nuclear Materials, 97 (1981) pp. 67-78.

38. H. R. Brager and J. L. Straalsund, previously unpublished data.

OVERVIEW OF THE SWELLING BEHAVIOR

OF 316 STAINLESS STEEL

F. A. Garner

Hanford Engineering Development Laboratory
Richland, WA 99352
USA

Summary

The austenitic stainless steel designated as AISI 316 is currently being used as the major structural material for fast breeder reactors in the United States, Britain and France. Efforts are now underway in each country to optimize the swelling resistance of this alloy for further application to both fission and fusion power generating devices. The optimization effort requires knowledge of the factors which control swelling in order that appropriate compositional and fabricational modifications can be made to the alloy specification. The swelling data for this alloy are reviewed and the conclusion is reached that optimization efforts must focus on the incubation or transient regime of swelling rather than the post-transient or "steady-state" regime. Attempts to reduce the swelling of this steel by solute modification have focused on elements such as phosphorus and titanium. It is shown that the action of these solutes is manifested only in their ability to extend the transient regime. It is also shown that irradiation at high helium/dpa ratios does not appear to change the conclusions of this study. Another important conclusion is that small differences in reactor environmental history can have a larger influence than either helium or solutes.

Introduction

The austenitic stainless steel designated as AISI 316 is currently being used in the cold-worked condition as the major structural material for fast breeder reactors in the United States, Britain and France. Due to the large amount of radiation experience on this steel it is also frequently considered to be a candidate for application in fusion reactors. Efforts are now underway in both the breeder and fusion materials communities to optimize the properties of this alloy for radiation service.

One major concern is the dimensional instability that develops as a result of the onset of void-induced swelling. Optimization efforts are of course directed toward the delay or reduction of swelling. This effort requires knowledge of the factors which control swelling in order that appropriate compositional and fabricational modifications can be made to the alloy specification. Second and even more important, it is essential that the materials community have a correct perception of the nature of swelling itself.

This latter consideration is most important when it is considered that many optimization efforts have been based on the implicit assumption that the steady-state swelling rate of austenitic steels is quite variable, depending strongly on temperature, displacement rate, composition, fabrication, stress and helium/dpa ratio. This variability offered the promise that the swelling rate could be reduced strongly by metallurgical solutions.

This paper reviews a large amount of recent data derived primarily from the U.S. Breeder Reactor Program and shows that solute-modified steels such as AISI 316 exhibit the same eventual behavior as developed by simple Fe-Cr-Ni ternary alloys. All austenitic alloys examined to date have been found to swell eventually at a post-transient swelling rate of $\sim$1%/dpa (1-3).

This conclusion previously has been obscured for a variety of reasons (1). First, the duration of the transient regime of swelling in AISI 316 is remarkably sensitive to minor differences in composition, thermal-mechanical treatment and particularly displacement rate (4-6), as shown in Figures 1 and 2. Second, the flux-temperature relationships in each fast reactor are different and there are strong gradients of displacement rate along structural components from which swelling data are extracted. Third, the transient swelling regime of AISI 316 is also quite sensitive to temperature history. Examples of this sensitivity are described in references 7-9. Fourth, swelling is also sensitive to applied stresses (10-12) found in operating reactor components but not necessarily in experimental subassemblies. Fifth, the accumulated strain at large exposures causes long reactor components to grow such that points far from the fixed end experience relatively large and time-dependent changes in position and environment (13). Each of these factors adds not only to the range of data scatter but also to the uncertainties associated with the description of environmental parameters and therefore to the difficulty of obtaining a correct perception of swelling.

Finally, most early experiments were conducted such that the data for the upper and lower temperatures of interest were obtained at the ends of the reactor core. Thus these data were obtained at lower displacement rates and lower total fluences, a condition which magnifies the influence of displacement rate on analysis of swelling data.

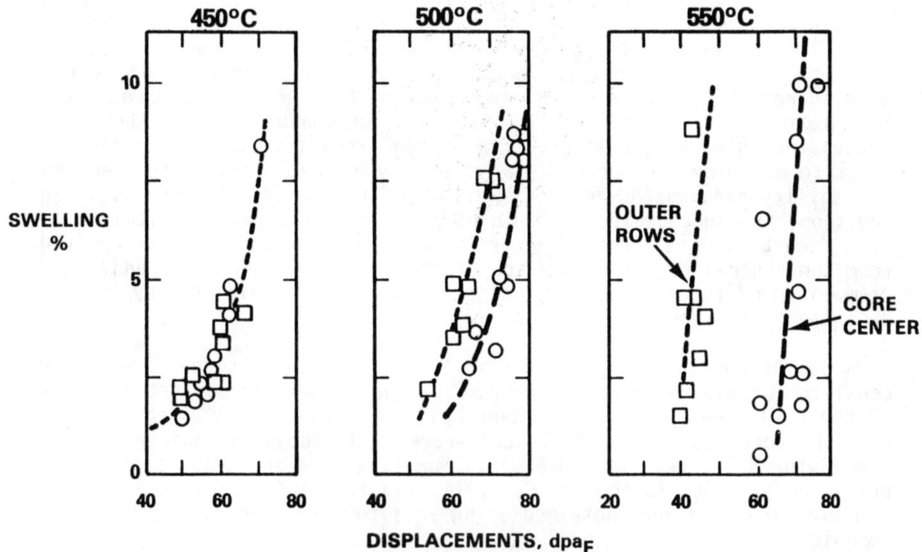

Figure 1 - The influence of temperature and displacement rate on the swelling of annealed AISI 316 in the Rapsodie reactor (5).

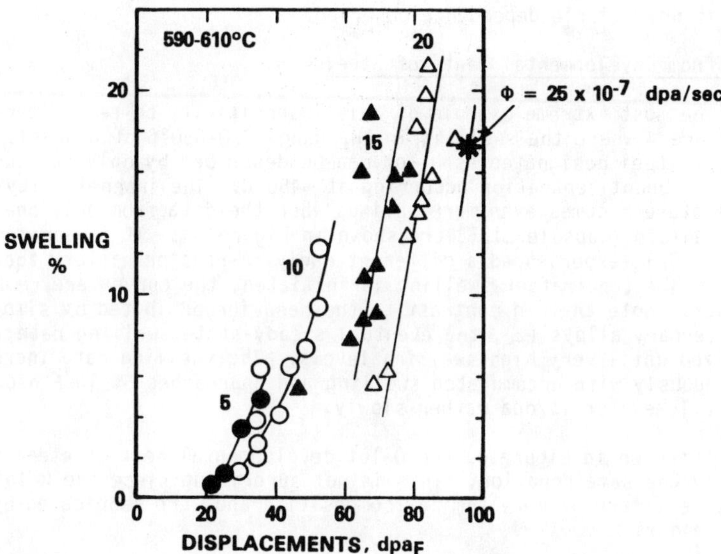

Figure 2 - The influence of displacement rate on swelling of cold-worked AISI 316 cladding at ~600°C in the Phenix reactor (6).

Results of Recent Experiments

All of the above considerations contributed to the early difficulty of forming a correct perception of the parametric dependency of swelling in AISI 316. In recent US experiments these factors were minimized. All data at any one temperature were obtained from one position in the core and were therefore developed at the same neutron flux and displacement rate. The specimens were periodically discharged for measurement and then reincapsulated into new subcapsules for further irradiation. This procedure tends to minimize the influence of reactor history, specimen-to-specimen variability and neutron-induced swelling of the irradiation vehicle. In addition, the subcapsules for each temperature were arranged within the core such that data for the higher irradiation temperatures were derived from the central region of the core rather than from the upper end. A comparison of the relative flux-temperature relationships for typical experiments is shown in Figure 3.

One of the major experiments in this series was designated RS-1 and consisted of several irradiation vehicles which were irradiated in Row 2 of the EBR-II fast reactor in Idaho Falls, Idaho. It contained twelve well-characterized heats of 20% cold-worked 316 tubing irradiated at nine temperatures between 370 and 650°C. Swelling data on these steels were extracted at peak fluences of 5, 9, 14, and 17 x 10^{22} n/cm^2 (E > 0.1 MeV). The swelling data were obtained using an automated immersion density technique.

The large amount of data obtained from this one experiment will not be covered in detail here. It is sufficient for our purpose to present only typical results showing the range of observed behavior.

The most surprising conclusion was that in contrast to our previous perception of the strong dependence of swelling on temperature, there was surprisingly little dependence observed.

Data from Developmental Heats of Steel

The most extreme example of this insensitivity to temperature is shown in Figure 4 where the swelling in the range 370-650°C of a developmental heat of steel designated as N-lot can be described by only two curves, with an abrupt separation occurring at ~480°C. The insensitivity to temperature becomes even more obvious when the data from only one irradiation vehicle (capsule B121) are shown in Figure 5a. It is apparent that capsule B210 experienced a different early in-reactor history than did capsule B211, but after swelling is initiated, the curves are remarkably similar. Note that in contrast to the behavior exhibited by simple solute-free ternary alloys (2), the eventual steady-state swelling rate is not realized until very high swelling levels. The swelling rate increases continuously with accumulated swelling and approaches 5%/10^{22} n/cm^2 (E > 0.1 MeV) or 1%/dpa rather slowly.

As shown in Figure 6, the O-lot developmental heat of steel exhibits exactly the same behavior. This is not surprising since the N-lot and O-lot heats are of very similar composition and were fabricated by the same vendor.

When another lot of developmental tubing designated R-lot is examined, we find the other extreme of the behavior exhibited by the steels in the RS-1 experiment. Note in Figure 7 that the post-transient behavior is essentially independent of temperature but the transient regime itself is

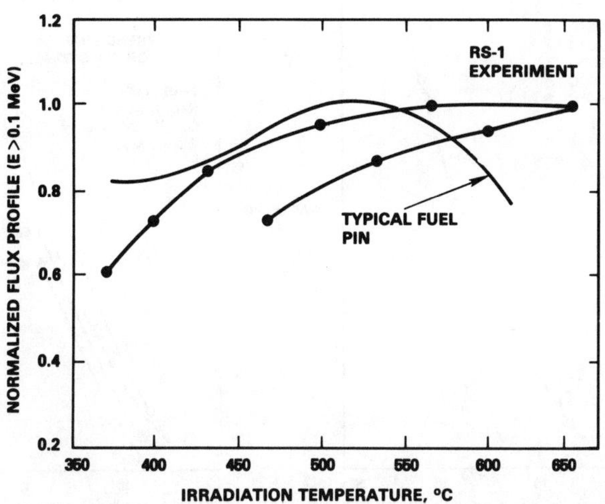

Figure 3 - Relative neutron flux/temperature relationships for the RS-1 experiment and a typical fuel pin.

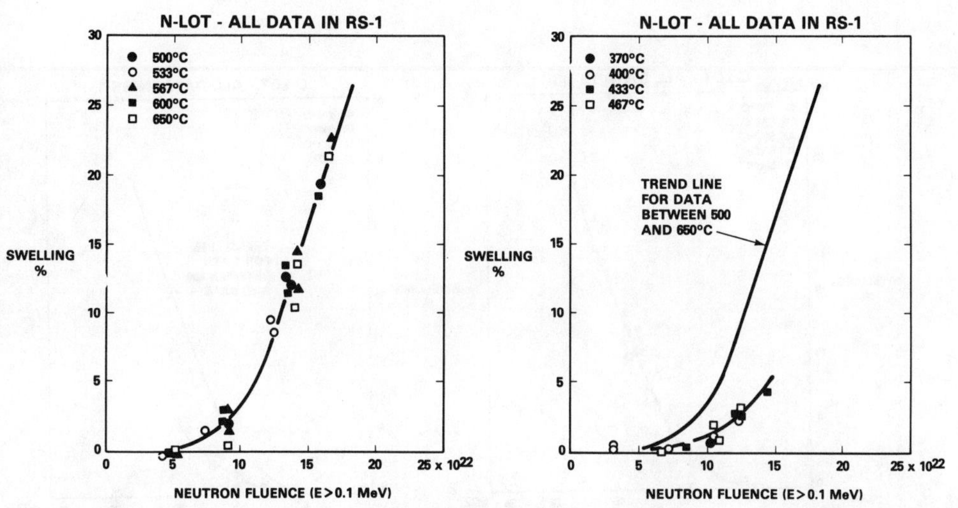

Figure 4 - Swelling data of N-lot from two capsules of the RS-1 experiment showing all data in the range 400-650°C.

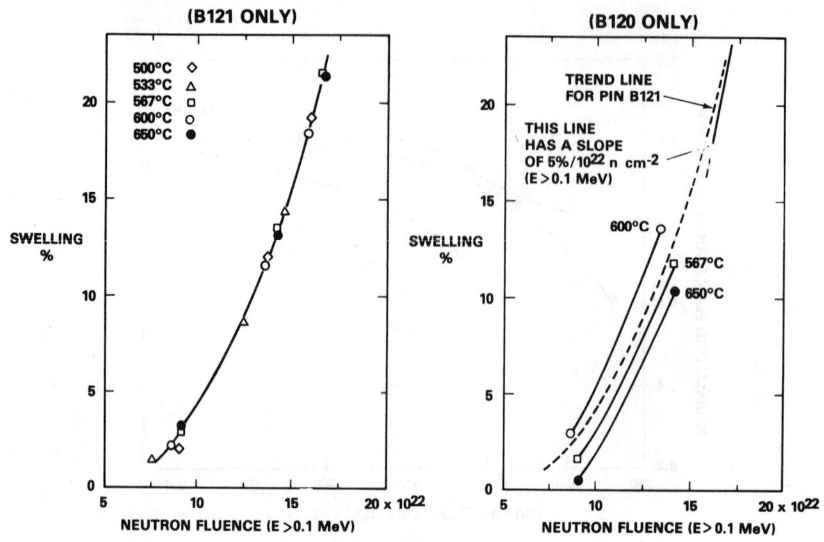

Figure 5 - N-lot swelling data from the RS-1 experiment, subdivided into data groups corresponding to the irradiation vehicle in which the irradiation occurred.

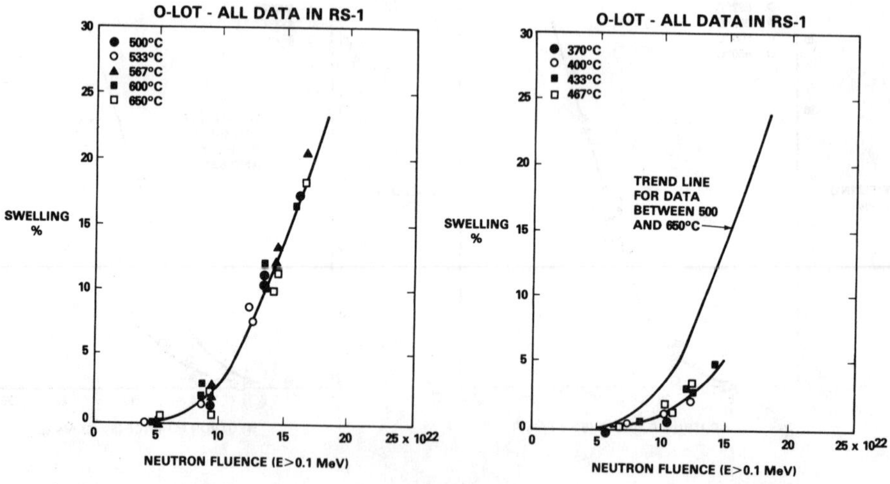

Figure 6 - O-lot swelling data from the RS-1 experiment.

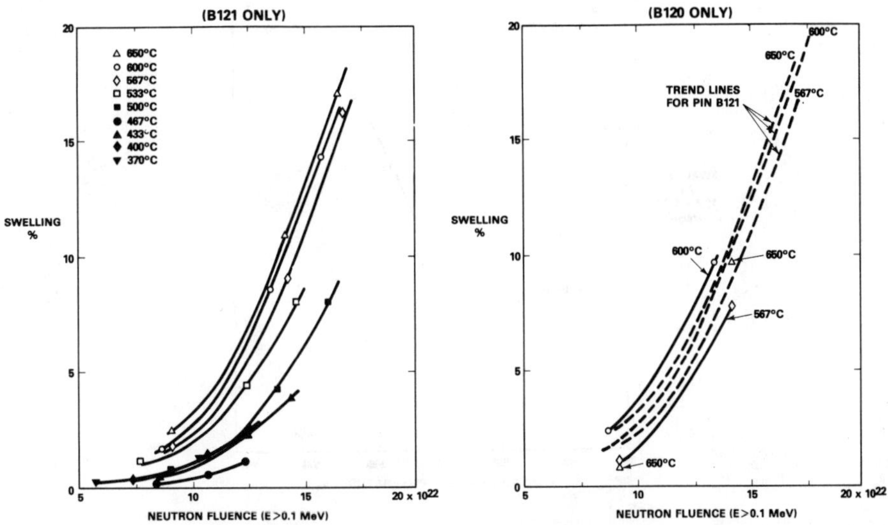

Figure 7 - Swelling of 20% cold-worked R-lot steel from the RS-1 experiment at 500-650°C.

very sensitive to temperature. In contrast to the behavior of the N-lot and O-lot heats, the temperature sensitivity of incubation was observed in both the B-120 and B-121 subcapsules. Note in Figure 7a that the largest incubation period is reached at 467°C and is shorter at both lower and higher temperatures. This maximum in incubation fluence is believed to be the origin of the "double-bump" swelling behavior shown in Figure 8 and is almost always observed in a more pronounced fashion in solution-annealed 316. It is occasionally observed in cold-worked AISI 316, particularly when the carbon level is relatively low (14). The double-bump swelling behavior will be covered in more depth in a later section.

Data from Prototypic FFTF Steels

Five separate heats of steel (designated CN-13, CN-17, BB, X and DD) were developed to meet the specifications required for construction of components in first core of the Fast Flux Test Facility (FFTF) in Richland, Washington. One of these heats, DD-lot, was rejected for first core service because its nitrogen content was above the specified limit. It was included in the experimental program, however.

Swelling Behavior of the BB-Lot of Steel

Figure 9 shows that BB-lot exhibits a plateau of void swelling with temperature which is not as broad as that seen in the N-lot and O-lot heats of developmental steels. It is obvious from Figure 9 that the primary difference in swelling response of these two heats of steel arises in the duration of the transient regime. BB-lot starts to swell earlier and is relatively insensitive to the differences that occur in the two irradiation

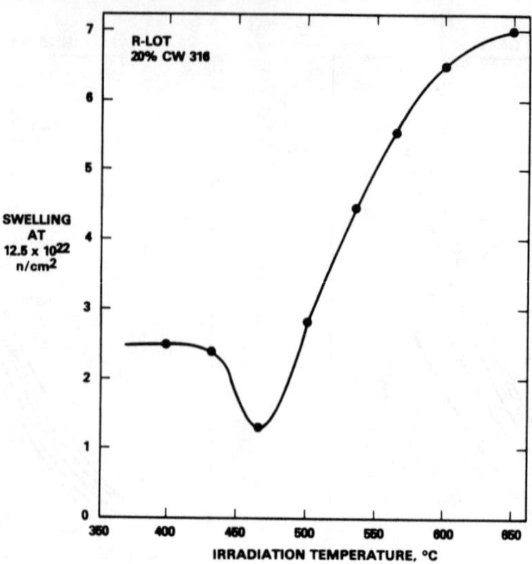

Figure 8 - Swelling of 20% cold-worked R-lot steel at 12.5 x 10^{22} n/cm^2 (E > 0.1 MeV) or 63 dpa.

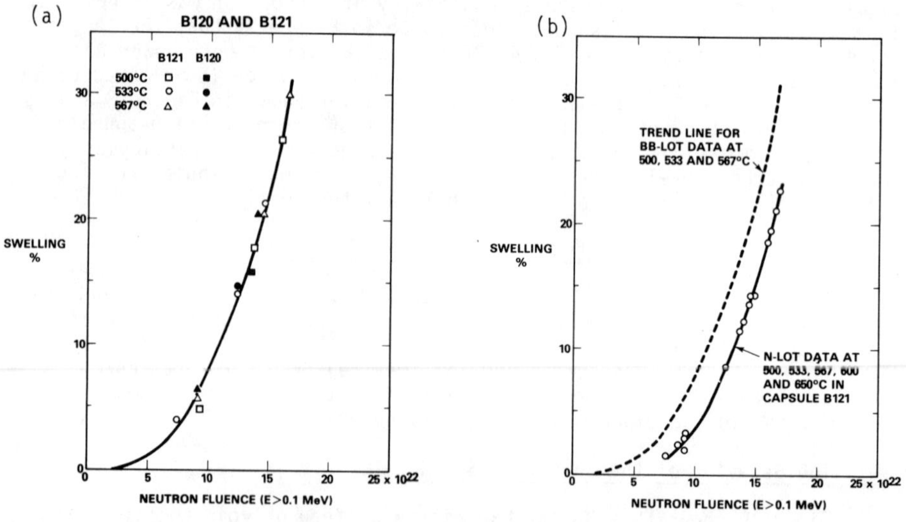

Figure 9 - (a) Swelling of 20% Cold-Worked BB-Lot (Heat 81583) at 500, 533 and 567°C in the RS-1 experiment, capsules B-120 and B-121. (b) Comparison of BB-lot and N-lot behavior in the RS-1 experiment.

capsules. As shown in Figure 10, differences in swelling due to differences in capsule history are found in BB-lot only above 600°C. The duration of the transient regime is also sensitive to temperature below 500°C.

Comparison of BB-Lot and Other First Core Steels

Figure 11 shows that in the range 400-567°C the swelling of BB, CN-13, CN-17 and X-lot heats in the B-121 capsule are essentially identical. Not only do these first core steels start to swell sooner than does N-lot, but their behavior below 480°C is not as abrupt as that of N-lot.

Figure 12 shows that X-lot experiences a progressively shorter transient regime at temperatures of 600°C and above. The CN-13, CN-17 and BB-lot steels behave in a very similar manner, however, and all could be described by one temperature-dependent swelling equation. While history-related effects of these heats in the various capsules are small at temperatures below 600°C, they become significant above this temperature, as shown in Figure 12b.

Swelling of DD-Lot

This heat swelled at slightly higher levels at all temperatures than was observed in the first core FFTF heats probably due to its higher nitrogen content. The higher swelling arose from a slightly shorter transient regime, but the curves for DD-lot are parallel with those of other heats. Note in Figure 13 that once again history-related effects occurred only at temperatures of 600 and 650°C.

Swelling of AISI 316 at Temperatures Greater Than 650°C

Early in the breeder program it was thought that swelling peaked between 550 and 600°C and declined quickly at temperatures above 600°C. The RS-1 experiment was therefore designed to put the 550-650°C temperature range in the peak flux regime of the EBR-II reactor but the experiment included no irradiation temperatures above 650°C. As can be seen from the many curves shown in this report the high-fluence swelling rate is independent of temperature in the range 500-650°C and shows no tendency to decline toward the upper end of that range.

How high in temperature does swelling of AISI 316 extend? At what temperature does the high fluence swelling rate begin to decline? While these questions cannot be answered definitively, there are some data available that indicate that swelling persists at least to temperatures in excess of 700°C.

Straalsund and coworkers (15) published some of the earliest high temperature data on both 316 and 304 stainless steels. They showed that voids formed in annealed 316 stainless steel irradiated to 3×10^{22} n/cm^2 ($E > 0.1$ MeV) at 740°C. No voids were found at this low fluence in 27% cold-worked AISI 316. It was also shown that swelling in 50% cold-worked 304 stainless steel persisted to ∿800°C.

Although the swelling-oriented subassemblies of the U.S. Breeder Program did not probe the temperature regime above 650°C, pressurized tubes used to study irradiation creep were irradiated at temperatures as high as 720°C. (Above this temperature the thermal creep rates are too high to warrant the service of AISI 316 in a breeder reactor). Two unstressed specimens of heat CN-13 irradiated at temperatures above 700°C have been examined by electron microscopy and the results are reported in detail

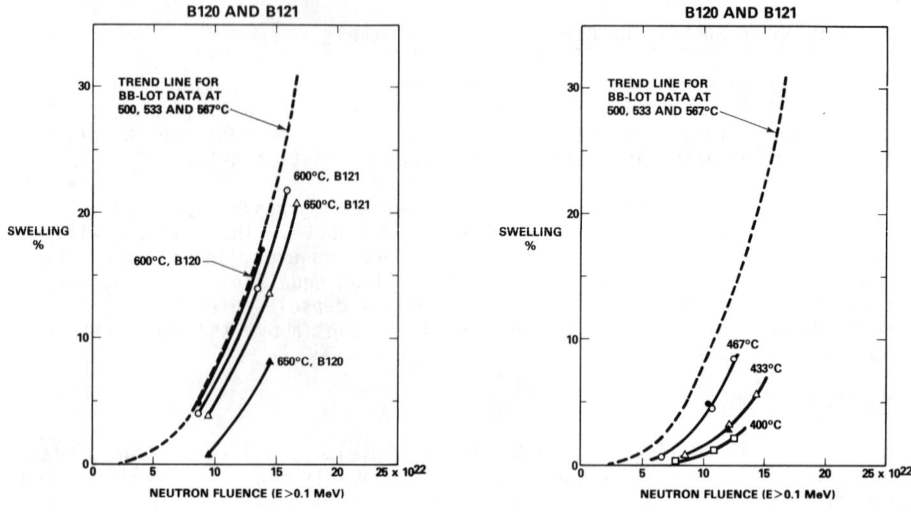

Figure 10 - Swelling of 20% cold-worked BB-lot at temperatures above and below the 500-567°C range.

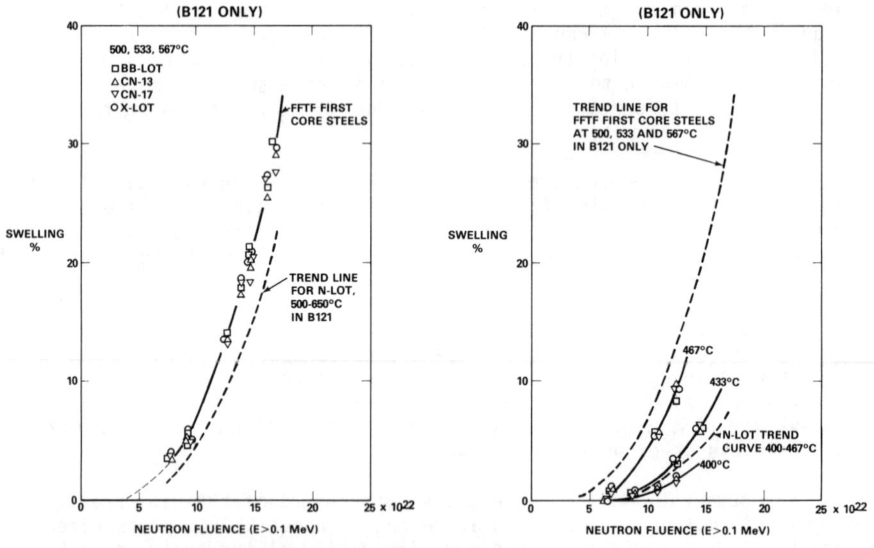

Figure 11 - Comparison of the swelling between 400 and 567°C of tubing lots BB, X, CN-13 and CN-17 in the B-121 capsule of the RS-1 experiment. The behavior of N-lot is shown for comparison.

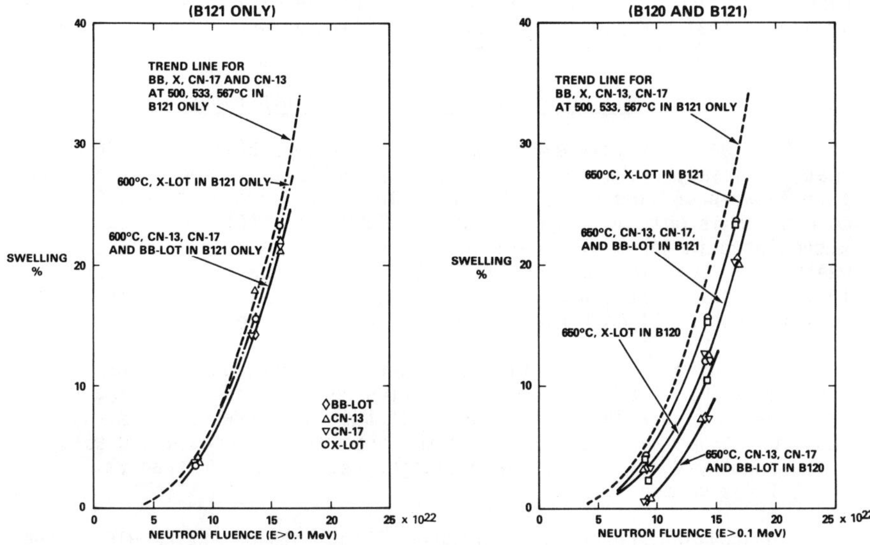

Figure 12 - Comparison of the swelling at 600 and 650°C of tubing lots BB, X, CN-13 and CN-17 in the B-121 and B-120 capsules.

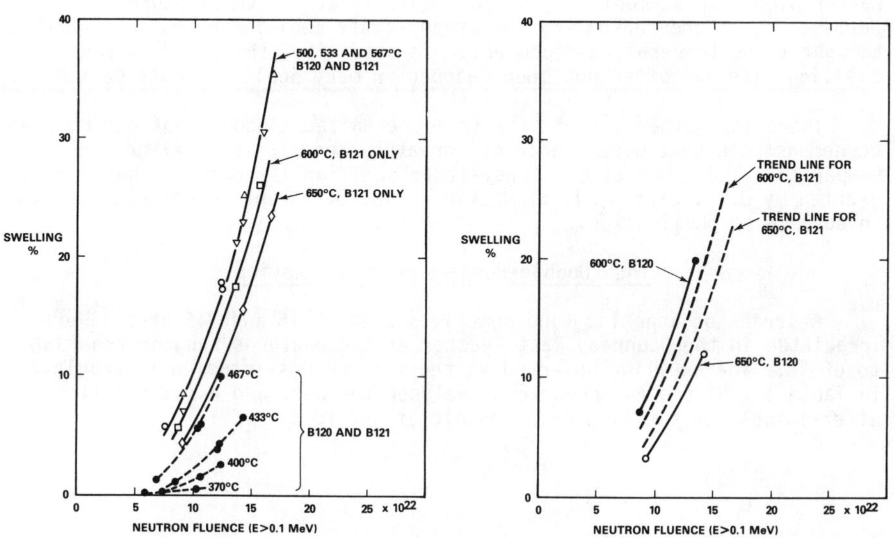

Figure 13 - Swelling of DD-Lot in the RS-1 experiment.

elsewhere (16). It was found that the swelling was comparable to that observed at 650°C for this heat of steel in the RS-1 experiment. This indicates that the plateau region of swelling extends to much higher temperatures than previously anticipated.

Swelling of AISI 316 at Lower Temperatures

For the relatively early swelling first core FFTF heats one can see that the steady-state swelling rate is essentially independent of temperature between 467 and 650°C. For the slower swelling heats (N, O, R) this conclusion is only defensible down to 500°C. Insufficient data have been accumulated in the RS-1 experiment at temperatures below 500°C to determine whether the swelling curves will continue to increase in slope at higher fluence. There are a number of reasons, however, to believe that the swelling rate will continue to increase with fluence.

First of all, the low fluence experiments are run in a portion of the core where the displacement cross section per neutron is falling rapidly as shown in Figure 14. This means that the curvature of the swelling behavior at 400 and 433°C shown in the preceding figures would actually be more pronounced if the data were plotted vs. dpa rather than vs. $E > 0.1$ MeV neutrons.

In addition, recently reported swelling data on 20% CW 316 in the AA-I experiment show that at 420°C the swelling rate is still increasing at 20.0×10^{22} n cm^{-2} ($E > 0.1$ MeV) and is almost at 1%/dpa (1,17). A similar behavior at low temperature has been observed in irradiations conducted in the Dounreay Fast Reactor in Scotland (18). Swelling has been observed at temperatures approaching 320°C in this reactor, which has a lower inlet temperature (270°C) compared to that of EBR-II (370°C).

It is obvious that transient curvature in swelling can persist to rather high displacement levels, particularly at low temperatures. This persistence is one reason why the steady-state swelling rate is frequently thought to be temperature-dependent. In actuality, the steady-state swelling rate has often not been reached in many published data sets.

There is another set of data from the United Kingdom that can be used to demonstrate that persistence of curvature tends to camouflage the temperature-independence of steady-state swelling. Permission has been granted by J. H. Gittus, J. S. Watkin and J. Standring to cite these data in advance of publication.

The "Double-Bump" Swelling Behavior

A series of annealed (and sometimes aged) M316 and 316 steels were irradiated in the Dounreay Fast Reactor at temperatures ranging from 350 to 600°C. The swelling observed at the maximum dose studied is tabulated in Table I. Gittus and coworkers analysed the data and assumed a linear-after-incubation swelling relationship of the form

$$\frac{\Delta V}{V_o}(\%) = A\,(D-D_o)\ ,$$

where D is the displacement dose (in Half-Nelson dpa) and A is the "steady-state" swelling rate in units of %/dpa. The values of A and D_o (the incubation dose) are also shown in Table I. A and D_o were considered to be material constants, varying for each material and with temperature.

Table I. Swelling of M316, 316 and 316L in the Dounreay Fast Reactor[†]

Material Condition	Irrad. Temp.	A	D_0	Number of Data Points	Standard Deviation	Swelling at Maximum Dose Studied	
	°C	% per dpa*	dpa*		$\frac{\Delta V}{V_o}$ %	$\frac{\Delta V}{V_o}$ %	D
M316, 1050°C, 1/2 h	595	0.4	21.6	6	1.0	10.06	45.7
	565	0.13	20.1	2	-	3.50	47.6
	535	0.19	24.3	2	-	3.70	43.8
	476	0.11	37.2	2	-	2.75	37.2
	430	0.26	20.5	3	0.02	3.47	34.0
	403	-	-	-	-	0	34.4
	380	-	-	-	-	0	23.0
M316, 1000°C,1 h + 850°C,24 h	595	0.48	21.5	6	2.5	15.40	47.2
	568	0.44	19.3	2	-	12.76	48.1
	538	-	-	-	-	0.9	11.9
	488	0.16	14.4	2	-	3.84	38.7
	430	0.41	20.4	3	0.01	5.56	33.8
	415	0.098	10.5	2	-	1.66	29.0
	403	0.075	26.3	3	0.03	0.6	34.8
	400	-	-	-	-	0	22.2
	360	-	-	-	-	0	21.8
M316, 1100°C, 1 h + 850°C, 24 h	595	0.52	23.0	4	3.0	13.30	44.2
	588	0.62	18.9	2	-	18.0	47.6
	576	0.49	20.9	2	-	13.45	48.5
	545	0.19	22.3	2	-	4.39	45.2
	499	0.15	14.8	2	-	3.82	40.0
	430	0.41	21.0	3	0.02	5.12	33.4
	422	0.091	12.4	2	-	1.64	30.4
	403	0.081	23.8	4	0.3	0.91	35.0
	350	-	-	-	-	0	20.6
M316, 1200°C, 1 h + 850°C, 24 h	595	0.43	21.4	4	0.8	10.7	44.9
	588	0.61	20.1	2	-	17.2	48.1
	580	0.49	20.2	2	-	14.2	48.8
	553	0.19	23.7	2	-	4.35	46.0
	507	0.16	18.2	2	-	3.88	41.1
	438	0.10	14.1	2	-	1.87	32.0
	430	0.43	21.8	3	0.31	4.61	33.0
	403	0.19	28.7	3	0.59	1.18	35.0
	350	-	-	-	-	0.1	19.2

†M316 and 316L are British designations for modified 316 and 316-low carbon.
*Dose given in Half-Nelson (N/2) displacements; 1 dpa (NRT, Austenitic) = 1.15 dpa (N/2 for Fe).

Table I Continued

Material Condition	Irrad. Temp.	A	D_0	Number of Data Points	Standard Deviation	Swelling at Maximum Dose Studied	
	°C	% per dpa*	dpa*		$\frac{\Delta V}{V_0}$ %	$\frac{\Delta V}{V_0}$ %	D
M316,							
1250°C, 1 h	595	0.42	22.2	4	0.4	9.90	45.6
+	588	0.68	23.4	2	-	17.1	48.4
850°C, 24 h	557	0.33	25.1	2	-	7.07	46.6
	518	0.14	17.9	2	-	3.45	42.1
	449	0.11	13.2	2	-	2.30	33.9
	430	0.46	22.6	3	0.2	4.51	32.8
	403	0.33	30.4	3	0.7	1.92	34.8
316,							
1070°C,	595	0.48	20.9	4	1.8	12.95	44.9
air cool	588	0.59	19.7	2	-	16.93	48.1
	568	0.33	19.0	2	-	9.78	48.1
	538	0.23	22.5	2	-	5.15	44.5
	488	0.09	13.2	2	-	2.3	38.7
	430	0.09	18.1	3	0.4	1.23	33.8
	415	0.06	15.4	2	-	0.79	29.0
	403	-	-	-	-	0	34.8
	350	-	-	-	-	0	18.2
316L*,							
*1050°C,	595	0.25	11.5	6	0.96	6.86	32.8
1/2 h	584	0.32	17.0	2	-	5.87	35.2
	557	0.19	6.9	2	-	5.12	33.7
	518	0.20	8.3	2	-	4.47	30.4
	430	0.20	8.5	5	0.05	4.65	31.8
	422	0.1	6.9	2	-	1.52	22.1
	403	0.26	11.4	3	0.02	1.33	16.6
	380	0.14	10.2	2	-	1.86	23.0

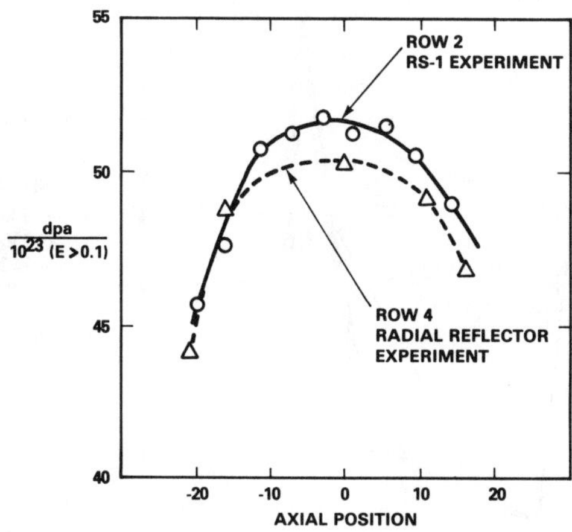

Figure 14 - Displacement Characteristics of EBR-II Neutrons as a Function of Core Position.

It is easy to see that linear-after-incubation descriptions of swelling in 300 series stainless steels yield underestimates of the swelling rate compared to that obtained in curvilinear descriptions. The error arising from the underestimate becomes smaller with accumulated swelling, however.

Five of the eight irradiated steels were derived from a single heat of M316 stainless steel (Fe-17.2Cr-13.7Ni-2.4Mo-1.85Mn-0.63Si-0.035C) and were irradiated in either the annealed condition (1050°C, 1/2 hr) or the annealed and aged condition, using four annealing temperatures and one aging condition. These temperatures are shown beside Figure 15. The four "steady-state" swelling rate curves shown in Figure 15 exhibit the double-bump behavior usually observed in the swelling of annealed AISI 316. The other steels listed in Table 1 also exhibit the same double-bump behavior.

The minimums observed in these curves at 475-525°C is thought to represent the longest incubation period for annealed steel and to be related to the temperature dependence of precipitate evolution(14). Note that these minimums are quite similar to that seen in Figure 8. There was a peaked neutron flux profile over the pin in which this experiment was conducted, however. This tends to pull the apparent swelling rate down at the extremities of the temperature range in which this experiment was conducted. The resultant fluence gradients can be normalized out of the data if either the swelling or swelling rate data are plotted against fluence. As seen in Table 1, however, only the maximum swelling values were reported by Gittus and coworkers.

A more instructive way to view the data for our purpose is to plot the swelling rate observed vs. the maximum swelling level in the data set from which that rate was derived. This allows us to test the hypothesis that the swelling rate increases with total swelling and that temperature is not an important variable. Note in Figure 16 that the swelling rate of all eight steels can be described by one curve for all temperatures above 475°C. It also increases with swelling level and is approaching a value

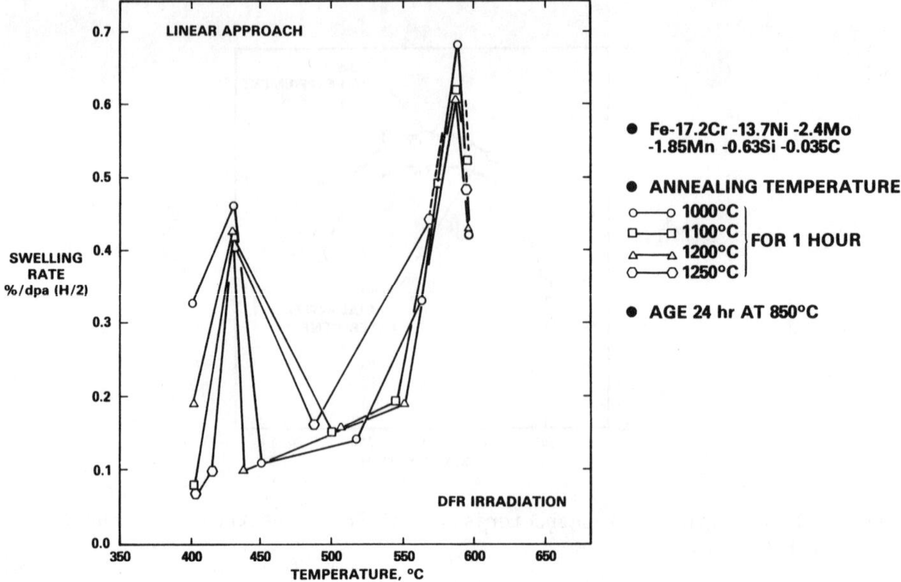

Figure 15 - "Double-bump" swelling rate profiles observed in annealed and aged M316 after irradiation in the Dounreay Fast Reactor, as observed by Gittus, Watkin and Standring. The swelling rates shown above are actually underestimates (of varying degree) arising from the premature assumption of swelling having reached the regime where it is linearly proportional to displacement dose.

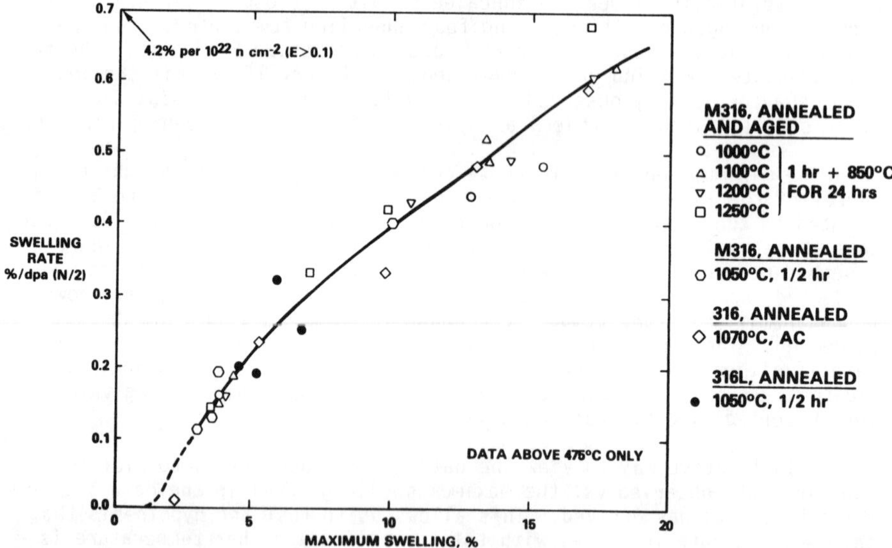

Figure 16 - Apparent swelling rate vs. maximum swelling in each data set for M316 and 316 steels from Table I, for temperatures greater than 475°C only.

of $\sim$5%/10^{22} n/cm^2 (E > 0.1 MeV). This is approximately 1%/dpa when converted from Half-Nelson dpa to the recommended NRT international system for calculating displacements.

A similar trend is observed for data below 475°C as shown in Figure 17. Even though the transients in swelling are longer, the post-transient rate of approach to steady state swelling is faster than that observed above 475°C. In effect, this analysis has demonstrated the relative temperature independence of swelling of annealed steels, but leads to the conclusion that there are two separate incubation regimes. Remember that a similar behavior was observed in the N-lot heat of 20% cold-worked AISI 316, as shown in Figure 4.

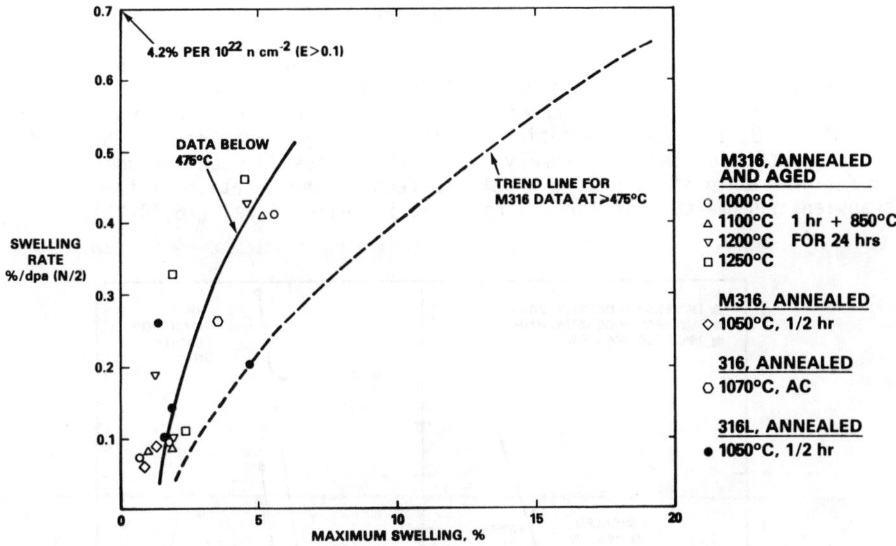

Figure 17 - Apparent swelling rate vs. maximum swelling in data set for M316 and 316 steels from Table I, for temperatures less than 475°C only.

It is obvious that the variation observed in the materials constant A (%/dpa) of Table I reflects more a consequence of prematurely assuming the linear behavior of swelling with dpa than any real dependence on either composition, thermal treatment or irradiation temperature.

Watkin noted in transmitting the data to the author that some caution should be exercised in interpreting the data from the Dounreay reactor. Some of the specimens, particularly those irradiated at higher temperatures, were subjected to changes in temperature due to rig malfunctions. He felt that this may have caused some of the higher swelling levels observed. He also noted that all specimens were in the same rig so that the relative temperature levels were maintained. Based on recent studies, however, it is now known that changes in temperature impact only the duration of the transient regime of swelling and exert no influence thereafter.[8] Since Figures 16 and 17 have been plotted in such a manner as to remove the influence of transient differences, there is no reason to discount the conclusion drawn from this analysis.

Effect of Large Levels of Helium on Swelling

Irradiation of AISI 316 in fusion environments will cause much larger amounts of solid transmutants and gases such as hydrogen and helium to be generated compared to that generated in fast reactors. While fusion reactor data are not yet available, irradiation in mixed-spectrum reactors such as HIFR (High Flux Isotope Reactor at Oak Ridge, Tennessee) also leads to large levels of helium.

The differences in non-gaseous transmutants in various neutron spectra can be substantial (19), but for AISI 316 in HFIR and EBR-II such differences were found to manifest themselves primarily in precipitate composition and not in the transient duration or the post-transient swelling rate (16,20-21). Considering the compositional insensitivity of the post-transient regime demonstrated in this study, such a result is not surprising, and is not expected to be altered in fusion environments.

Another consideration is that of the helium/dpa ratio. As shown in Figure 18, the very large difference in helium/dpa ratio between HFIR and EBR-II did not lead to a difference in the post-transient swelling rate of annealed AISI 316. A compilation of other helium effects studies has led to the conclusion that relatively small differences in starting composition or fabrication history have a greater effect on the duration of the transient regime than do large variations in helium level (16,20-22).

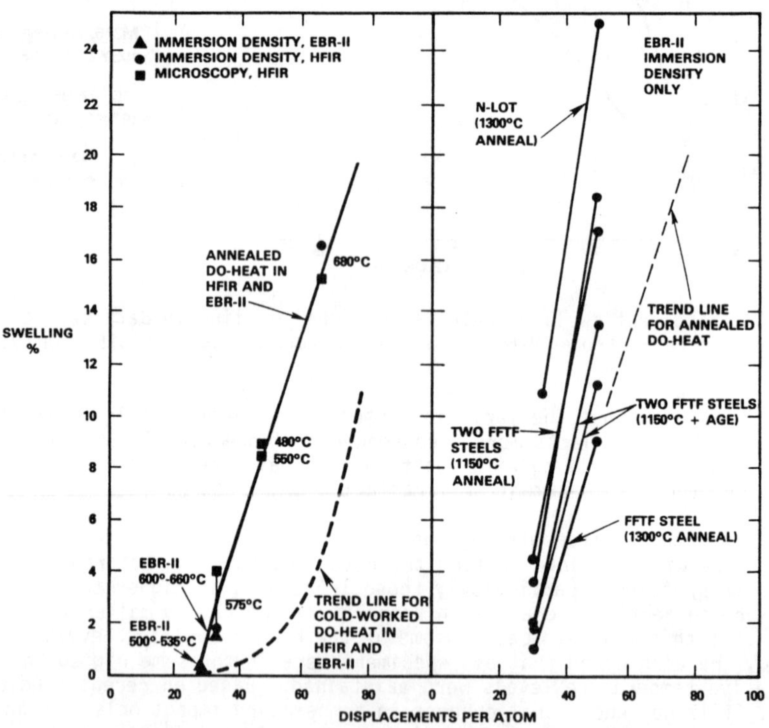

Figure 18 - Comparison of several heats of annealed AISI 316 irradiated in HFIR and EBR-II, showing relative insensitivity of post-transient swelling to temperature and neutron spectra (1,16).

Solute Modification of AISI 316

The question is often raised whether the modification of austenitic steels by solute additions can forestall the inevitability of reaching the ∿1%/dpa swelling rate exhibited by pure Fe-Ni-Cr ternary alloys and various 300 series stainless steels (1-3,13). Although it is known from numerous studies that titanium additions tend to suppress swelling of austenitic alloys, it has not been shown previously whether the benefit of adding titanium and other elements results in a permanent reduction in the maximum swelling rate or just a delay in the onset of swelling. In a recent design study on the Starfire Tokamak fusion plant it was stated that the assumed swelling of the titanium-modified alloy designated PCA (Prime Candidate Alloy for the fusion Path A alloy series) was one-tenth that of AISI 316 (23). This optimistic assumption implies that solute additions, particularly of titanium, suppress the steady-state swelling rate rather than just extend the incubation period.

The most relevant fission reactor data is that derived from irradiations of PCA and 316+Ti in the high He/dpa environment of HFIR, (24,25) but these data are too sparce to determine whether the inherent post-transient swelling rate of austenitics (∿1%/dpa) can be modified by titanium or other elements. If one ignores the possibly synergistic effects of the helium and solutes, however, there are relevant data from the U.S. Breeder Reactor Program that can be used to address this question.

Three sets of EBR-II data will be considered. The first set will be used to establish the relative behavior of Ti-modified 316 and unmodified 316, and also to determine the influence of cold-work on this relationship. Using the insight gained from this analysis two other data sets will be analyzed for the effect of more extensive compositional variations.

Swelling of LS-1

The alloy LS-1 was developed at Oak Ridge National Laboratory and is one of the earliest titanium-modified alloys irradiated in the U.S. Breeder Program. Consequently this alloy has one of the highest levels of reactor exposure. Its composition in wt.% is Fe-13Ni-16Cr-1.8Mo-1.0Mn-0.9Si-0.10Ti-0.05C. In addition to the titanium, this alloy has about twice the normal silicon level used in AISI 316 in the U.S. Breeder Program. Silicon has a particularly pronounced influence on delaying the onset of swelling (26-28).

Figure 19a shows a comparison of the swelling behavior of 20% cold-worked LS-1 with the temperature-independent behavior of 20% cold-worked N-lot AISI 316. As shown earlier, N-lot, which contains very little titanium, does not exhibit the temperature sensitivity of incubation observed in many heats of AISI 316. It is therefore used as a standard "template" curve with which to compare the development of swelling in other alloys. Note that for swelling levels >5% the LS-1 curves tend to parallel the N-lot curve. The transient behavior of LS-1 appears to be quite sensitive to temperature however. This sensitivity was observed in other heats of AISI 316 and is therefore not entirely a consequence of titanium addition.

Figure 19b shows that the annealed condition of LS-1 at 510°C is also swelling at a rate comparable to that of N-lot but with a shorter transient period. Note that the dotted lines drawn in Figure 19b indicate that the N-lot steel at comparable voidage levels swells at the same average rate as that determined for LS-1 at 510°C. Figure 19c shows that annealed LS-1 at other irradiation temperatures also approaches a steady-state swelling

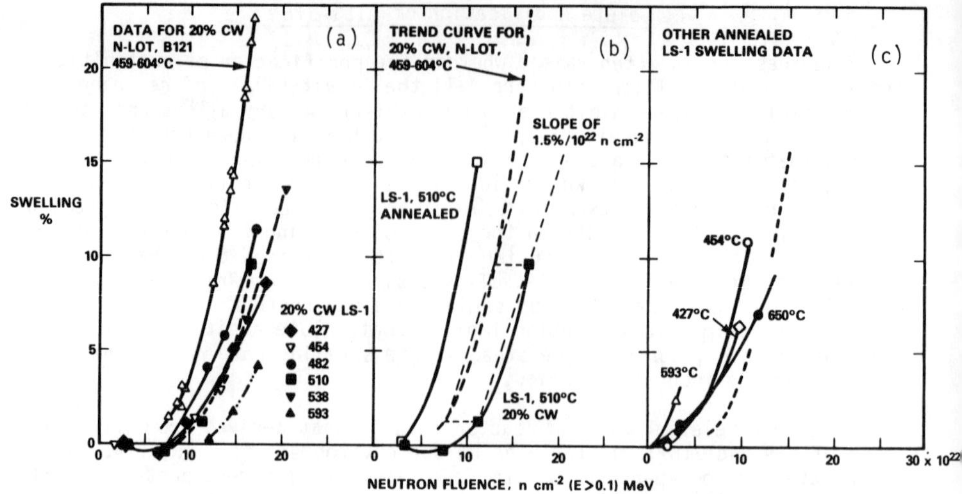

Figure 19 - (a) Comparison of the swelling behavior in EBR-II of N-Lot AISI 316 and LS-1 at various temperatures. Both alloys are in the 20% cold-worked condition. (b) Comparison of annealed and cold-worked LS-1 swelling behavior at 510°C. Note that linear extrapolation of LS-1 data gives the same result as would be obtained for N-lot at comparable swelling levels. (c) Swelling of annealed LS-1 at other temperatures.

rate comparable to that of the N-lot steel. It therefore appears that the compositional differences between LS-1 and AISI 316 only affect the duration of the transient regime of swelling, which is also sensitive to the starting thermal-mechanical state of the alloy. In addition the post-transient swelling rate of Ti-modified alloys also appears to be relatively insensitive to irradiation temperature over a very broad temperature range.

The similarity of post-transient behavior in annealed and cold-worked alloys thus suggests the validity of using data on annealed steels to forecast the influence of solute additions on the swelling of cold-worked alloys. This approach will be used in the following section to study the synergistic effects of titanium and other elements.

Swelling of a Series of Titanium-Modified Alloys

In another experiment, the general effect of titanium on swelling can be assessed in synergism with that of other compositional changes. The MV-III experiment involved the irradiation in EBR-II of a large number of compositionally-modified 316-base alloys in both the cold-worked and annealed conditions (29). Density change data are available for these alloys to exposure levels as large as $\sim 15 \times 10^{22}$ n/cm^2 (E > 0.1 MeV) or ∿75 dpa.

In general, the predominant effect of titanium additions to both cold-worked and annealed alloys is to extend the duration of the transient regime, but there are occasional exceptions. Figure 20 shows density change data at 540°C for nine series of annealed alloys, each with a different base composition and each having variations in titanium. Note that with few exceptions the addition of titanium results only in a shift of

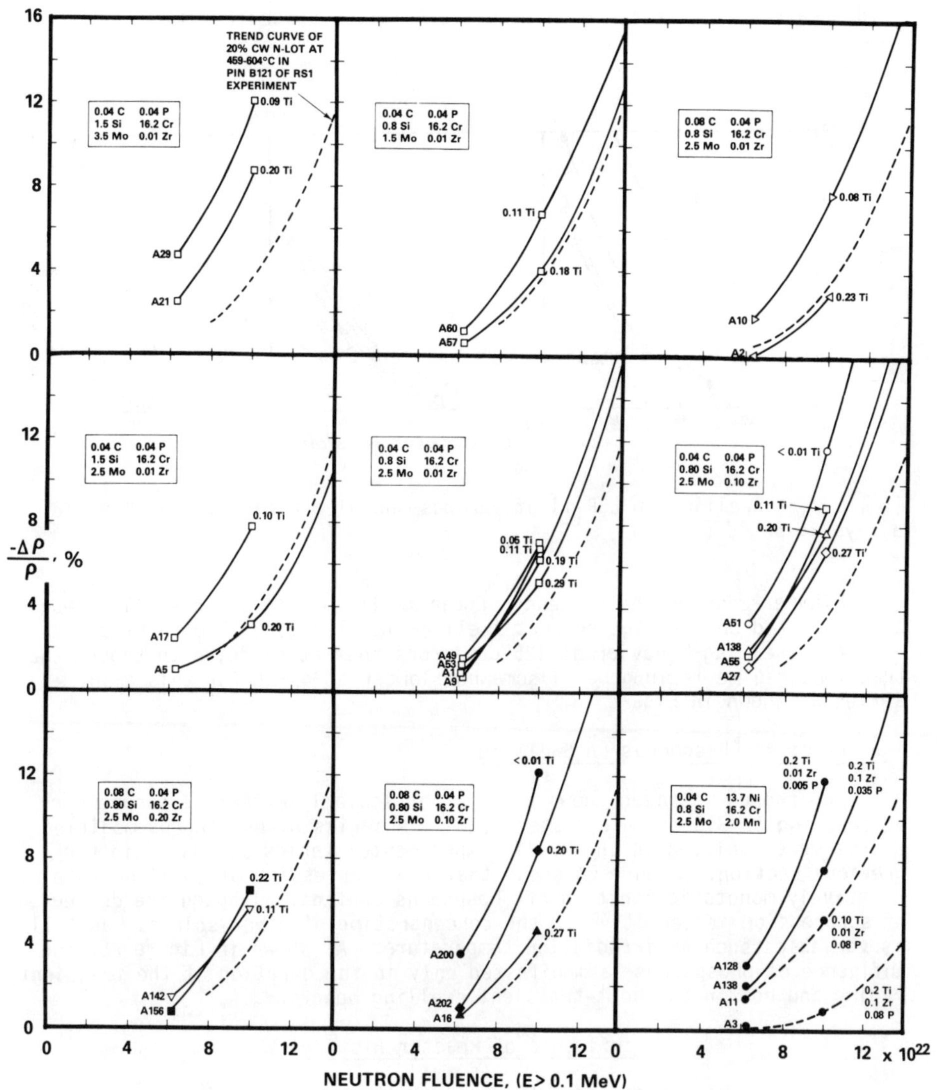

Figure 20 - Influence of titanium on swelling of various compositionally-modified 316-type alloys in the annealed condition at 540°C. Alloy designations and compositions are shown for each alloy series. Note that percent density change $\Delta\rho/\rho o$ is plotted in these curves instead of swelling and that measured density changes in excess of 16% lie off-scale for many of these curves.

the swelling curve to higher fluence. If these data are replotted to show the influence of other solutes a similar conclusion can be drawn, namely that solute additions to AISI 316 affect primarily the duration of the transient regime. For some combinations of solutes the transient regime is shortened rather than extended, however. As shown in Figure 21 the swelling curves continuously accelerate in swelling rate and approach the 1%/dpa rate intrinsic to the austenitic Fe-Ni-Cr system (1-3).

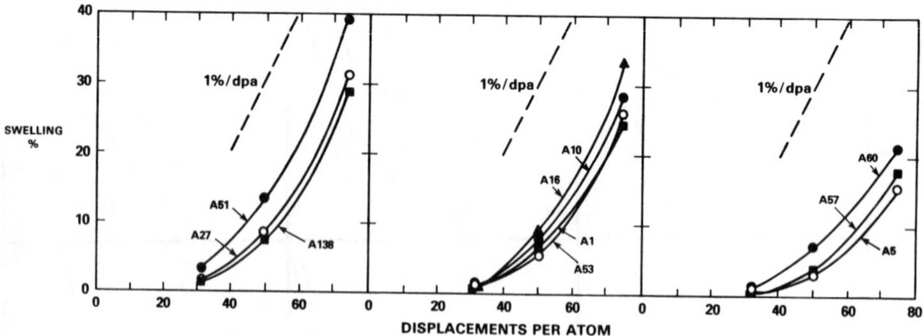

Figure 21 - Swelling in EBR-II of various annealed titanium-modified 316 alloys at 540°C (29).

Figure 22 shows the tendency of the swelling rate in this alloy system to be related only to the current swelling level. Note also in Figure 22 that the swelling behavior at 425°C mirrors that at 540°C, even though the transient regime is in general somewhat longer at 425°C for this alloy series as shown in Figure 23.

The Effect of Phosphorus on Swelling

Phosphorus has been shown to be a particularly effective suppressor of swelling in 316 stainless steel (30). A series of phosphorus-modified alloys were included in the MV-III experimental series discussed in the previous section. Figure 24 shows that the suppression of swelling is a relatively monotonic function of phosphorus content, although the degree of suppression is sensitive to the concentration of other solutes, as well as variables such as irradiation temperature. As shown in Figure 25, the influence of phosphorus is manifested only in the duration of the transient regime and not in the post-transient swelling behavior.

Influence of Reactor History

It should be emphasized that environmental history effects on the transient regime can overwhelm the influence of either compositional or fabricational modifications. Makenas has recently shown that fuel pins constructed from the fourth core FFTF steels indeed developed a relatively temperature-independent swelling behavior for temperatures above 475°C (Figure 26) and exhibit an increasing transient duration below 475°C (31). He also showed that differences occurred in transient duration compared to that of first core steels in the RS-1 experiment but the differences were relatively small.

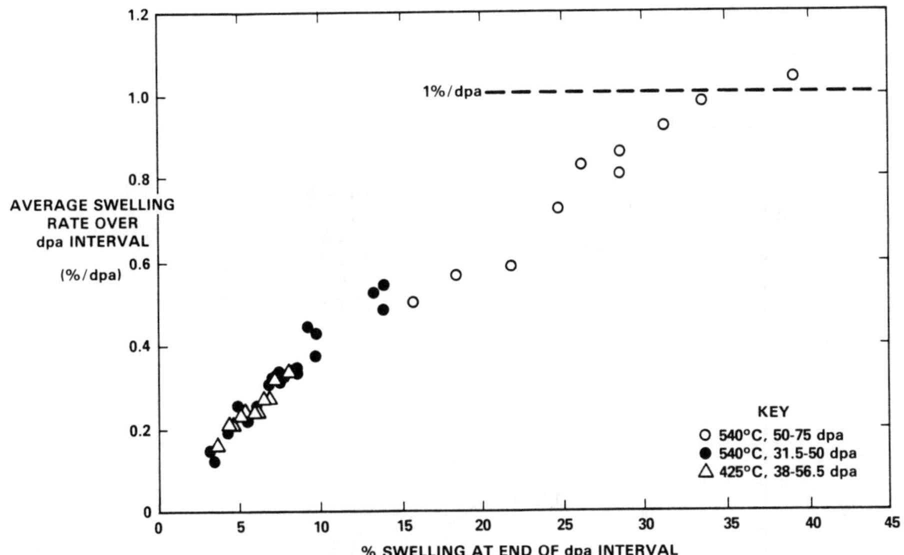

Figure 22 - Average swelling rate observed in annealed Ti-modified steels at 425 and 540°C, showing similar approach at both temperatures to a swelling rate of ∿1%/dpa. The rates are determined from the data in figures 20, 21 and 23.

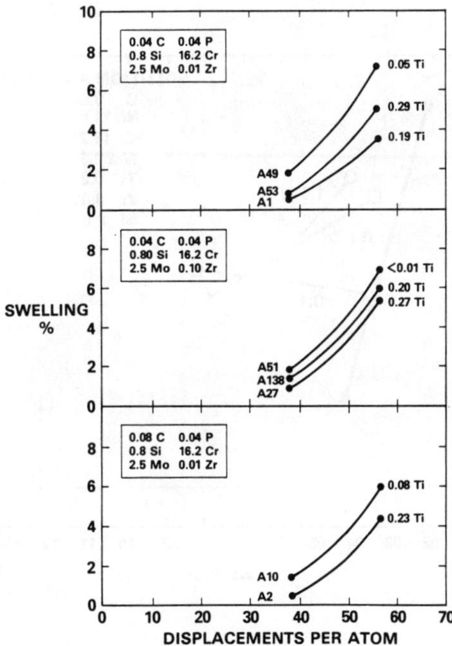

Figure 23 - Swelling of various titanium-modified AISI 316 steels at 425°C in EBR-II.

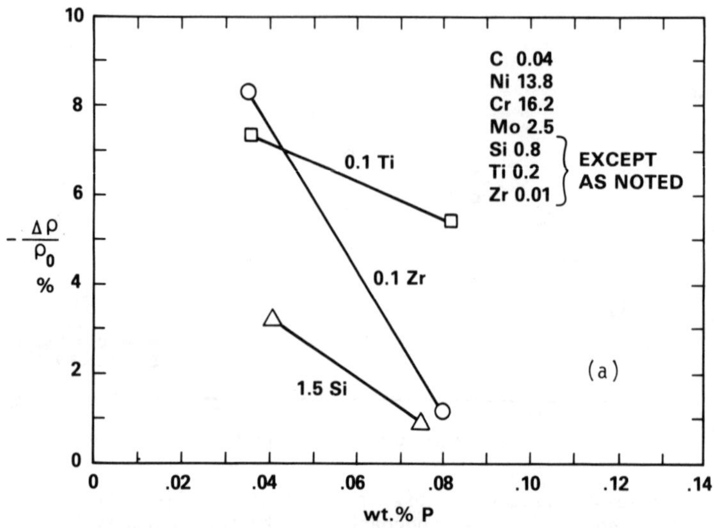

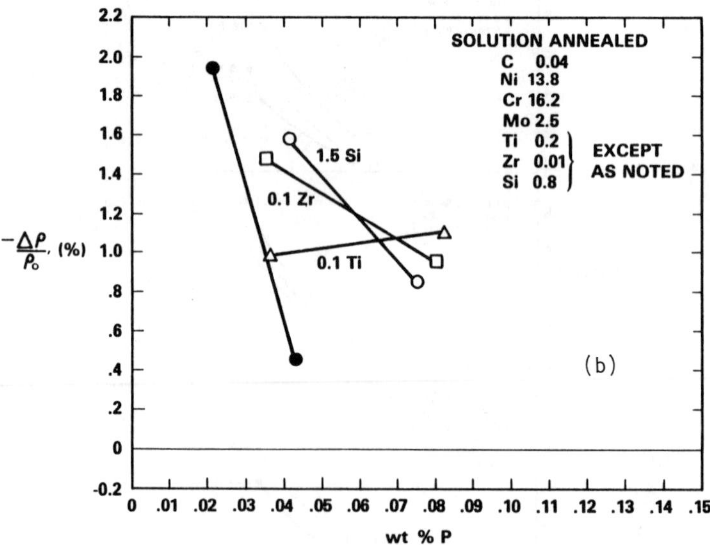

Figure 24 - Influence of phosphorus on density change of annealed AISI 316 steels in EBR-II (a) 540°C, 10.0 x 10^{22} n cm^{-2} (E > 0.1 MeV) or ∿50 dpa and (b) 450°C, 7.6 x 10^{22} n cm^{-2} (E > 0.1 MeV) or ∿38 dpa.

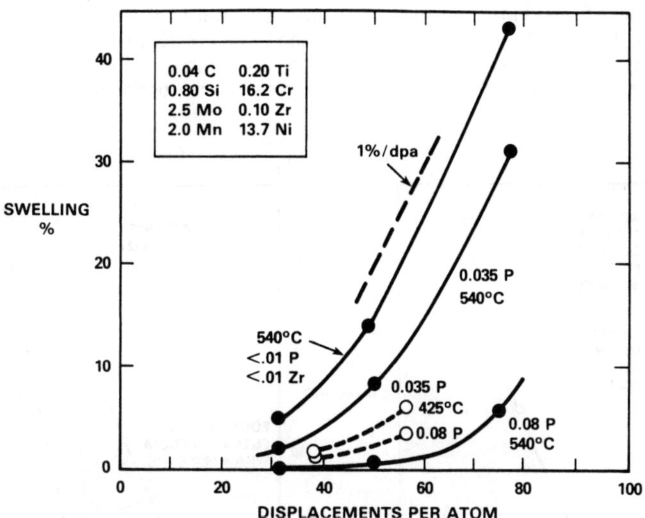

Figure 25 - Effect of phorphorus and irradiation temperature on swelling of annealed AISI 316.

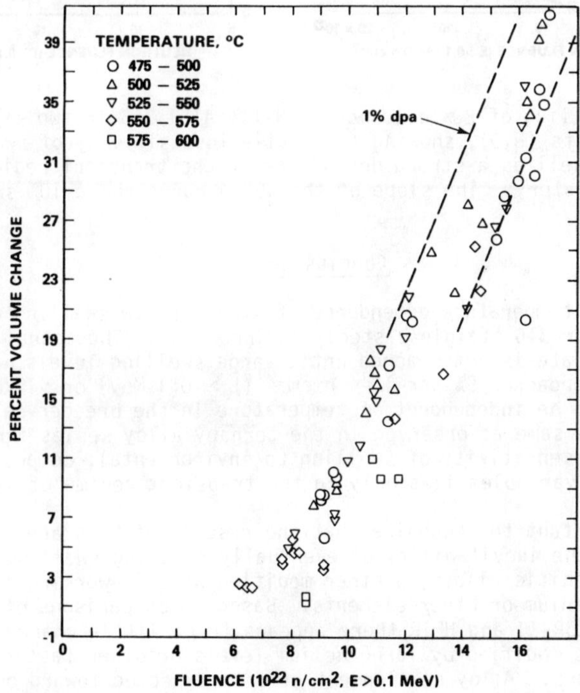

Figure 26 - Swelling of fuel pin cladding at temperatures above 475°C, as reported by Makenas (31).

In another fuel pin experiment designated as WSA-4 (9), however, the N-lot heat of steel was employed as shown in Figure 27. Note that the transient regime of swelling above 475°C was substantially increased relative to that of the RS-1 experiment, such that no temperature dependence of swelling was observed between 404 and 604°C. Currently we are unable to predict the occurrence of such shifts in transient behavior.

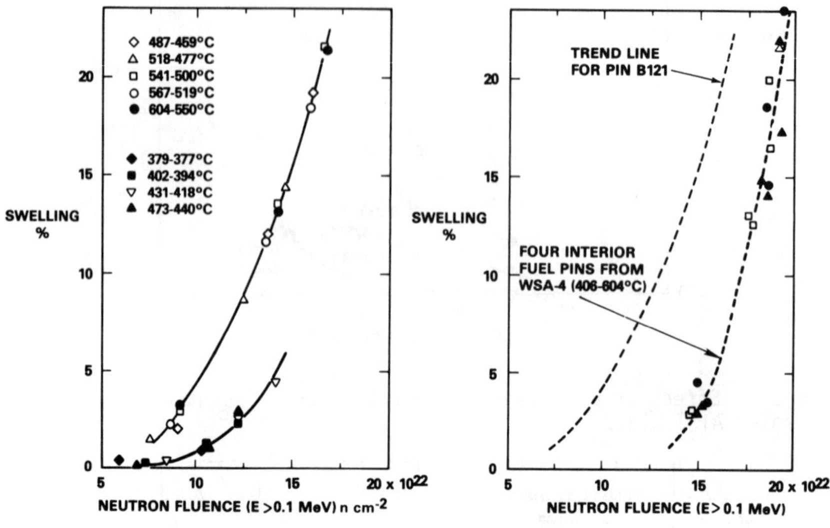

Figure 27 - Swelling of 20% cold-worked N-lot AISI 316 in two separate EBR-II experiments (8,9), showing a relative insensitivity of swelling to temperature as well as a strong dependence of the transient regime on environmental history. The slope of the WSA-4 curve above 10% swelling is ∿1%/dpa.

Conclusions

The strong temperature dependence of steady-state swelling rate commonly assumed for 316 stainless steels is incorrect. The actual steady-state swelling rate is not reached until large swelling levels have occurred and approaches 5% per 10^{22} n cm^{-2} (E > 0.1 MeV) or ∿1%/dpa. This value appears to be independent of temperature in the breeder-relevant range and is the same as observed in the ternary alloy series irradiated in EBR-II. The sensitivity of swelling to environmental, compositional and fabrication variables lies only in the transient regime of swelling.

It is important to recognize that the results of this and other studies signal the inevitability of eventually reaching swelling rates of ∿1%/dpa in austenitic alloys, whether modified by cold-working or by addition of titanium or other elements. Based on comparisons of swelling of AISI 316 in EBR-II and HFIR there appears to be little chance that such behavior will be modified by large helium levels or other factors relevant to fusion reactors. Alloy development efforts directed toward optimizing the swelling resistance of austenitic alloys for fusion environments therefore should not be directed toward modification of the swelling rate but toward the maximum extension of the duration of the transient regime.

It is very important to realize that this steel is relatively unstable with respect to the onset of void growth. Therefore, small differences in reactor history, some of which are perhaps beyond our control, can have a larger influence on swelling than can be realized through careful control of either composition or fabrication variables.

Acknowledgments

This work was initiated in the U.S. Breeder Reactor Program and completed under the auspices of the U.S. Magnetic Fusion Energy Program, both funded by the U.S. Department of Energy.

References

1. F. A. Garner, "Recent Insights on the Swelling and Creep of Irradiated Austenitic Alloys," Journal of Nuclear Materials, 122 & 123 (1984) pp. 459-471.

2. F. A. Garner and H. R. Brager, "Dependence of Neutron-Induced Swelling On Composition In Iron-Based Austenitic Alloys," these proceedings.

3. F. A. Garner, "Application of High Fluence Fast Reactor Data to Fusion-Relevant Materials Problems," to be published in Proceedings of First International Conference on Fusion Reactor Materials, Tokyo, Japan, December 3-6, 1985.

4. J. L. Seran and J. M. Dupouy, "The Swelling of Solution Annealed 316 Cladding in Rapsodie and Phenix," pp. 5-16 in Effects of Radiation on Materials: Eleventh Conference, ASTM STP 782, H.R. Brager and J. S. Perrin, eds.; American Society for Testing and Materials, Philadelphia, PA, 1982.

5. J. L. Seran and J. M. Dupouy, "Effect of Time and Dose Rate on the Swelling of 316 Cladding in Phenix, pp. 25-28 in Vol. I of Proceedings of Conference on Dimensional Stability and Mechanical Behavior of Irradiated Metals and Alloys, Brighton, England, British Nuclear Energy Society, 1983.

6. J. L. Boutard, G. Brun, J. Lehmann, J. L. Seran and J. M. Dupouy, pp. 137-144 in Proceedings of Conference on Irradiation Behavior of Metallic Materials for Fast Reactor Core Components, Ajaccio, Corsica, June 4-8, 1979.

7. F. A. Garner, E. R. Gilbert, D. S. Gelles and J. P. Foster, pp. 698-712 in Effects of Radiation on Materials: Tenth Conference, ASTM STP 725, D. Kramer, H. R. Brager and J. S. Perrin, eds.; American Society for Testing and Materials, Philadelphia, PA, 1981.

8. W. J. S. Yang and F. A. Garner, "Relationship Between Phase Development and Swelling of AISI 316 During Temperature Changes", pp. 186-206 in reference 4.

9. A. Boltax, J. P. Foster and U. P. Nayak, "High Dose Stainless Steel Swelling Data on Interior and Peripheral Oxide Fuel Pins," pp. 21-24 in Vol. I of reference 5.

10. F. A. Garner, E. R. Gilbert and D. L. Porter, "Stress-Enhanced Swelling of Metals During Irradiation," pp. 680-697 in reference 7.

11. K. Ehrlich, "Irradiation Creep and Interrelation with Swelling in Austenitic Steels," Journal of Nuclear Materials, 100 (1981) pp. 149-166.

12. T. Lauritzen, W. L. Bell, J. M. Rosa and S. Vaidyanathan, "Some Observations on the Effect of Stress on Irradiation-Induced Swelling in AISI 316," to be published in Effects of Radiation on Materials: Twelfth Conference, F. A. Garner and J. S. Perrin, eds.; American Society for Testing and Materials, Philadelphia, PA, 1984.

13. F. A. Garner and D. L. Porter, "A Reassessment of the Swelling Behavior of AISI 304 Stainless Steel," pp. 41-44 in Vol. II of reference 5.

14. F. A. Garner, "The Microchemical Evolution of Irradiated Stainless Steels," pp. 165-189 in Phase Stability During Irradiation, J. R. Holland, D. I. Potter and L. K. Mansur, eds.; AIME, New York, NY, 1980.

15. J. L. Straalsund, H. R. Brager and J. J. Holmes, "Effects of Cold Work on Void Formation in Austenitic Stainless Steel," pp. 142-155 in Radiation-Induced Voids in Metals, AEC Symposium Series 26, J. W. Corbett and L. C. Ianniello, eds.; 1972.

16. H. R. Brager and F. A. Garner, "Microstructural and Microchemical Comparison of AISI 316 Irradiated in HFIR and EBR-II," Journal of Nuclear Materials, 117 (1983), pp. 159-176.

17. D. S. Gelles, "Swelling in Several Commercial Alloys Irradiated to Very High Neutron Fluence," Journal of Nuclear Materials, 122 & 123 (1984), pp. 207-213.

18. C. Brown, R. M. Sharpe, E. J. Fulton and C. Cawthorne, "Void Swelling Behavior of U.K. Steels," pp. 63-67 in Vol. I of reference 5.

19. J. F. Bates, F. A. Garner and F. M. Mann, "The Effect of Solid Transmutation Products on Swelling in 316 Stainless Steel," Journal of Nuclear Materials, 103 & 104 (1981), pp. 999-1004.

20. H. R. Brager and F. A. Garner, "Comparison of the Swelling and the Microstructural/Microchemical Evolution of AISI 316 Irradiated in EBR-II and HFIR," Journal of Nuclear Materials, 103 & 104 (1981), pp. 993-998.

21. H. R. Brager and F. A. Garner, "Influence of Neutron Spectra on the Radiation-Induced Evolution of AISI 316," Journal of Nuclear Materials, 108 & 109 (1982) pp. 347-358.

22. F. A. Garner, "Impact of the Injected Interstitial on the Correlation of Changed Particle and Neutron-Induced Radiation Damage," Journal of Nuclear Materials, 117 (1983), pp. 177-197.

23. "Starfire - A Commercial Tokamak Fusion Power Plant Study," Argonne National Laboratory Report ANL/FPP-80-1, Vol. 1, pp. 10-41.

24. P. J. Maziasz and E. E. Bloom, "Comparison of Titanium-Modified and Standard Type 316 Stainless Steel Irradiated in HFIR: Swelling and Microstructure," pp. 40-53 in ADIP Quarterly Progress Report, DOE/ET-0058/1, August 1978.

25. P. J. Maziasz and D. N. Braski, "Swelling and Microstructural Development in Path A PCA and Type 316 Stainless Steel Irradiated in HFIR to about 22 dpa," pp. 44-66 in ADIP Semiannual Progress Report, DOE/ER-0045/9, September, 1982.

26. F. A. Garner and W. G. Wolfer, "Factors Which Determine the Swelling Behavior of Austenitic Alloys," Journal of Nuclear Materials, 122 and 123 (1984) pp. 201-206.

27. F. A. Garner and W. G. Wolfer, "The Effect of Solute Additions on Void Nucleation," Journal of Nuclear Materials, 103 (1981) pp. 143-150.

28. B. Esmailzadeh, A. S. Kumar and F. A. Garner, "The Influence of Silicon on Void Nucleation in Irradiated Alloys," to be published in reference 3.

29. F. A. Garner, H. R. Brager and R. J. Puigh, "Swelling of Titanium-Modified AISI 316 Alloys," to be published in reference 3.

30. F. A. Garner and H. R. Brager, "The Role of Phosphorus in the Swelling and Creep of Austenitic Alloys," to be published in reference 3.

31. B. J. Makenas, "Swelling of 316 Stainless Steel Cold-Worked Fuel Pin Claddings and Ducts," to be published in reference 12.

THE ORIGIN OF THE LARGE RESISTANCE TO VOID SWELLING OBSERVED IN Fe-35.5Ni-7.5Cr

H. R. Brager and F. A. Garner

Hanford Engineering Development Laboratory
Richland, WA 99352

Abstract

While it is well known that increases in the nickel content of austenitic steels delay swelling, less attention has been paid to the fact that decreases in the chromium level from the usual 15-18% range also suppress swelling. Neutron irradiation studies show the longest postponement of swelling in Fe-Ni-Cr alloys occuring at ∿35 wt% nickel and 7.5% chromium. The causes of the suppression and the subsequent loss of swelling resistance have been sought using electron microscopy and quantitative x-ray microanalysis. Swelling in this alloy is preceded by a substantial densification arising from large radiation-induced spinodal-like micro-oscillations in composition which appear to destroy the resistance to void nucleation, such that void nucleation is favored in the nickel-poor chromium-rich regions. Alloy optimization efforts now concentrate on solute additions which are designed to delay the microsegregation process.

Introduction

Several recent studies have shown that the eventual fate in a fast reactor environment of all austenitic iron-based alloys is to swell at a rate of ∿1%/dpa.(1-4) Therefore there does not appear to be any potential for development of an alloy optimized for swelling resistance that arises as a consequence of a reduced post-transient swelling rate. The development of such an alloy must be based on attempts to extend the transient regime of swelling and thereby delay the onset of significant swelling.

While cold-working and various solute additions are well established as means of extending the duration of the transient regime, it is reasonable to expect that the development effort should start at a composition of solvent atoms (Fe,Cr,Ni) that exhibits the greatest intrinsic swelling resistance. This paper addresses the early results of an effort designed to identify the Fe-Cr-Ni compositional regime exhibiting the greatest resistance to swelling and to discern the processes involved in the loss of such resistance at high neutron exposure. Once the mechanisms of swelling initiation are understood, attempts will then be made to further postpone their action via solute additions and/or thermal-mechanical treatment.

While it is well known that increases in nickel content of austenitic steels tend to delay swelling (1,5), it has also been shown that decreases in chromium level from the usual 15-18% range also suppress swelling (1,2,6,7). Figure 1 clearly shows the benefit of a simultaneous reduction in chromium and an increase in nickel, providing the nickel content does not exceed ∿45%.

It is unrealistic to design alloys in which chromium is completely removed, however, since it is necessary to provide corrosion resistance. There are also advantages in reducing the nickel to the lowest level compatible with the goal of enhanced swelling resistance, since nickel is relatively expensive and possesses a larger cross section for neutron capture than does either iron or chromium.

Recent neutron irradiation studies have shown that the largest postponement of swelling in simple Fe-Ni-Cr ternary alloys occurs at ∿35 wt% nickel and 7.5% chromium, although no lower levels of chromium were included in the experimental series(2). Figure 2 shows a compilation of swelling data for this alloy in the solution-annealed condition over the temperature range of 400 to 650°C to fluences as large as 16-22 x 10^{22} n/cm^2 (E > 0.1 MeV) or 90-110 dpa. Not only is the swelling very low, even when compared to a cold-worked solute-laden alloy such as AISI 316, but many of the data indicate that a net decrease in lattice parameter has occurred. The resulting densification was as large as 0.9% as indicated by the 593°C datum highlighted with an arrow in Figure 2. The thermal equilibrium diagram for the Fe-Ni-Cr system at 600°C does not display any miscibility gaps or precipitate phases at this composition, and the observed densification was therefore rather surprising.

As shown in Figure 3, there exists a maximum in the lattice parameter at Fe-40Ni-0Cr at room temperature(8) and a corresponding minimum in the density(9). Thus the Fe-35.5Ni-7.5Cr alloy lies very close to this minimum and radiation-induced segregation within the alloy might lead to an average increase in the density. An experimental program to investigate this possibility was therefore initiated.

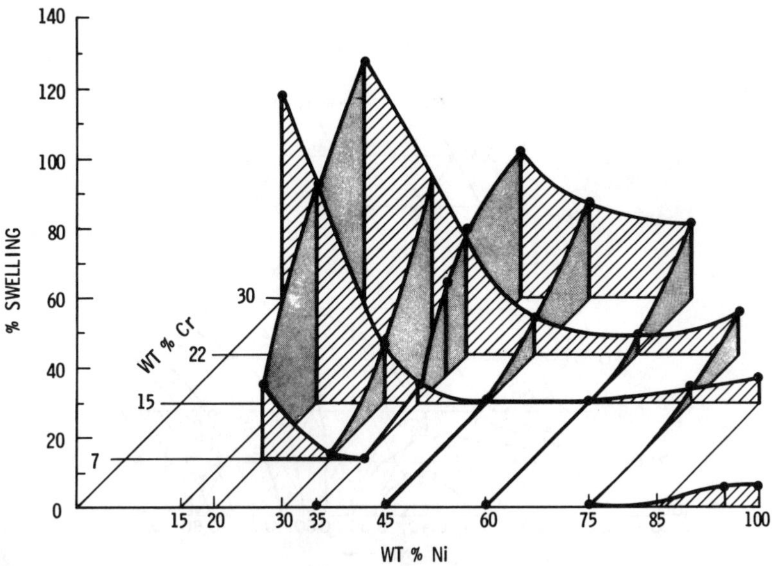

Figure 1 - Swelling-induced in Fe-Cr-Ni alloys by irradiation with 5 MeV Ni^+ ions to 116 dpa at 675°C.(5,6)

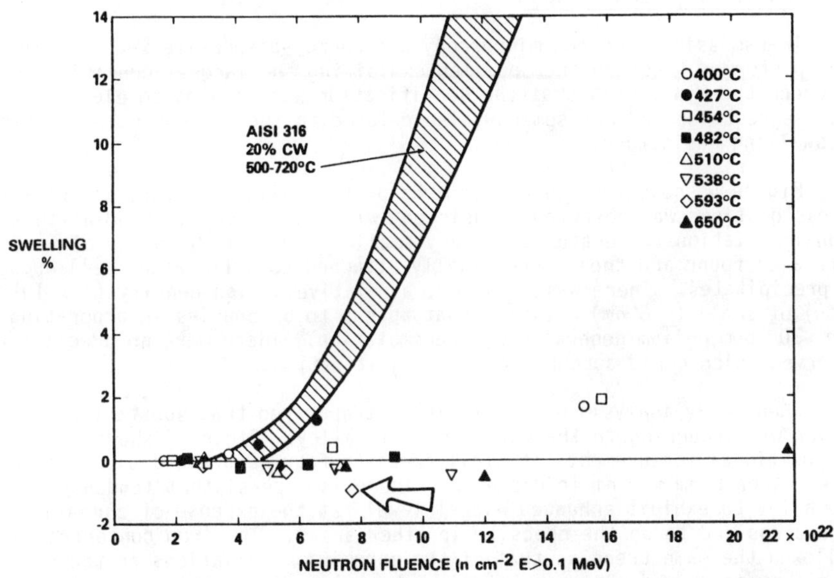

Figure 2 - Comparison of the swelling of annealed Fe-35.5Ni-7.5Cr (2) and cold-worked AISI 316 (1,3), both irradiated in EBR-II.

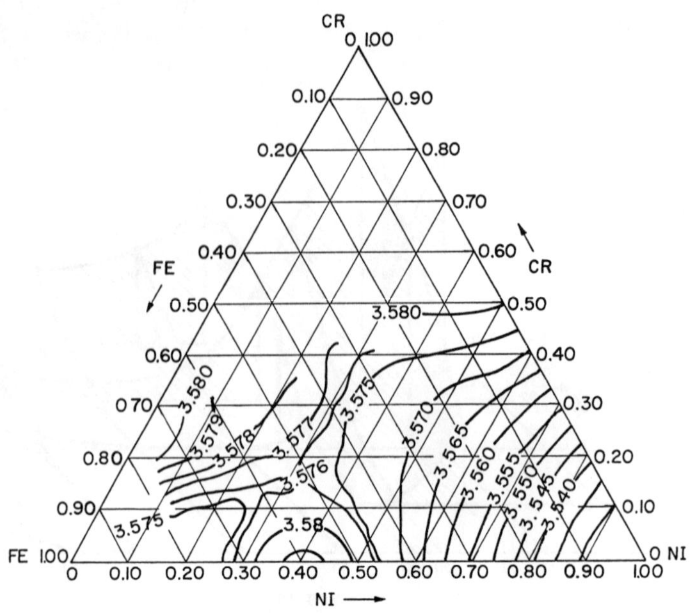

Figure 3 - Lattice parameter in Å for austenite phase Fe-Cr-Ni alloys at room temperature.(8)

Examination of Neutron-Irradiated Specimens

Specimen at 593°C and 38 dpa

Transmission electron microscopy and energy-dispersive x-ray analysis were performed first on the specimen exhibiting the largest densification to check the hypothesis that the densification was related to elemental microsegregation and was somehow also related to the subsequent breakdown in swelling resistance.

Figure 4 shows typical micrographs of this alloy. A very low density of dislocations was observed, consistent with that usually found at this high irradiation temperature. A very small ($\leq 10^{12}$ cm^{-3}) number of voids were also found and these were usually attached to a likewise small number of precipitates. There was, however, a relatively high density (1×10^{16} cm^{-3}) of small ($\leq$ 5 nm) cavities that appear to be bubbles incorporating the ∿30 appm helium generated by transmutation. There were no precipitates observed which could account for the densification.

When x-ray analysis was performed, it appeared that substantial segregation was occurring in the matrix of this alloy. Figure 5 shows ∿100 compositional measurements of small areas in four separate grains, using an electron beam ∿30 nm in diameter. There is a persistent tendency of the alloy to exhibit enhanced nickel levels at the expense of chromium in some areas and to do the opposite in other areas. The iron concentration followed the same trend as that of the chromium. Variations in the nickel level betwen 25 and 53% were found which could easily lead to a net densification of the alloy.

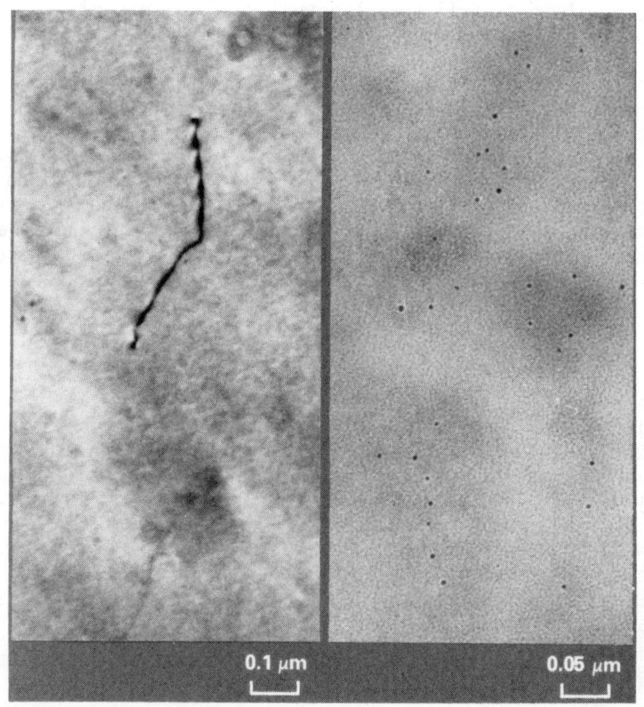

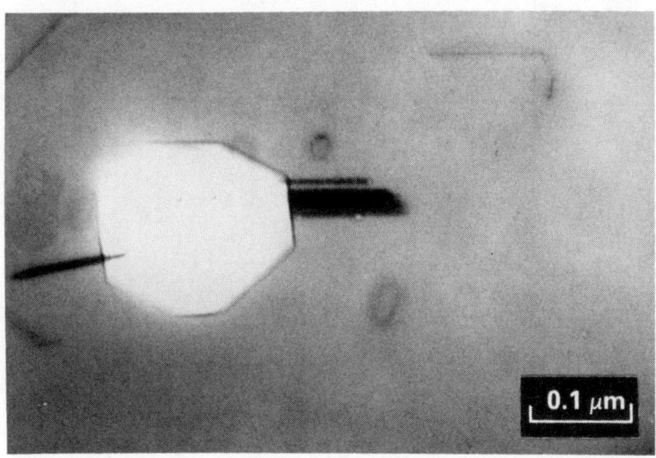

Figure 4 - Micrographs of Fe-35.5Ni-7.5Cr after irradiation to 7.6 x 10^{22} n/cm^2 (E > 0.1 MeV) at 593°C, showing a very low density of dislocations, an absence of Frank dislocation loops and a high density ($\sim 10^{16}$ cm^{-3}) of small cavities assumed to be helium bubbles. The void in the lower micrograph is a representative example of one of the small density of voids found in this specimens.

It is important to note, however, that these measurements are averaged over a finite volume of foil so the actual range of nickel segregation is probably more extensive than it appears in Figure 5.

It was therefore decided to measure the spatial variation of the matrix composition at fixed intervals along traverses across thin (∿50 nm) sections of foil. Figures 6 and 7 show the results of a typical traverse for both irradiated and unirradiated specimens, using 50 nm sampling intervals and a beam width of ∿20-40 nm. While no large compositional variations are seen in the unirradiated specimen, rather large and periodic variations are seen in the nickel concentration in the irradiated specimen. Note also that whenever the nickel level was above that of the bulk average concentration, the iron level tended to be below its bulk-averaged concentration. Both the iron and chromium profiles appear to be mirror images of the nickel profile. The traverses shown in Figures 6 and 7 used dislocations as starting points.

When a grain boundary or a rare precipitate or void were used as a marker (Figures 8-10), one observes a perturbation related to the surface of the marker and then the development of the compositional micro-oscillations far from the marker's surface. These oscillations exhibit a periodicity on the order of several hundred nanometers. The micro-oscillations could not be correlated with any components of the microstructure which currently existed in the foil. To verify that the micro-oscillations were reproducible, measurements on many of the traces were taken on more than one occasion, as illustrated using different symbols in Figures 6 and 8-10.

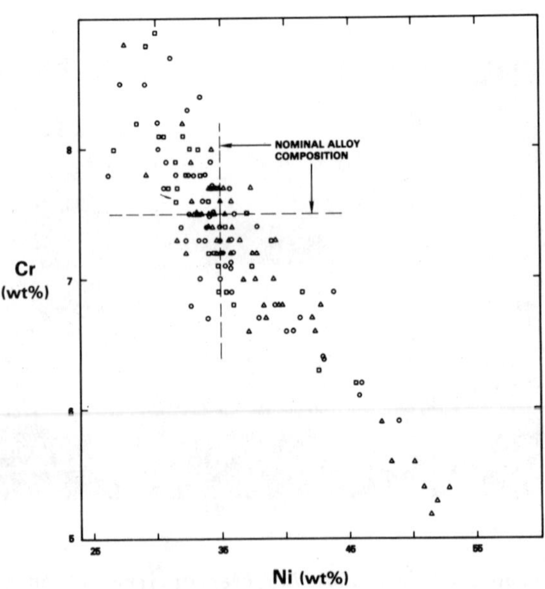

Figure 5 - Relationship between local nickel and chromium levels, as measured in random small areas ∿30 nm in size. The three types of symbols denote three separate sets of measurements.

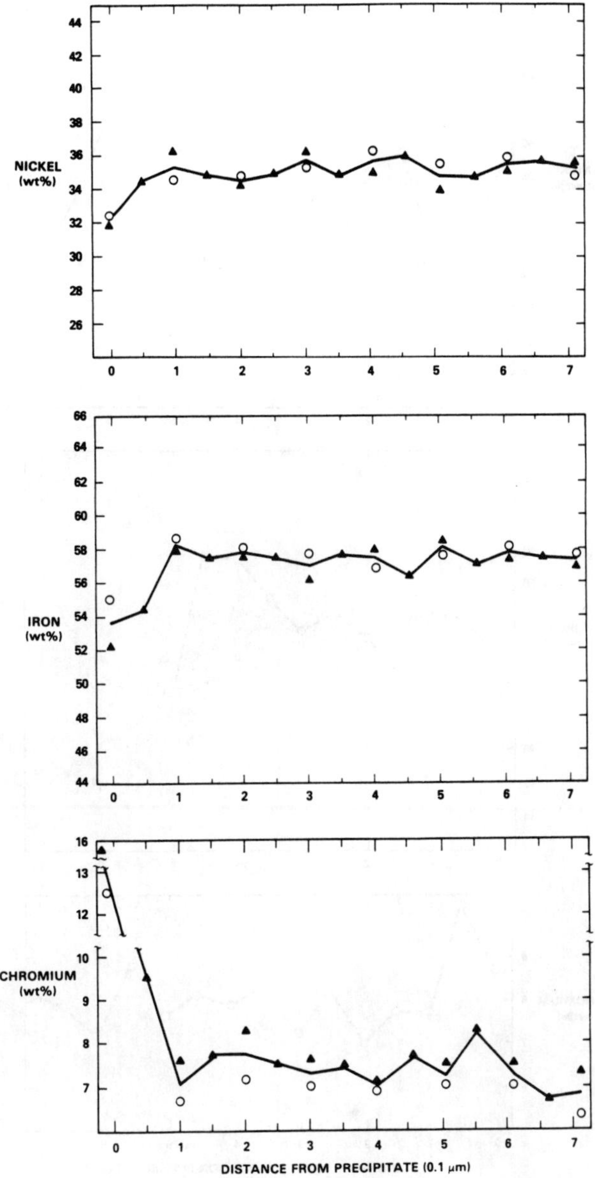

Figure 6 - Compositional variations observed in a thin foil traverse on one side of a precipitate-inclusion in unirradiated Fe-35.5Ni-7.5Cr. Open circle and closed triangle symbols represent alloy concentrations measured on two independent scans to illustrate the reproducibility in the spatial variation of the alloy composition.

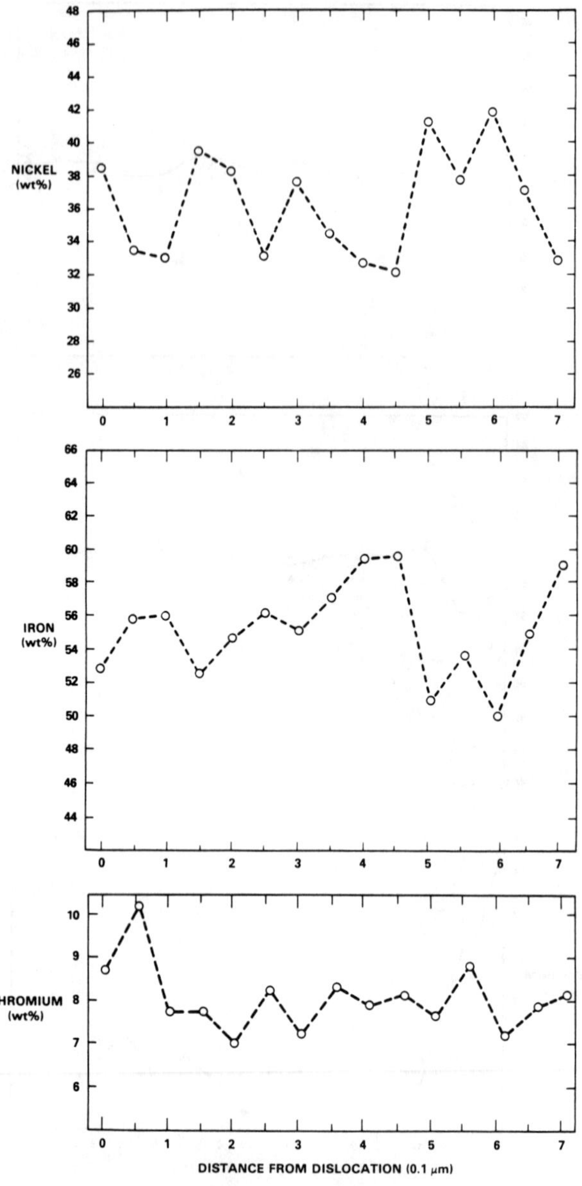

Figure 7 - Thin foil compositional traverse in irradiated Fe-35.5Ni-7.5Cr using a dislocation as a location marker (593°C, 38 dpa).

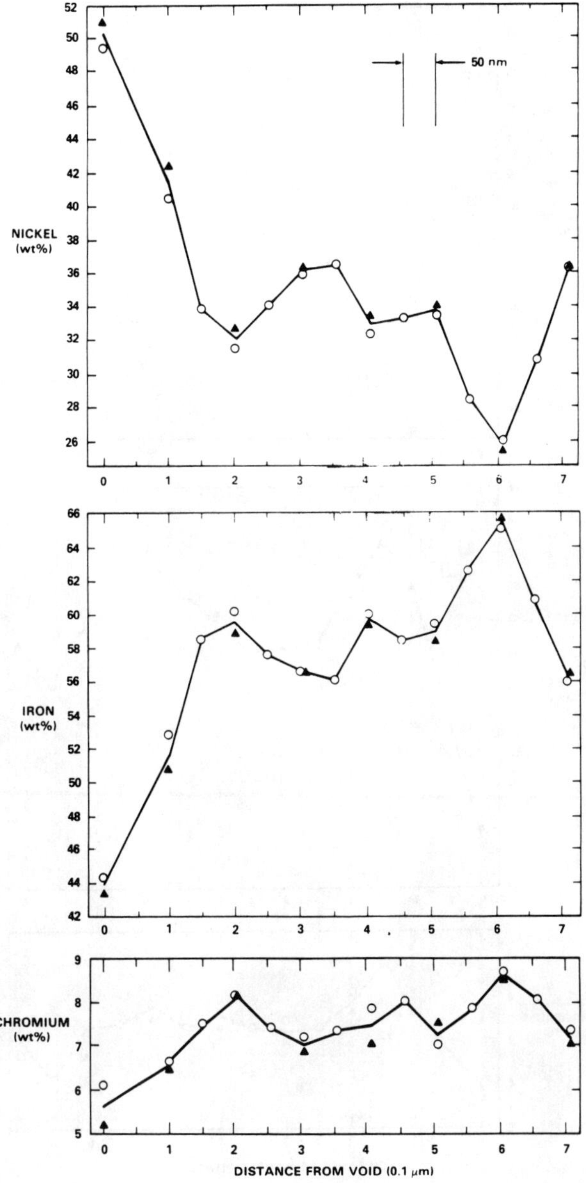

Figure 8 - Thin foil compositional traverse in irradiated Fe-35.5Ni-7.5Cr using a void as marker. (593°C, 38 dpa)

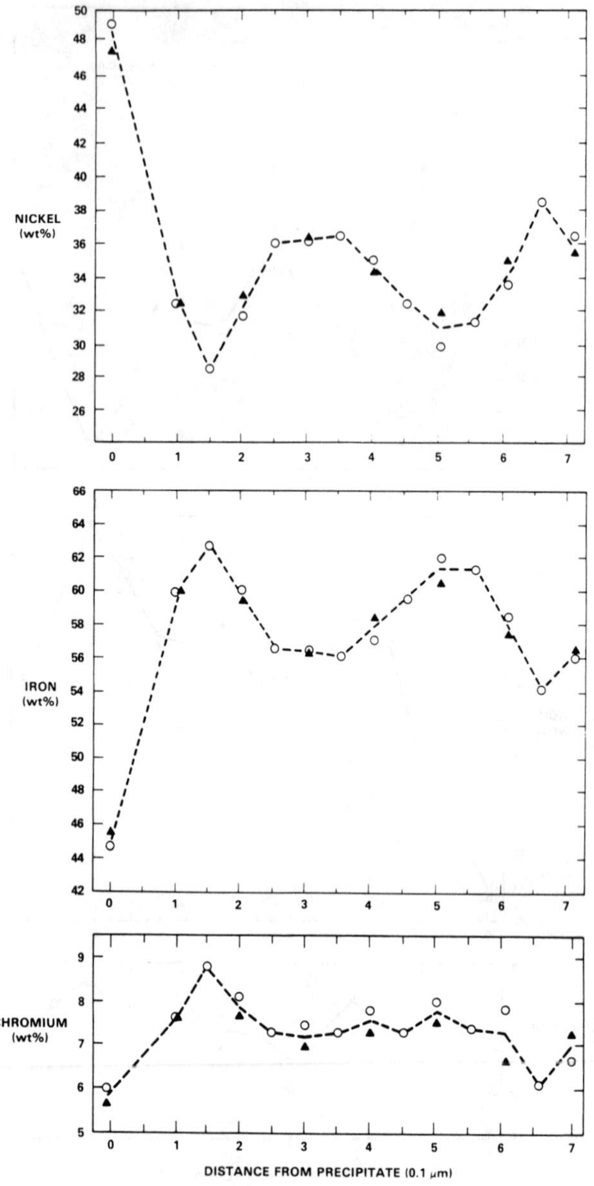

Figure 9 - Thin foil composition traverse in neutron irradiated Fe-35.5Ni-7.5Cr using a precipitate as a location marker. (593°C, 38 dpa)

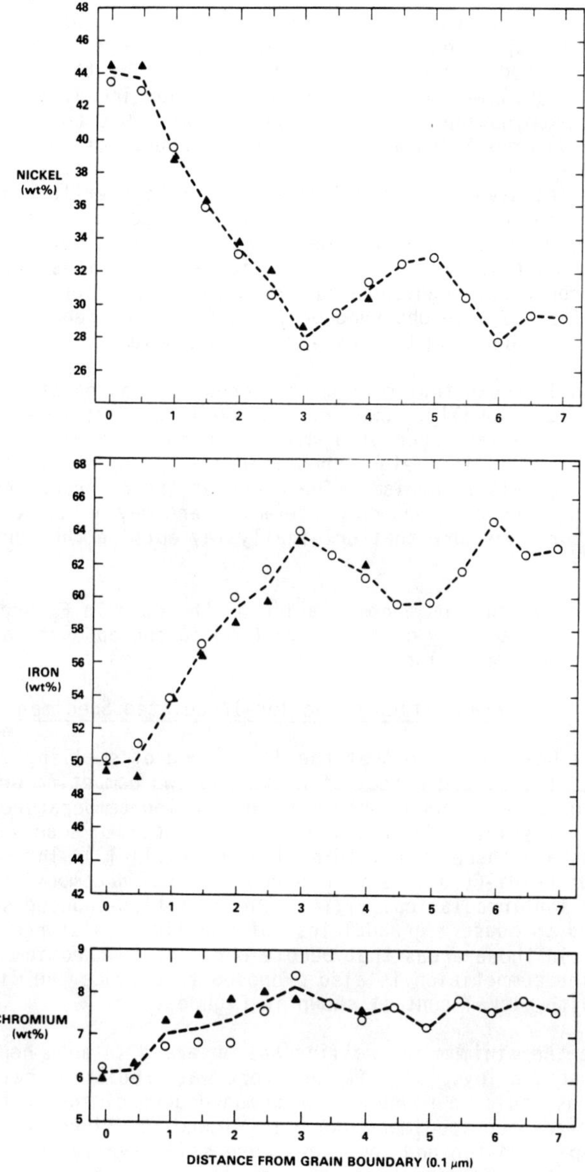

Figure 10 - Thin foil traverse in neutron irradiated Fe-35.5Ni-7.5Cr using a grain boundary as marker. (593°C, 38 dpa)

Specimen at 550°C and 12 dpa

Figure 2 shows that the densification process does not appear to develop until exposures of 2 to 3 x 10^{22} n/cm^2 (E > 0.1 MeV) or 10-15 dpa have been reached. In order to study the onset of the micro-oscillation process a Fe-35.5Ni-7.5Cr specimen from another irradiation experiment was chosen for examination. It was irradiated at 550°C to 2.5 x 10^{22} n/cm^2 (E > 0.1 MeV) or ∿12 dpa and had a density change of only -0.14% (10).

As anticipated, the void density at this temperature and exposure level was very low (<10^{12} cm^{-3}) and the dislocation density was on the order of 10^{-9} cm^{-2}. No Frank loops or precipitates were observed. Figures 11 and 12 show that, on the average, micro-oscillations exist, but they are less pronounced than those observed in the earlier study at higher exposure. As was also observed previously, nickel tends to flow in the opposite direction from that of iron and chromium.

Figure 13 shows that much more pronounced ranges of variation were also found occasionally, however. The dominant feature in Run #7 of Figure 13 is the influence of a small (∿50 nm) void which strongly alters the composition in its neighborhood, presumably by the action of the inverse-Kirkendall mechanism. The cause of the far-reaching gradient shown in run #8 of Figure 13 was not determined and may be associated with a microstructural feature that originally lay outside the current foil volume.

Figure 14 contains a compilation of the data in Figures 11-13 showing once again the tendency of nickel to flow in the opposite direction from that of chromium and also iron.

Examination of an Ion-Irradiated Specimen

It has been proposed that the dependence of swelling on nickel content arises from the strong effect of nickel on two competing processes, both of which increase in importance with increasing temperature. The first is the effect of solvent composition on the effective vacancy diffusion coefficient and thereby on void nucleation.(1,11-13) The second is the tendency of Fe-Ni-Cr alloys to undergo spinodal decomposition during either neutron or ion irradiation.(11,14) The radiation-induced spinodal process is proposed to cause a gradual loss of swelling resistance with voids nucleating in those areas that become enriched in chromium and depleted in nickel. The competition is also proposed to produce the minimum in swelling with nickel content shown in Figure 15.

Since the minimum in swelling was observed in both neutron (2) and ion-irradiated alloys,(5,6) it therefore was predicted that such micro-oscillations would be found in ion-bombarded specimens such as the 5 MeV Ni^+ ion-irradiated specimens shown in Figure 1. Since the Fe-Cr-Ni specimens of Johnston and coworkers were still available in the post-irradiation thinned condition, the specimens were forwarded to these authors for examination.

The regions available for examination were 850-1050 nm from the original foil suface, which corresponds to the peak displacement region examined by Johnston and coworkers. The procedures used in the analysis were identical to those used for the neutron-irradiated specimens. To date, one specimen has been examined, that of Fe-35.0Ni-7.0Cr irradiated to 117 dpa at 625°C. Very little swelling (< 0.1%) was found by Johnston to have occurred in this specimen.

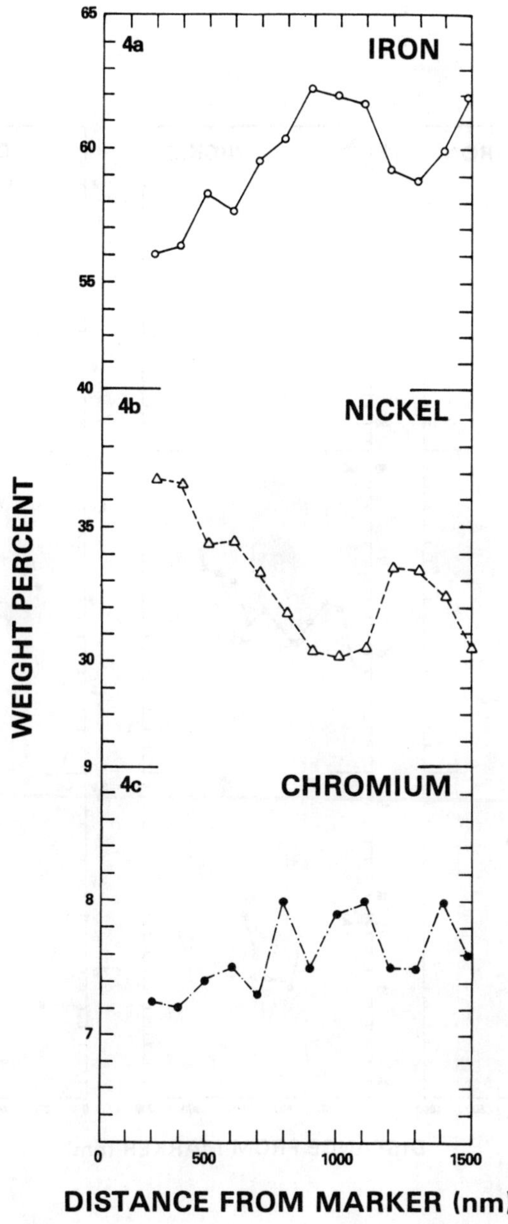

Figure 11 - Compositional traverse of Fe-35.5Ni-7.5Cr alloy after irradiation in EBR-II at 550°C and 2.5 x 10^{22} n/cm^2 (E > 0.1 MeV). The number in the left corner of each graph is the traverse identification number referred to in the text.

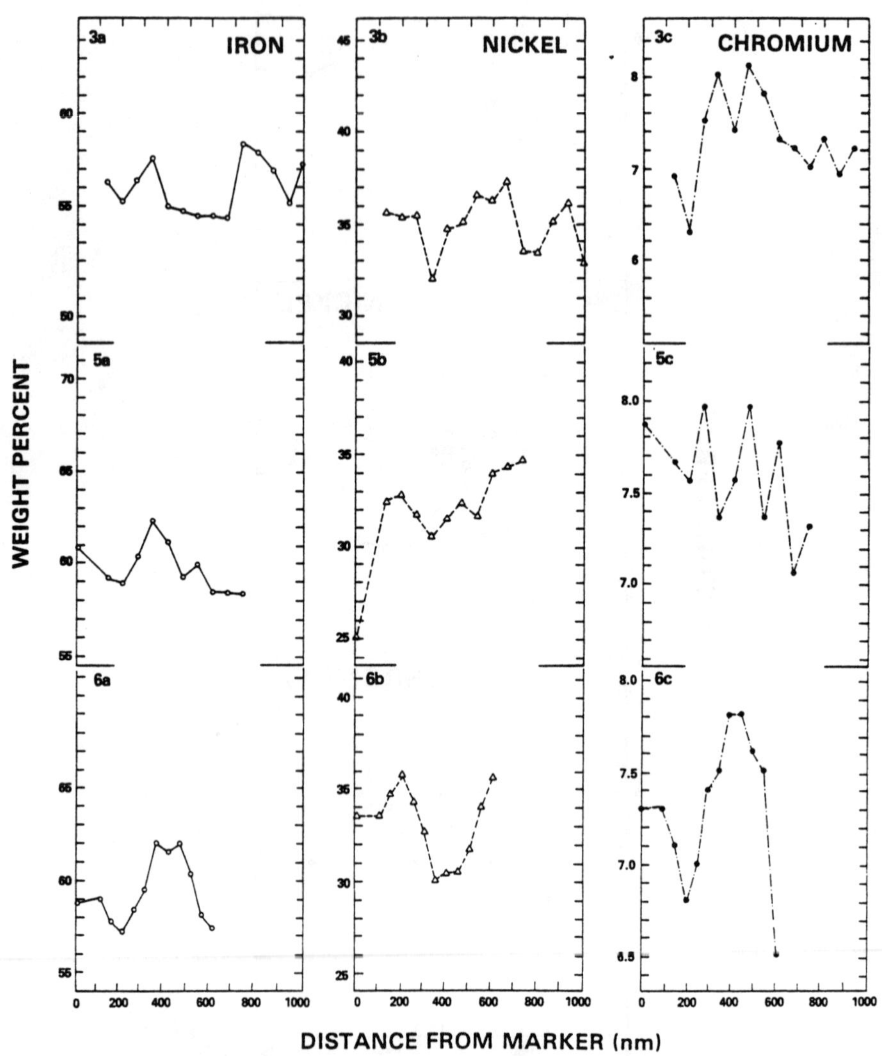

Figure 12 - Compositional traverses of Fe-35.5Ni-7.5Cr after irradiation in EBR-II at 550°C and 2.5 x 10^{22} n/cm^2 (E > 0.1 MeV).

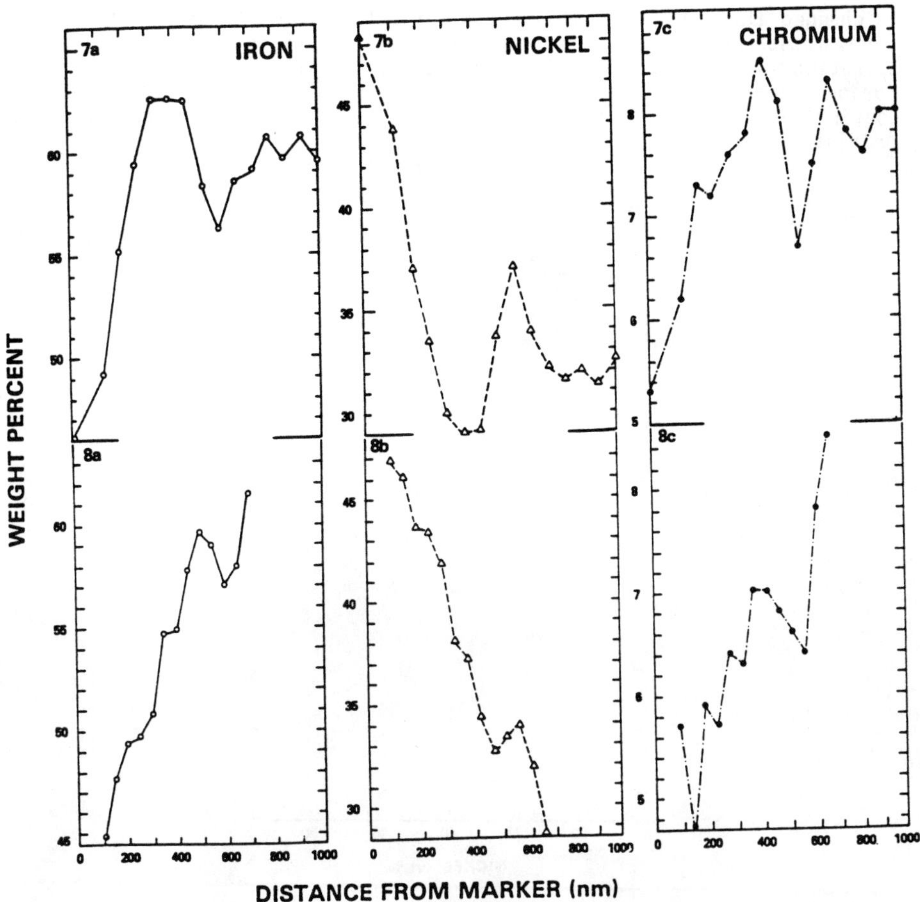

Figure 13 - Additional composition traverses of Fe-35.5Ni-7.5Cr after irradiation in EBR-II at 550°C and 2.5 x 10^{22} n/cm^2 (E > 0.1 Mev).

Although there were very few microstructural features observed within suitably thin areas of the specimen, compositional traces were obtained starting at features which appeared to be surface contamination particles. Figure 16 shows four such sets of measurements (15). Note that the period of the oscillations is on the order of 200-400 nm and that once again the chromium and iron traces tend to be mirror images of that of nickel.

In Figure 17 each of the chromium and nickel levels shown in Figure 16 has been plotted to show that they indeed segregate in opposite directions. Similar data points for the (593°C, 38 dpa) neutron-irradiated specimen are also shown. Note that in addition to the similar trend of the two groups of data that there is an offset between the two groups. Although a small offset arises from the slight difference in composition in the two alloys, it is thought that the offset arises primarily from an ion-induced modification of the average composition at the depth examined. Johnston, (16) as well as others,(17-18) have shown that the elemental distribution is changed along the ion path in Fe-Ni-Cr alloys. This is due to the combined influence of the foil surface and the gradient in displacement rate, operating in conjunction with the inverse Kirkendall effect and other diffusion mechanisms. Large-area compositional scans (compiled in Table 1) confirm that the average composition at this depth was shifted from the original level.

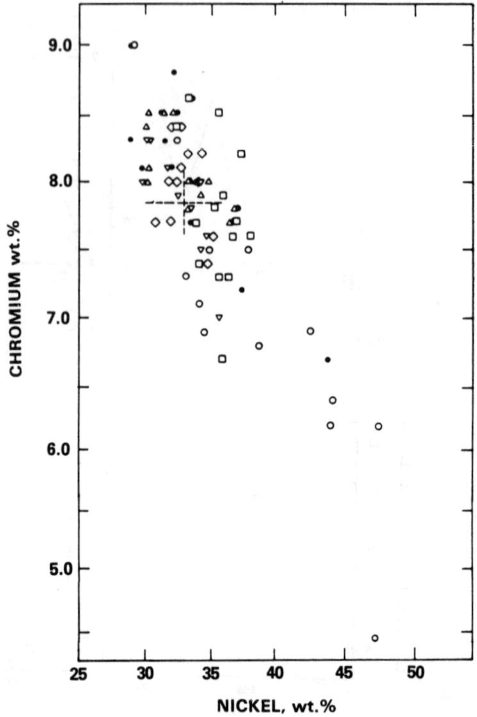

Figure 14 - Compilation of radiation-induced decomposition data for Fe-35.5Ni-7.5Cr after irradiation in EBR-II to 2.5 x 10^{22} n/cm^2 (E > 0.1 MeV) at 550°C.

Table I. Large Area Measurements Of Average Composition In An Ion-Bombarded Specimen of Fe-35.0Ni-7.0Cr

Area	Cr (wt %)	Ni (wt %)
#1	6.1	33.9
#2	5.8	33.6
#3	6.3	34.5
#4	6.6	34.8
#5	6.6	33.4
#6	6.3	33.9
Average	6.3	34.0

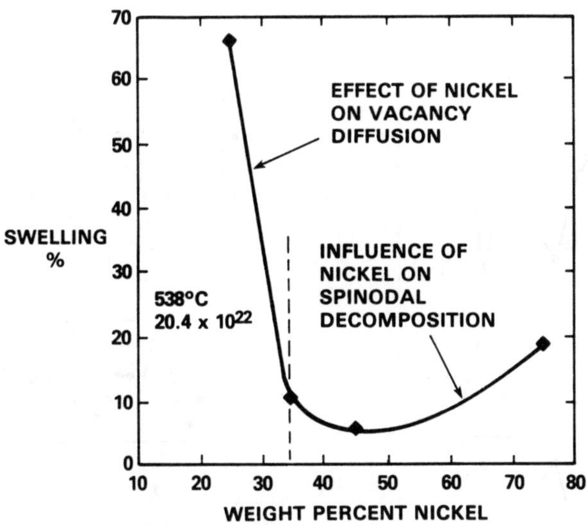

Figure 15 - Swelling of Fe-15Cr-XNi alloys after irradiation in EBR-II to 102 dpa at 538°C, showing two proposed roles of nickel competing to produce the minimum swelling observed in this and other experiments.

It is considered particularly significant that the period of the oscillation increases progressively with irradiation temperature but is not particularly sensitive to the displacement rate, which in the ion bombardment experiment was four orders of magnitude greater than that in the neutron irradiation.

Discussion

Each of the three specimens investigated in this study has several features in common. First, they all exhibit micro-oscillations in composition that exhibit periods of 100-400 nm, increasing with irradiation temperature. The period seems to be more sensitive to temperature than displacement rate. Second, for the most part the micro-oscillations do not appear to be related to currently existing microstructural components. Third, the compositional fluctuations of chromium and iron appear to be almost mirrior images of those of nickel. Finally, the range of nickel compositions appears to be from ~25% to ~50%.

The oscillations can arise from three possible sources. Two of these are perturbations due to microstructural sinks either outside of the foil volume or formed earlier within the foil volume but which no longer exist. The first of these is exemplified by voids and grain boundaries, both of which are known to segregate nickel at the expense of iron and chromium, but the concentrations of these microstructural features are judged to be too low to create the oscillations. The second type of perturbation might arise from dislocation loops which are also known to segregate nickel. Perhaps the loop causes the compositional perturbation and later unfaults and then the resulting dislocation moves away. The highest density of loops found at these temperatures is much too small to yield the observed scale of oscillations, however.

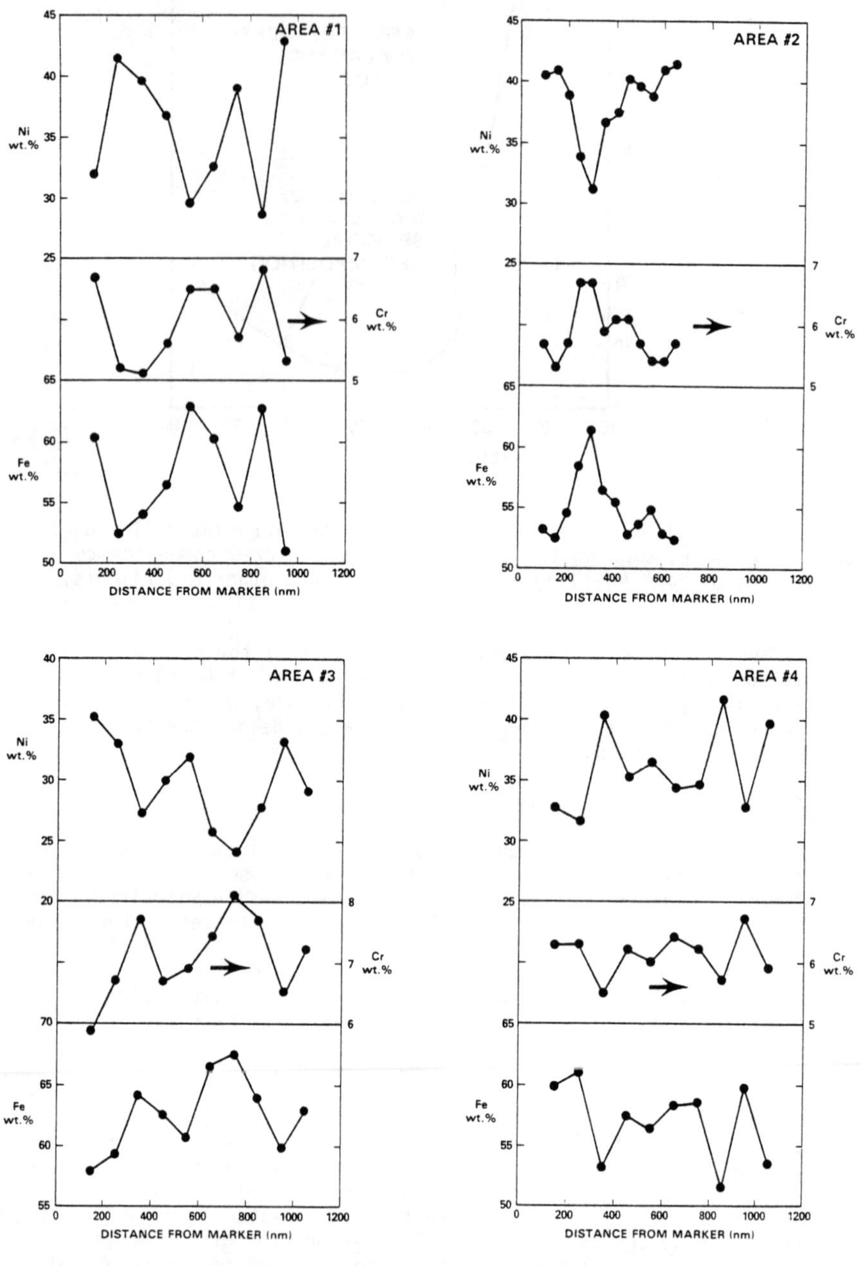

Figure 16 - Compositional oscillations observed in Fe-35.0Ni-7.0Cr irradiated to 117 dpa with 5 MeV Ni^+ ions at 625°C (15).

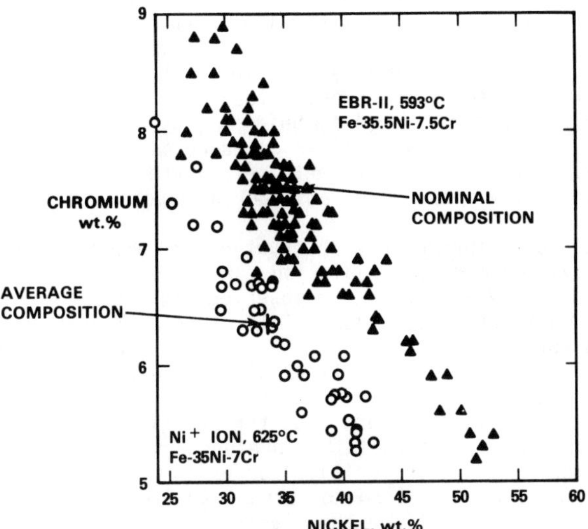

Figure 17 - Compilation of ion-induced decomposition data from figure 16. The neutron-induced composition data from figure 5 is also shown for comparison.

The third candidate mechanism does not require either pre-existing or radiation-induced sinks. Assume for the moment that spontaneous spinodal-type oscillations occur that are created or perhaps merely accelerated by irradiation. Although no actual miscibility gap has been observed in this compositional regime in thermally-aged Fe-Ni or Fe-Ni-Cr alloys, the existence of concentration inhomogeneities on a small scale has frequently been suggested in the scientific literature for this alloy system. It has even been suggested that the true equilibrium diagram for the Fe-Ni system contains a number of previously unobserved miscibility gaps and comprises the γ-disordered phase, γ'-FeNi ordered phase, γ''-$FeNi_3$ ordered phase and γ'''-Fe_3Ni ordered phase.(19) This assumption seems to be partially confirmed by the observation of an FeNi ordered phase in slow cooling meteorites formed billions of years ago.(20) If this assumption is correct then alloys in the composition range occupied by the Fe-35.5Ni-7.5Cr alloy do not exist in the true equilibrium state, but the kinetics of transformation to the ordered state are too slow to be observable in human experience.

This compositional regime has received a large amount of scientific attention since the low swelling regime happens to coincide with the compositional range referred to as Invar alloys. It is well known that many physical properties, such as thermal expansion, magnetic properties, elastic moduli, lattice parameter and excess free energy exhibit rather marked variations with relatively small changes in composition. The Invar regime is frequently referred to as the anomalous property regime.

Many researchers have therefore used charged particle or neutron irradiation to speed up the kinetics of transformation and test the stability of Invar alloys. The earliest of these studies are summarized in Reference 21. Their major conclusions were that the order-disorder transformation is accelerated by irradiation and is a vacancy-dominated process, although interstitials produced by irradiation may also participate. The anomalous Invar properties usually disappear after irradiation and both short-range and long-range ordering have subsequently been observed. The degree of ordering is sensitive to the composition, temperature and displacement rate. Most evidence of ordering has been inferred from indirect observational techniques (Mossbauer, induced magnetic anisotropy, resistivity, nuclear gamma resonance, neutron scattering, lattice parameter measurements, etc.) and comparison with similar studies on easily ordered systems such as CuAu.

There has been one other direct observation of compositional segregation in Fe-Ni alloys induced by irradiation. Penisson and Bourret irradiated Fe-50Ni with 1.0 MeV electrons and observed high densities of ordered micro-domains $\leq$50 nm in size using dark field electron microscopy.(22) The micro-domains formed only when the irradiation was conducted below the critical ordering temperature, which is ∿320°C for Fe-50Ni.(23)

It is not unreasonable to assume that the ordering process requires first a segregation process and that this precursor segregation process does not cease abruptly above the critical ordering temperature. In the disordered state, however, one cannot observe with electrons the existence of compositional oscillations since the electron structure factors of the three solvent atoms (Fe,Ni,Cr) are quite similar. Therefore the presence of micro-oscillations at reactor-relevant temperatures can only be observed using x-ray measurements or by the use of techniques which measure the radiation-induced modification of various anomalous Invar properties.

It has been shown by Pauleve and coworkers(23) that two overlapping types of ordering (FeNi, $FeNi_3$) develop in irradiated Fe-Ni alloys in the range of 37-70% nickel, as shown in Figure 18. If we ignore the chromium in our alloy for a moment we can see from Figure 18 that for Fe-35.5Ni our entire range of irradiation temperature (400-650°C) lies above the critical ordering temperatures of both phases. The addition of chromium has been shown to reduce the tendency toward short-range order(24,25) but not suppress it totally. Chromium additions also tend to reduce the degree of anomaly in some properties such as Young's modulus.(26) The addition of chromium also introduces the possibility of ordered phases involving chromium(27-29), such as Ni_2Cr which forms thermally. Wahi recently reported that precipitates of $Cr_{38}Fe_{11}Ni$ are formed in thin foils of Fe-40Ni-13Cr which were thermally-annealed in vacuum at 700-900°C, although such precipitates did not form in bulk specimens.(30)

The effect of chromium, then, is such that one would expect to see similar compositional micro-oscillations even more easily in Fe-35.5Ni than in Fe-35.5Ni-7.5Cr. There has been one indirect measurement that supports this conclusion. Chamberod and coworkers(19) irradiated with electrons a series of alloys with composition between Fe-28Ni and Fe-50Ni at temperatures of 80, 250 and 400°C. The last temperature lies above the critical ordering temperature for this composition range. As shown in Figure 19, the anomalous peak in lattice parameter in the Invar range disappears at the lower two irradiation temperatures. In effect, the Invar alloys densify just as did the Fe-35.5Ni-7.5Cr alloy of this study. Note that the maximum change in lattice parameter is ∿0.3%, corresponding to a 0.9% change in density. Remember that the same value was observed in the neutron-irradiated specimen at 593°C and 38 dpa.

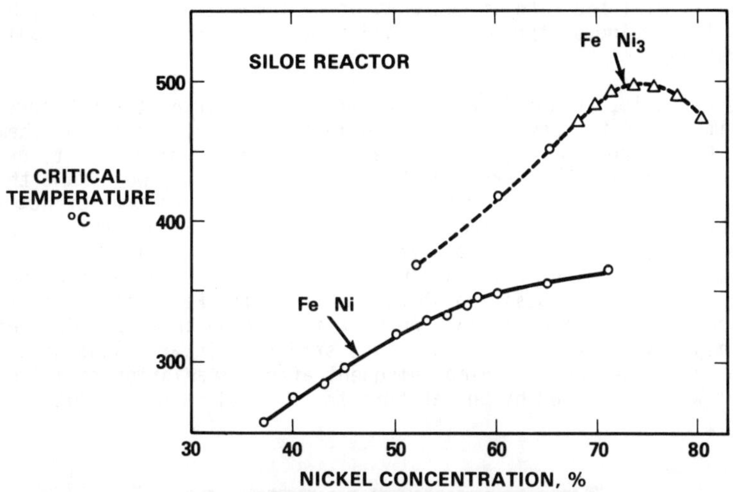

Figure 18 - Critical temperatures for the order-disorder transitions of NiFe and Ni_3Fe in the Fe-XNi alloy system.(23)

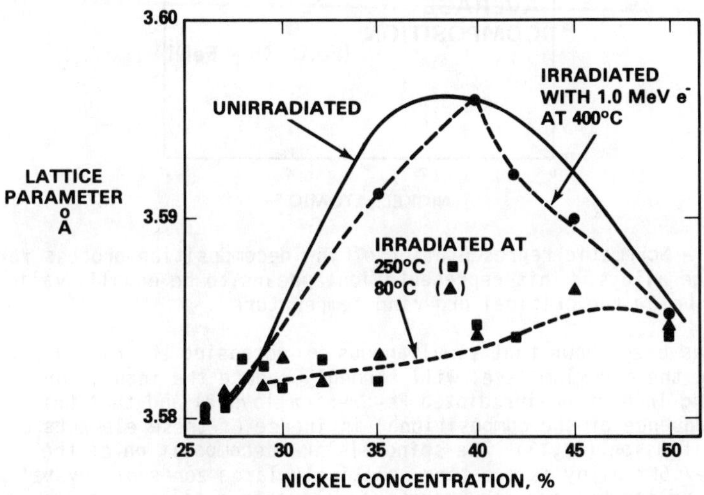

Figure 19 - Lattice parameters of Fe-XNi alloys measured at room temperature before and after electron irradiation.(19)

One can also see in Figure 19, however, that at 400°C the anomaly in lattice parameter is also beginning to disappear, even though the irradiation was conducted above the critical ordering temperature. This supports our hypothesis that the precursor segregation process continues well above the critical order-disorder temperature. It is not unreasonable to expect that not only will the rate of segregation be temperature-dependent, but also that the period of the micro-oscillations will be strongly sensitive to temperature.

References 19, 20 and 23 all describe one feature of the decomposition that is similar to the results presented in this paper. As shown schematically in Figure 20, each study found a tendency of Invar alloys to decompose toward Fe_3Ni and FeNi (25 and 50% nickel), providing we assume that the chromium in Fe-35.5Ni-7.5Cr substitutes for iron. Whether or not this decomposition occurs via a spinodal process is not yet completely clear and additional studies are in progress to discern the mechanism. In recent years, however, a number of authors have independently reached the conclusion that spinodal decomposition should occur in the Fe-Ni system in the Invar compositional range.(31-33) Another indication that spinodal decomposition may have occurred was found in a series of Invar alloys which displayed an unexpected hardening component after irradiation to 12 dpa at 450°C that was attributed by the authors to spinodal hardening.(10)

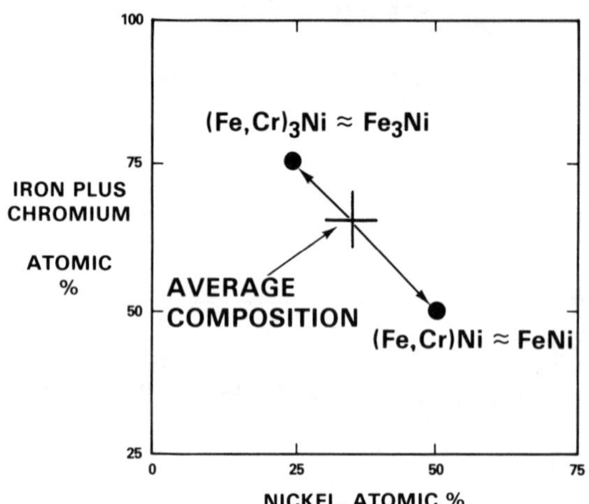

Figure 20 - Schematic representation of the decomposition process for Invar range alloys. This representation appears to be equally valid both above and below the critical ordering temperature.

It has been shown that simultaneously decreasing the nickel and increasing the chromium level will strongly reduce the incubation period of swelling in neutron-irradiated Fe-Cr-Ni alloys(2) and that this occurs as a consequence of the compositional influence of these elements on vacancy diffusion.(12,13) The spinodal-like decomposition of the Fe-35.5Ni-7.5Cr alloy is creating relatively large zones of crystal with this kind of composition and therefore a lowered swelling resistance, and it is reasonable to expect that voids will nucleate first in these regions as shown in Figure 21. Once voids nucleate and grow to reasonably large sizes they begin to segregate nickel at their surfaces via the inverse

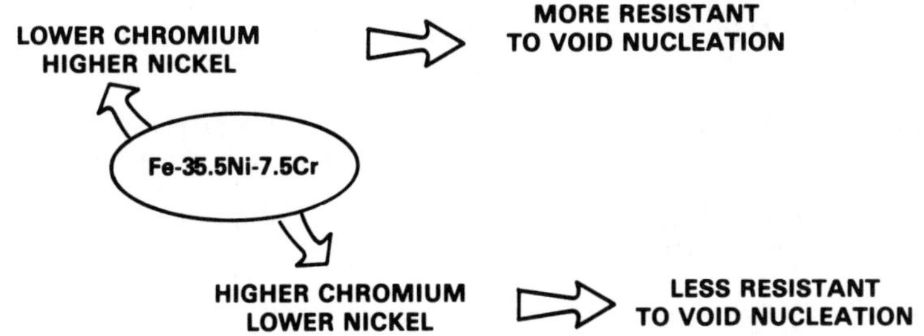

Figure 21 - Schematic representation of the effect of radiation-induced spinodal decomposition on void nucleation.

Kirkendall mechanism and reject chromium back into the matrix.(14) Each void, once formed, thus reduces the nickel content of the surrounding matrix in a manner which makes the nucleation of the next void easier. This sets into motion a positive feedback process of enhanced void nucleation. While nickel-rich precipitates do not form in these alloys the voids themselves become the reservoir for additional nickel removal. It appears, however, that the radiation-induced spinodal process may be responsible for the initial destruction of the swelling resistance.

Conclusions

The simple homogeneous austenite region shown on phase diagrams for Fe-Ni-Cr alloys near the Invar compositional range appears not be the true equilibrium state. When the results of this investigation are compared with those of other studies it appears that radiation seems to accelerate the evolution toward stability, particularly if the irradiation occurs below the critical order-disorder temperature. Even above the critical temperature a spinodal-like decomposition occurs during irradiation. Tnis produces compositional micro-oscillations with periods on the order of hundreds of nanometers in the range 550-625°C.

The compositional oscillations are proposed to destroy the swelling resistance that results from increased nickel and decreased chromium levels, both of which act to reduce void nucleation in a homogeneous alloy. Research is now underway to develop ways to delay the onset of the radiation-induced spinodal decomposition by solute additions and thermal-mechanical treatments. If this effort is successful, an austenitic alloy optimized for swelling resistance may be developed from the Invar class of alloys.

Acknowledgments

This work was supported by the Magnetic Fusion Energy Program of the U.S. Department of Energy under contract DE-AC06-76FF02170. The authors are indebted to T. Lauritzen of General Electric Company for supplying the ion-bombarded specimen.

References

1. F. A. Garner, "Recent Insights on the Swelling and Creep of Irradiated Austenitic Alloys," Journal of Nuclear Materials, 122 & 123 (1984) pp. 459-471.

2. F. A. Garner and H. R. Brager, "Dependence of Neutron-Induced Swelling on Composition in Iron-Based Austenitic Alloys," these proceedings.

3. F. A. Garner, "Overview of the Swelling Behavior of 316 Stainless Steel," these proceedings.

4. F. A. Garner and D. L. Porter, "A Reassessment of the Swelling Behavior of AISI 304 Stainless Steel," pp. 41-44 in Dimensional Stability and Mechanical Behavior of Irradiated Metals and Alloys, Vol. 1, British Nuclear Energy Society, London, 1983.

5. W. G. Johnston, T. Lauritzen, J. H. Rosolowski, and A. M. Turkalo, "The Effect of Metallurgical Variables on Void Swelling," pp. 227-266 in Radiation Damage in Metals, N. L. Peterson and S. D. Harkness, eds.; ASM, Cleveland, OH, 1976.

6. W. G. Johnston, J. H. Rosolowski, A. M. Turkalo, and T. Lauritzen, "An Experimental Survey of Swelling in Commercial Fe-Ni-Cr Alloys Bombarded with 5 MeV Ni^{+} Ions," Journal of Nuclear Materials, 54 (1974) pp. 24-40.

7. D. Gulden and K. Ehrlich, "Void Swelling and Microstructural Development in Experimental Alloys," pp. 13-16 in reference 4.

8. W. R. Rees, B. D. Burns and A. T. Cook, "Constitution of Iron-Nickel-Chromium Alloys at 650-800°C," Journal of the Iron and Steel Institute, 162 (1949) pp. 325-330.

9. W. J. Tomlinson and A. J. Andrews, "Densities of FCC Nickel-Iron Alloys," Metal Science, 12 (5) (1978) pp. 263-264.

10. H. R. Brager, F. A. Garner and M. L. Hamilton, "The Influence of Composition on the Microstructural Evolution and Mechanical Properties of Irradiated Fe-Cr-Ni Ternaries," to be published in Proceedings of First International Conference on Fusion Reactor Materials, Tokyo, Japan, December 3-6, 1985.

11. F. A. Garner, "Application of High Fluence Fast Reactor Data to Fusion-Relevant Materials Problems," ibid.

12. F. A. Garner and W. G. Wolfer, "Factors Which Determine the Swelling Behavior of Austenitic Stainless Steels," Journal of Nuclear Materials, 122 & 123 (1984) pp. 201-206.

13. B. Esmailzadeh and A. Kumar, "Influence of Composition on Steady-State Void Nucleation in Irradiated Metals," to be published in the Proceedings of ASTM 12th International Symposium on Effects of Irradiation on Materials, ASTM STP 870, F. A. Garner and J. S. Perrin, eds.; Williamsburg, VA., June 1984.

14. H. R. Brager and F. A. Garner, "Microsegregation Observed in Fe-35.5Ni-7.5Cr Irradiated in EBR-II," to be published in reference 13.

15. H. R. Brager, F. A. Garner and T. Lauritzen, "Compositional Micro-Oscillations in Ion-Bombarded Fe-35.0Ni-7.0Cr," Damage Analysis and Fundamental Studies Quarterly Progress Report DOE/ER-0046/19, (November 1984) pp. 56-61.

16. W. G. Johnston, W. G. Morris and A. M. Turkalo, "Excess Sub-Surface Swelling Produced by Ni^+ Ion Bombardment," pp. 421-430 in Proceedings of International Conference on Radiation Effects on Breeder Reactor Structural Materials, M. L. Bleiberg and J. W. Bennett, eds.; AIME, New York, NY, 1977.

17. N. Q. Lam, "Compositional Changes in Fe-Cr-Ni Alloys Under Proton Bombardment at Elevated Temperatures," Journal of Nuclear Materials, 117 (1983) pp. 106-112.

18. A. D. Marwick, R. C. Pillar and M. E. Horten, "Radiation-Induced Segregation in Fe-Ni-Cr Alloys," pp. 11-15 in reference 4.

19. A. Chamberod, J. Laugier and J. M. Penisson, "Electron Irradiation Effects on Iron-Nickel Invar Alloys," Journal of Magnetism and Magnetic Materials, 10 (1979) pp. 139-144.

20. J. F. Peterson, A. Aydin and J. M. Knudsen, "Mossbauer Spectroscopy of an Ordered Phase (Superstructure) of FeNi in an Iron Meteorite," Physics Letters, 62A (1977) pp. 192-194.

21. D. Dautrebbe, "Radiation Effects in Magnetic Materials" pp.1-151 in Studies in Radiation Effects in Solids, Vol. 3, Gordon and Breach, G. J. Dienes, ed.; 1966.

22. J. M. Penisson and A. Bourret, "Mise en ordre de l'alliage Fe Ni équiatomique sous irradiation," Bulletin d'Informations Scientifiques et Techniques, 207 (1975) pp. 59-63.

23. J. Pauleve, A. Chamberod, K. Krebs and A. Marchand, "Transformations ordre-désorde et propriétés magnétiques des alliages fer-nickel," IEEE Transactions Magnetics, MAG-2, (3) (1966) pp. 475-478.

24. K. N. Binnatov, A. A. Katsnel'son and Yu. L. Rodionov, "Influence of Chromium on the Atomic Ordering of Nickel-Iron Alloys," Fizika Metallov I Metallovedenie, 39, (5) (1975) pp. 1021-1025.

25. V. I. Goman'kov, I. M Puzei, and E. I. Mal'tsev, "Effect of Alloying Elements on the Superstructure of Ni_3Fe," Soviet Physics-Doklady, 15, (9) (1971) pp. 874-876.

26. V. M. Kalinin and V. A. Kornyakov, "Young's Modulus of Alloyed Iron-Nickel Invars," Fizika Metallov I Metallovedenie, 51 (5) (1981) pp. 1110-1113.

27. H. Heidsiek, R. Scheffel and K. Lucke, " The Influence of Quenched-In and Thermal Vacancies Upon Short-Range Order Formation in a Ni-11.4% Cr Alloy," Journal De Physique, Colloque C7, Supplement No. 12, 38, (1977) p. C7-174.

28. V. I. Goman'kov and N. I. Nogin, "Concentration Structure Transition $Ni_3Fe \rightarrow Ni_2Cr$ in the Ternary System Nickel-Iron-Chromium," Fizika Metallov I Metallovedenie, 52, (3) (1981) pp. 665-667.

29. E. Z. Vintaikin and G. G. Urushadze, "Formation of Long-Range Order in Nickel Chromium Alloys," Soviet Physics-Doklady, 14 (1) (1969) pp. 84-87.

30. R. P. Wahi, "Precipitation in Thin Foils of an Fe-Ni-Cr Alloy," Scripta Metallurgica, 14, (11) (1980) pp. 1229-1232.

31. J. R. C. Guimaraes, J. Danon, R. B. Scorzelli and I. Souza Azevedo, "Phase Stability in Iron-Nickel Alloys," Journal of Physics F: Metal Physics, 10 (1980) pp. L197-L202.

32. P. L. Rossiter and P. J. Lawrence, "Phase Transformations on Fe-Ni Invar Alloys," Philosophical Magazine A, 49 (4) (1984) pp. 535-546.

33. Y. Tanji, Y. Nakagawa, Y. Saito, K. Nishimura and K. Nakatsuka, "Anomalous Thermodynamic Properties of Iron-Nickel (F.C.C.) Alloys," Physica Status Solidi A, 56 (1979) pp. 513-519.

AN EVALUATION OF STEELS AND WELDS PRODUCED OVERSEAS FOR 288°C RADIATION EMBRITTLEMENT RESISTANCE

J. R. Hawthorne

Materials Engineering Associates, Inc.
Lanham, Maryland 20706
USA

Summary

Eight reactor pressure vessel steels and welds produced overseas were irradiated at 288°C to $\sim 2 \times 10^{19}$ n/cm^2, E > 1.0 MeV for assessments of relative notch ductility and dynamic fracture toughness change. Notch ductility and fracture toughness were determined, respectively, by Charpy-V and fatigue precracked Charpy-V test methods. The objectives of this study were to increase our knowledge of the metallurgical requirements for radiation resistance, and concurrently, to assess the adequacy of new specifications of the American Society For Testing and Materials and the American Welding Society for the production outside the USA of radiation resistant steels and welds.

The radiation resistance of each of the eight materials was found to correspond well to the trend of radiation embrittlement resistance reported for USA-produced steels and welds having similar low copper and phosphorus contents. The experimental results fully support application of Nuclear Regulatory Commission Regulatory Guide 1.99 to foreign steels either in use or contemplated for USA reactor vessels.

Introduction

The International Working Group on Reliability of Reactor Pressure Components (IWG-RRPC), sponsored by the International Atomic Energy Agency (IAEA) is embarked on a coordinated study of the radiation embrittlement behavior of pressure vessel steels (1). A primary goal of recent studies was to assess the radiation behavior of improved steels produced in various countries. The intent was to demonstrate that: (a) a careful specification of reactor steels can eliminate the problem of potential failure due to neutron irradiation effects, and (b) knowledge has advanced to the point where steel manufacture and welding technology can routinely produce steel vessels exhibiting a high resistance to radiation. The study encompassed plates, forgings and welds produced by the Federal Republic of Germany, France and Japan and also included a reference steel plate produced in the United States.

The U.S. Nuclear Regulatory Commission (NRC) has supported the IWG-RRPC effort since its inception in 1977, in the form of studies sponsored at the Naval Research Laboratory (NRL) and later at Materials Engineering Associates, Inc. (MEA). Its interest in forwarding the multi-laboratory, international program stems from its objective of furthering knowledge on the metallurgical requirements for radiation resistant steels. Specific research interests of NRL and MEA included the determination of irradiation characteristics of overseas steel production, the relative radiation resistance of USA vs. non-USA produced steels, and the correlation of Charpy-V (C_V) notch ductility and fracture toughness property changes produced by irradiation. The results were expected to be of significant value to the NRC's Regulatory Guide 1.99 (2) in applications of foreign steels in USA vessels and to construction codes.

Current technology for improved steels is founded in great measure on earlier research conducted at NRL (3-6). The cited research, prompted by observations of significant variability in the irradiation response among structural steels, systematically explored the metallurgical causes of this variability. Suspect variables included the identity and quantity of major alloying constituents and residual elements, steelmaking practices, microstructure, and gaseous impurity content, particularly oxygen, hydrogen and nitrogen. Working with commercially-produced melts and special laboratory melts, a strong dependence of irradiation response on residual element content was disclosed (see Figure 1). The determination proved to be a key breakthrough in the understanding of variable radiation embrittlement sensitivity.

Building on this finding, subsequent investigations of specific impurity element contributions employing split laboratory melts, revealed a particularly detrimental effect of copper and phosphorus on resistance to radiation embrittlement as shown in Figure 2. It was found that the mechanism of increased radiation embrittlement sensitivity to copper is not the same as that for phosphorus. Moreover, it was established that synergisms can and do exist between various alloying elements and the copper content and also between different impurity elements as will be discussed later in this report. A second important determination arising from the early laboratory tests was that a minimum specified content of residual elements (e.g., Cu, P, S, Sb, Sn, As) can markedly improve the resistance to radiation embrittlement at power reactor service temperatures (~288°C). In other words, a low residual element content is essential for achieving a low sensitivity to radiation embrittlement in pressure vessel and welds.

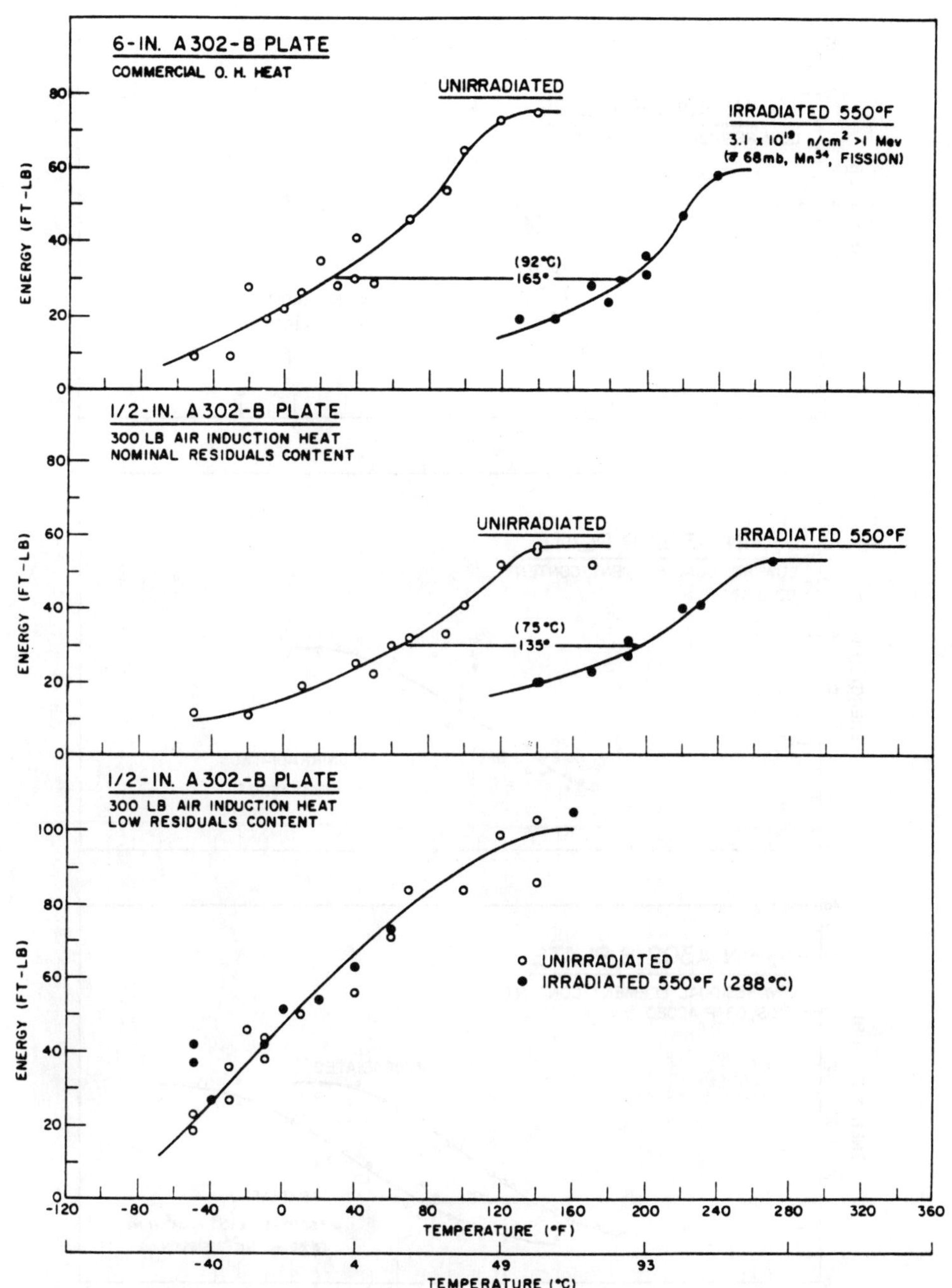

Figure 1 - Comparison of radiation embrittlement sensitivities of one large commercial melt and two air induction melts of A302-B steel with nominal and low residual element contents, based on Charpy-V notch ductility.

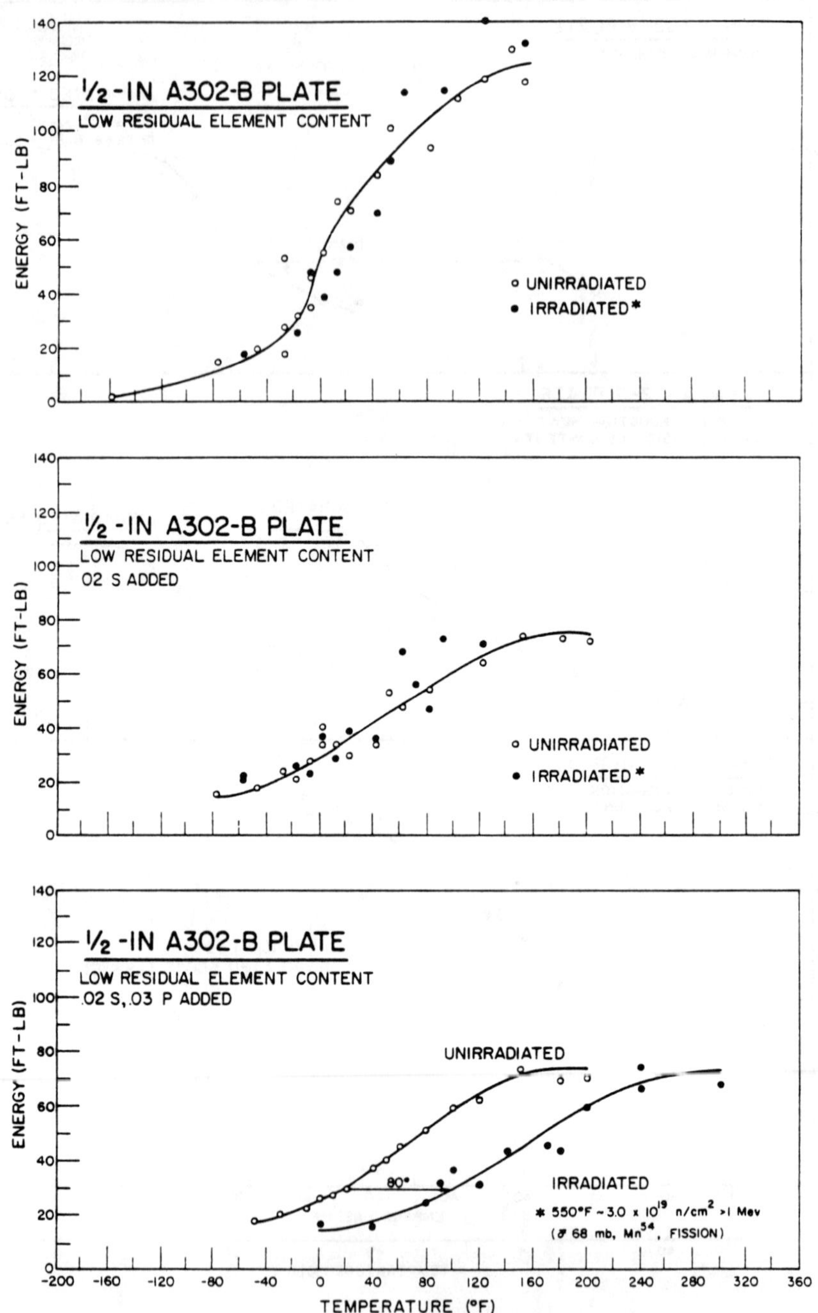

Figure 2 - Notch ductility test results for three plates from a 3-way split melt of A302-B steel showing the detrimental effect of phosphorous impurities on radiation resistance.

Recent efforts, including those of the present study, have centered on the transfer of these laboratory findings to commercial practice. The focus of these efforts is directed toward assessments of the radiation-induced behavior of plates, forgings, and weld deposits from commercial production and applying these assessments to the development of new specifications and guidelines. Studies of improved USA steels and welds yielded results shown in Figure 3 (7). Irradiation test results for three series of materials are depicted: non-improved production steel (Series 1: > 0.15% Cu), improved production steels (Series 2: 0.10% Cu maximum with 0.012% P maximum) and optimum steelmaking practice (Series 3: < 0.06% Cu). Figure 3 shows that a benefit to radiation resistance clearly results from a low copper content. The data also indicate a comparable behavior of Series 2 and Series 3 materials, showing that the use of the "best" steelmaking practice, with its attendant costs, is not necessary.

The experimental investigation reported here represents an extended parallel study and qualification of improved radiation resistance for overseas steel and weld production. The underlying concern was that the use of raw material from sources other than those employed for USA steel production would introduce different impurity element concentrations (or ratios) with a subsequent impact on the radiation sensitivity.

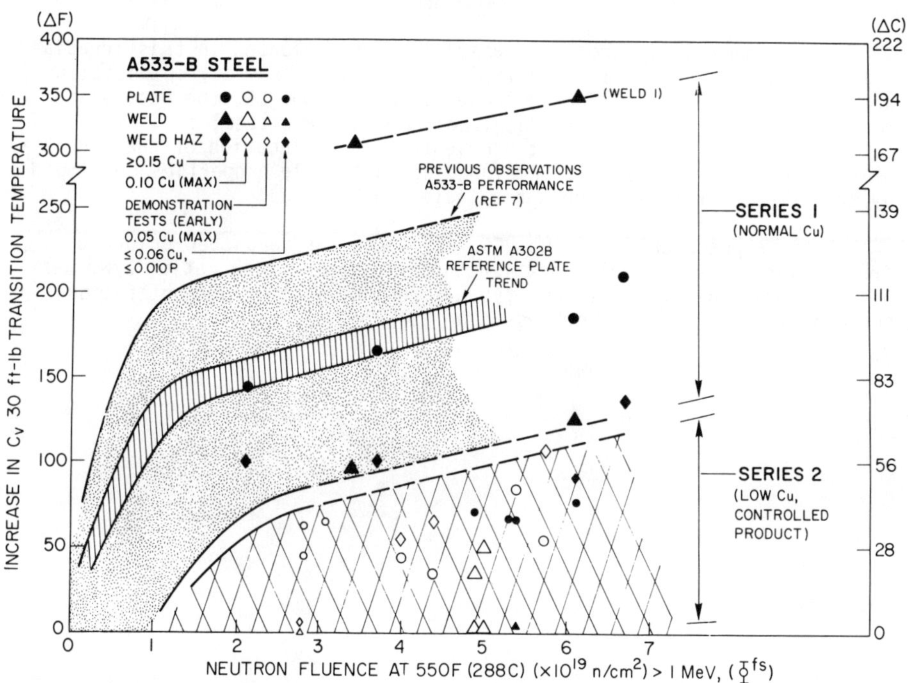

Figure 3 - Summary of Charpy-V 41 Joules (30 ft-lb) transition temperature elevations with 288°C irradiation observed for three series of materials representing increasing levels of copper content control. A clear benefit to radiation embrittlement resistance by a low copper content (0.10% Cu maximum) is evident.

Approach

The determination of relative radiation resistance centers on the comparison of results of preirradiation and postirradiation notch ductility and dynamic fracture toughness tests. The notch ductility tests employed the standard ASTM Charpy-V (C_V) specimen, type A (8). The fracture toughness (K_J) tests employed a fatigue precracked Charpy-V (PCC_V) specimen and J-integral assessment procedures. K_J values were based on the energy absorbed to maximum load corrected for specimen and test machine compliance. The test procedure, developed by the Electric Power Research Institute (EPRI), is described in Reference 9. It should be noted that J-integral values based on maximum load imply an absence of stable (rising load) crack extension; thus, the reported values for upper shelf temperatures may overestimate the value of K_J at crack initiation for the relatively high toughness materials investigated in this study.

Irradiations of these steels were conducted at 288°C in the 2 megawatt UBR pool reactor of the Nuclear Science and Technology Facility at the State University of New York at Buffalo. This temperature is comparable to reactor vessel operating temperatures. Typically, the C_V and PCC_V specimens of an individual material were commingled in the irradiation assembly to provide an exact match in neutron fluence and exposure temperature histories.

Materials

Nine commercially produced materials were evaluated in this irradiation series. The materials are identified by their source, composition and heat treatment condition in Table I. Plate code LG, the base plate material for weld code JW, was derived from the same steel melt as plate code 107. Table II indicates the orientation and location of the specimens in the source material. Preirradiation tensile properties developed by this study for these steels are listed in Table III.

Fatigue precracking of the PCC_V specimens was conducted prior to irradiation. The nominal crack length to specimen width ratio, a/W, was 0.5. The maximum stress intensity, K_f, for the last 0.76 mm of fatigue crack growth was equal to or less than 22 $MPa\sqrt{m}$.

TABLE I. Test Materials

Material/ Producer	Codes	Thickness (mm)	Chemical Composition (wt.%) C	Mn	Si	P	S	Ni	Cr	Mo	Cu	Other	Heat Treatment
HSST A533-B Plate 03 (Lukens Steels)	3 MU, 3 NE	305	.20	1.26	.25	.011	.018	.56	.10	.45	.12		1
Japan A533-B Plate (Nippon Steel)	107	251	.18	1.48	.22	.007	.007	.66	.20	.57	.01		2
Japan A508-3 Forging (Japan Steel)	212, 213	302	.18	1.35	.27	.007	.005	.76	.12	.49	.04		3
Japan S/A Weld (Mitsubishi)	502, 503	248	.07	1.20	.32	.008	.003	.89	.06	.50	.04		4
Japan A533-B Plate (Parent Plate for Weld 502)	LG	245	(Not Available)										5
A533-B Class 1 (MARREL)	FP	310	.23	1.44	.23	.007	.002	.65	.03	.51	.03	.02 Al .004 Sn .022 As	6
A508 Class 3 (FRAMATOME)	FF	230	.15	1.37	.25	.009	.008	.69	.24	.47	.07	.02 Co <.01 V	7
S/A Weld* (FRAMATOME)	FW	220	(1) .056	1.37	.54	.015	.011	.56	.05	.52	.05	.02 Co .01 V	8
			(2) .075	1.60	.38	.011	.008	.73	.09	.55	.06	.02 Co .01 V	
S/A Weld (Thyssen-Maschinenbau)	GW	250	.075	1.48	.16	.011	.009	.93	.02	.62	.03	.013 Al <.01 Co <.01 V	9

Heat Treatments:
1. 843 to 899°C - 4 hr, WQ; 649 to 677°C - 4 hr. AC; 607 to 636°C - 20 hr, FC.
2. 880°C - 8 hr, WQ; 660°C - 6 hr, AC; 620°C - 26 hr, FC
3. 870 to 900°C - 6.4 hr, WQ; 635 to 645°C - 7.8 hr, AC.
4. (PWHT) - 615°C - 26 hr, FC.
5. Not known.
6. 900°C - 7 hr, 40 min, WQ; 640°C - 10 hr 35, AC; 610° - 24 hr, FC at 40°C/hr
7. 865/880°C - 3 hr, WQ; 630/650°C - 5 hr 30, AC; 615°C - 8 hr, AC.
8. 617°C - 8 hr, FC at 30°C/hr.
9. 610°C - 20 hr, FC.

* (1) 2.5 mm analysis.
(2) 4.0 mm analysis.

TABLE II. Specimen Location/Orientation in Test Materials

Material	Code	Specimen Locations (Thickness Direction)	Specimen Orientation (Long Dimension)
A533-B Plate (HSST 03)	3 MU, 3 NE	3/4T*(4 layers)+	Transverse to Rolling Direction (TL)
Japan A533-B Plate	JP 107	1/4T (4 layers)	Transverse
Japan A508-3 Forging	JF 212, JF 213	1/4T (4 layers)	Transverse
Japan S/A Weld	JW 502, JW 503	Through Weld Thickness (Major Groove-6 layers)	Perpendicular to Welding Direction
Japan A533-B Plate	LG (JW-BP)	1/4T, 3/4T (2 layers each)	Transverse
France A533-B Plate	FP	1/4T (4 layers)	Transverse
France A508-3 Forging	FF	1/4T (3 layers)	Transverse
France S/A Weld	FW	Through Weld Thickness (Major Groove-6 layers)	Perpendicular to Welding Direction
FRG S/A Weld	GW	1/4T (2 layers)	Perpendicular to Welding Direction

+ Four layers spanning the quarter thickness plane

* 3/4T designated a location 3/4 of the way through the plate thickness.

TABLE III. Tensile Properties at 24°C

Material	Identification Codes	Yield Strength* (0.2% Offset) (MPa)	Tensile Strength (MPa)	Reduction in Area (%)	Total Elongation (in 25.4 mm) (%)
Plate	3 MU, 3 NE	464	628	65.0	25.3
Plate	JP 107	442	589	72.6	24.7
Forging	JF 212, 213	462	597	70.7	26.2
Weld	JW 502, 503	530	619	73.0	24.8
Plate	LG (JW-BP)	444	581	72.6	28.8
Plate	FP	486	632	66.7	25.8
Forging	FF	501	626	70.2	25.6
Weld	FW	508+	607	71.1	24.9
Weld	GW	529	583	74.6	28.2

* Average of duplicate tests (6.4 mm gage diameter x 25.4 mm gage length)

+ 5.74 mm diameter x 25.4 mm gage length specimens

TABLE IV. Test Material Irradiations

Experiment No.	Compartment No.	Material Codes	Fluence Φ^{cs} n cm^{-2} (E > 1.0 MeV)
UBR-23	1	3 MU	2.1×10^{19}
	2	JW 502	2.3×10^{19}
	3	JF 212	2.0×10^{19}
		LG (JW-BP)	1.9×10^{19}
UBR-24	1	JP 107	2.3×10^{19}
	2	JF 212	2.5×10^{19}
UBR-29	1	FF	2.0×10^{19}
	2	FW	2.5×10^{19}
	3	FP	2.1×10^{19}
		3 MU	2.1×10^{19}
UBR-30	3	GW	2.2×10^{19}

Irradiation Experiments

The large number of test materials necessitated the use of four irradiation assemblies. The contents of individual assemblies and the neutron fluences received by each assembly are listed in Table IV. The neutron fluences received by the specimens were determined from iron wire dosimeters placed within each array of specimens. The neutron spectrum for irradiation locations within the fuel lattice of the UBR reactor have been calculated by the Hanford Engineering Development Laboratory (10). For the specific in-core positions used, the calculated spectrum fluence (ϕ^{cs}) and the fluence based on an assumed fission spectrum (ϕ^{fs}) have the relation

$$\phi^{cs} = 1.22\,\phi^{fs}\ (E > 1.0\ \text{MeV}). \tag{1}$$

Also, the calculations show that the calculated spectrum fluence for neutron energies greater than 1.0 MeV and the calculated spectrum fluence for neutron energies greater than 0.1 MeV have the relation

$$\phi^{cs}\ (E > 0.1\ \text{MeV}) = 2.91\,\phi^{cs}\ (E > 1.0\ \text{MeV}). \tag{2}$$

Results

The C_V notch ductility determinations for the materials produced overseas are presented in Figures 4 through 11. Plate code LG shown in Figure 7 was the base plate material used for weld code JW. The C_V results for the USA-produced reference plate (A533-B steel, HSST Plate 03) are illustrated in Figure 12. Fracture toughness K_J determinations for the materials are presented in Figures 13 through 21. The results from both test methods are summarized for comparison in Tables V and VI. The choice of the C_V 41 Joules and the K_J 100 MPa$\sqrt{m}$ temperatures to index the transition temperature was arbitrary. We note that the C_V 41 Joules temperature is also used as a convenient index by the Nuclear Regulatory Commission Regulatory Guide 1.99 and the Code of Federal Regulations (10 CFR 50) (11) for radiation effects projections and transition temperature comparisons.

TABLE V. Preirradiation and Postirradiation Charpy-V Notch Ductility Properties

Material	Fluence* (ϕ^{CS})	C_v 41 Joules (°C)			Transition Temperature C_v 68 Joules (°C)			C_v 0.9 mm (°C)			Upper Shelf Energy (J)		
		Initial	Irrad.	Change	Initial	Irrad.	Change	Initial	Irrad.	Change	Initial	Irrad.	Change
HSST 03	2.1	-1	43	44	29	76	47	24	80	56	138	136	~0
JP 107	2.3	-46	-29	17	-35	-21	14	-38	-21	17	>270	212	≥58
JF 212	2.5	-63	-32	31	-54	-21	33	-54	-21	33	~233	~218	~15
JW 502	2.3	-42	-23	19	-29	-12	17	-34	-7	28	184	184	~0
LG (JW-BP)	1.9	-37	-23	14	-32	-15	17	-34	-15	19	>270	228	≥37
FP	2.1	-18	-4	14	7	24	17	-1	18	19	165	165	~0
FF	2.0	~-1	-7	(-)6	~7	2	(-)5	7	-1	(-)8	200	197	~0
FW (Layers 1-3)	2.5	-54	-46	8	-34	-26	8	-40	-32	8	220	195	25
FW (Layers 4-6)	2.5	-54	-23	31	-34	2	36	-40	2	42	193	168	25
GW	2.2	-62	-34	28	-51	-23	28	-57	-26	31	>270	210	≥60

* in units of 10^{19} n/cm^2, E > 1.0 MeV

TABLE VI. Preirradiation and Postirradiation

Dynamic Fracture Toughness (K_J) Properties

PCC_V Test Method

Material Codes	Fluence* (ϕ^{CS})	K_J 100 $MPa\sqrt{m}$ (°C) Initial	Irrad.	Change	Upper Shelf ($MPa\sqrt{m}$)+ Initial	Irrad.	Change
HTSS 03	2.1	33	77	44	265	260	∿0
JP 107	2.3	-1	16	17	335	335	∿0
JF 212	2.5	2	35	33	355	325	∿0
JW 502	2.3	-17	16	33	320	320	∿0
LG(JW-BP)	1.9	-1	16	17	335	325	∿0
FP	2.1	13	32	19	315	320	∿0
FF	2.0	4	10	6	315	285	>30
FW (Layers 1-3)	2.5	-34	-3	31	315°	330	∿0°
FW (Layers 4-6)	2.5	-34	11	45	315	325	∿0
GW	2.2	-34	2	36	330	295	35

* 10^{19} n/cm^2, E > 1 MeV

\+ Lowest value of upper shelf test data

° Estimated

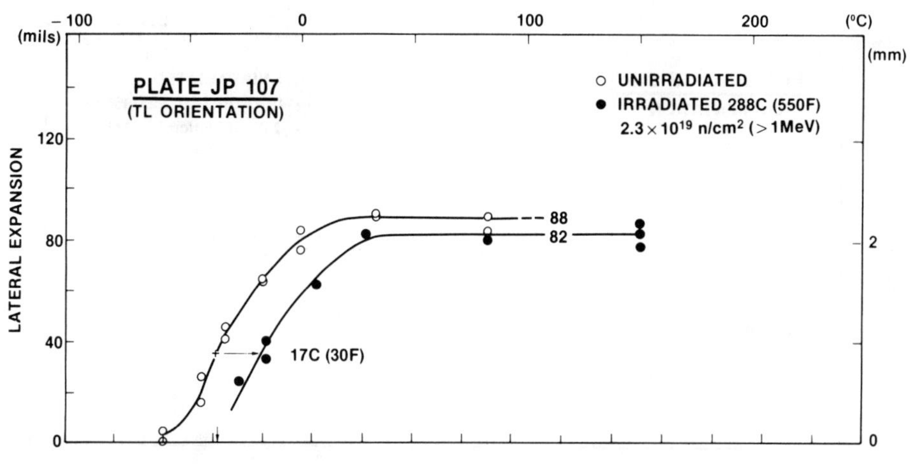

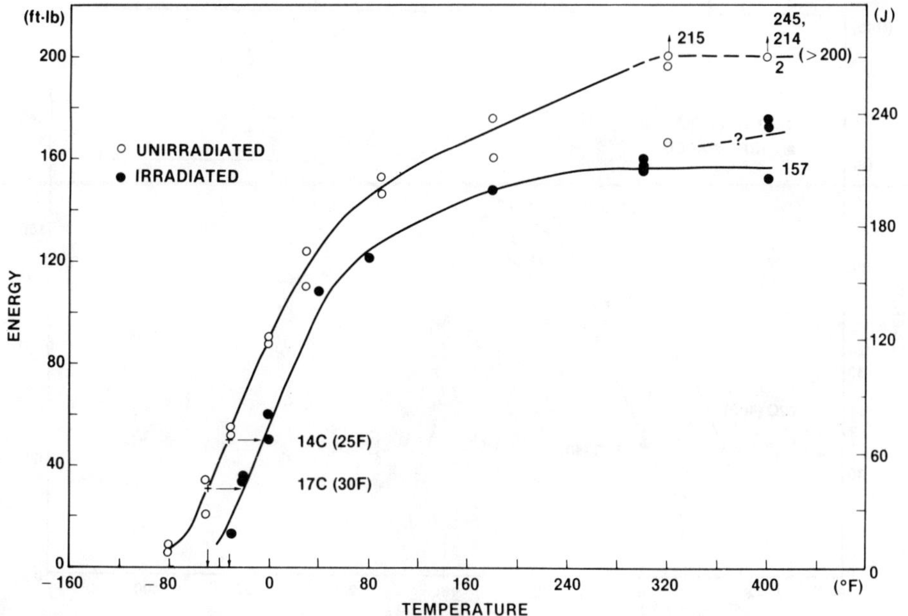

Figure 4 - Charpy-V notch ductility of the A533-B Class 1 steel plate, Code JP 107, supplied by Japan. In this Figure and in Figures 5 through 12 the upper graph relates specimen lateral expansion and temperature while the lower graph relates specimen energy absorption and temperature.

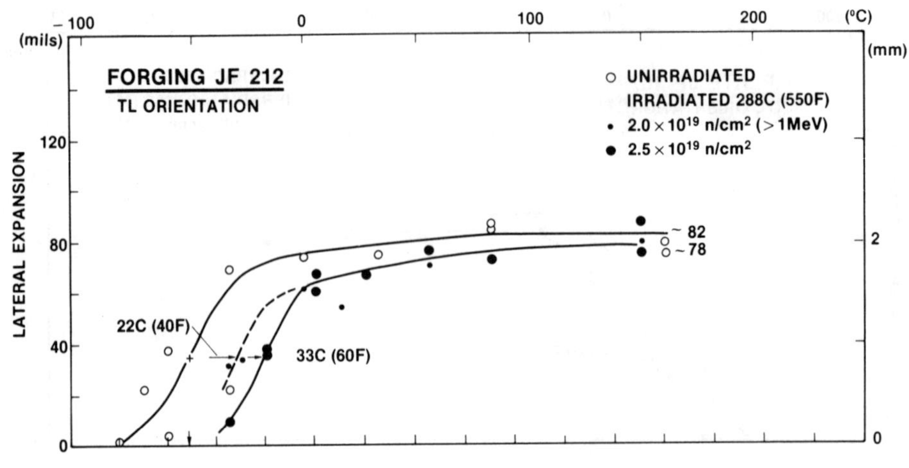

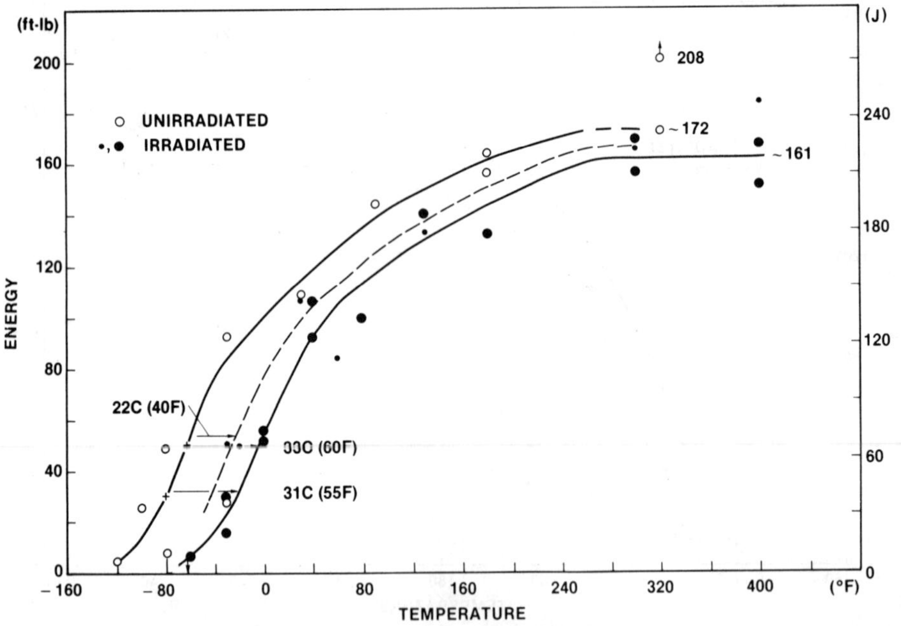

Figure 5 - Charpy-V notch ductility of the A508 Class 3 steel forging, Code JF 212, supplied by Japan.

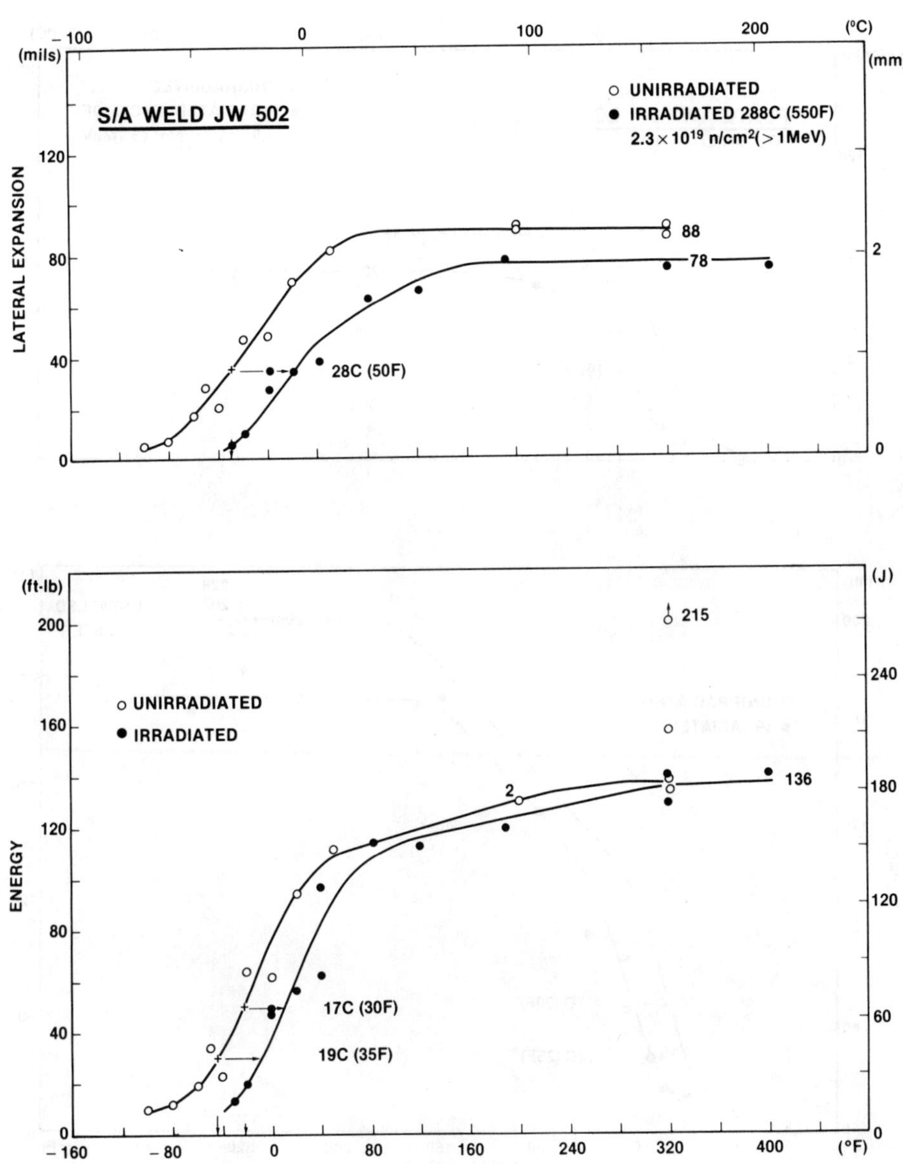

Figure 6 - Charpy-V notch ductility of the submerged arc weld, Code JW 502, supplied by Japan.

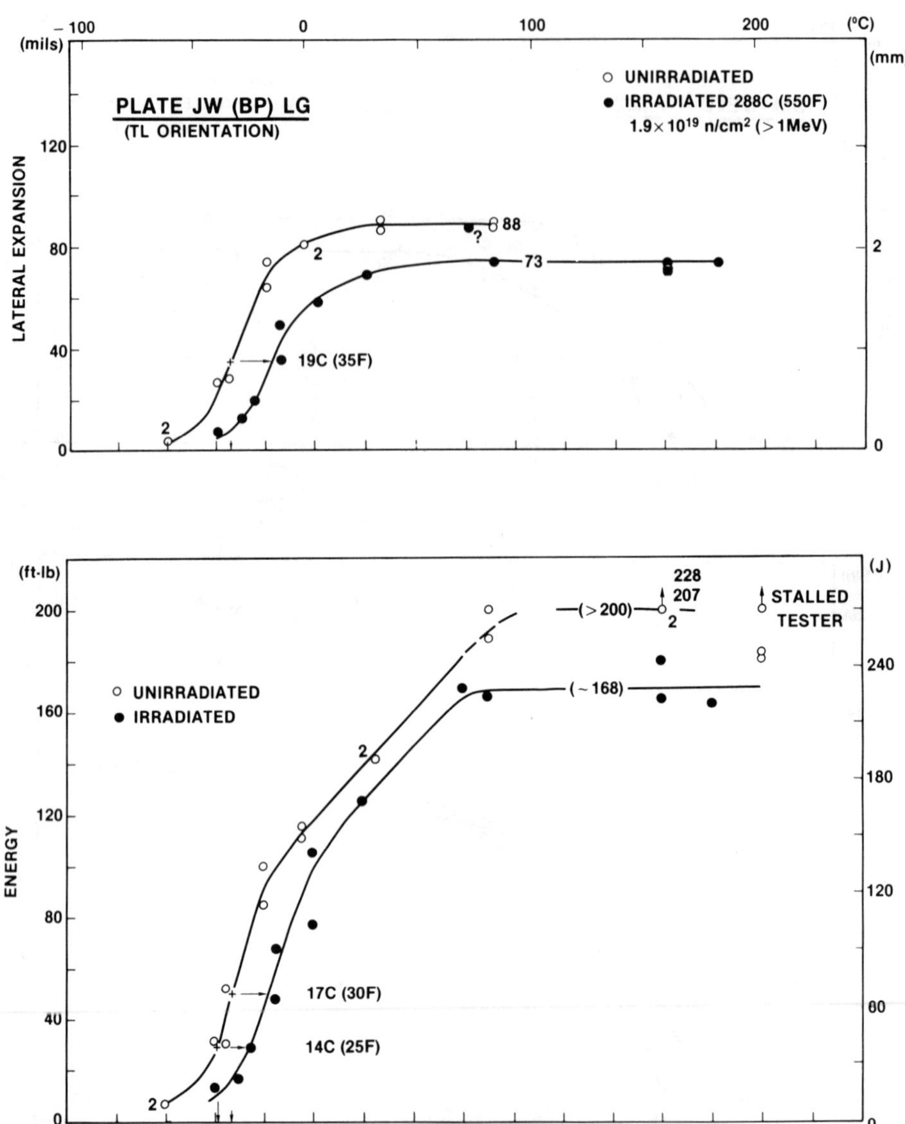

Figure 7 - Charpy-V notch ductility of the A533-B steel base plate, Code LG, for the weld Code JW 502.

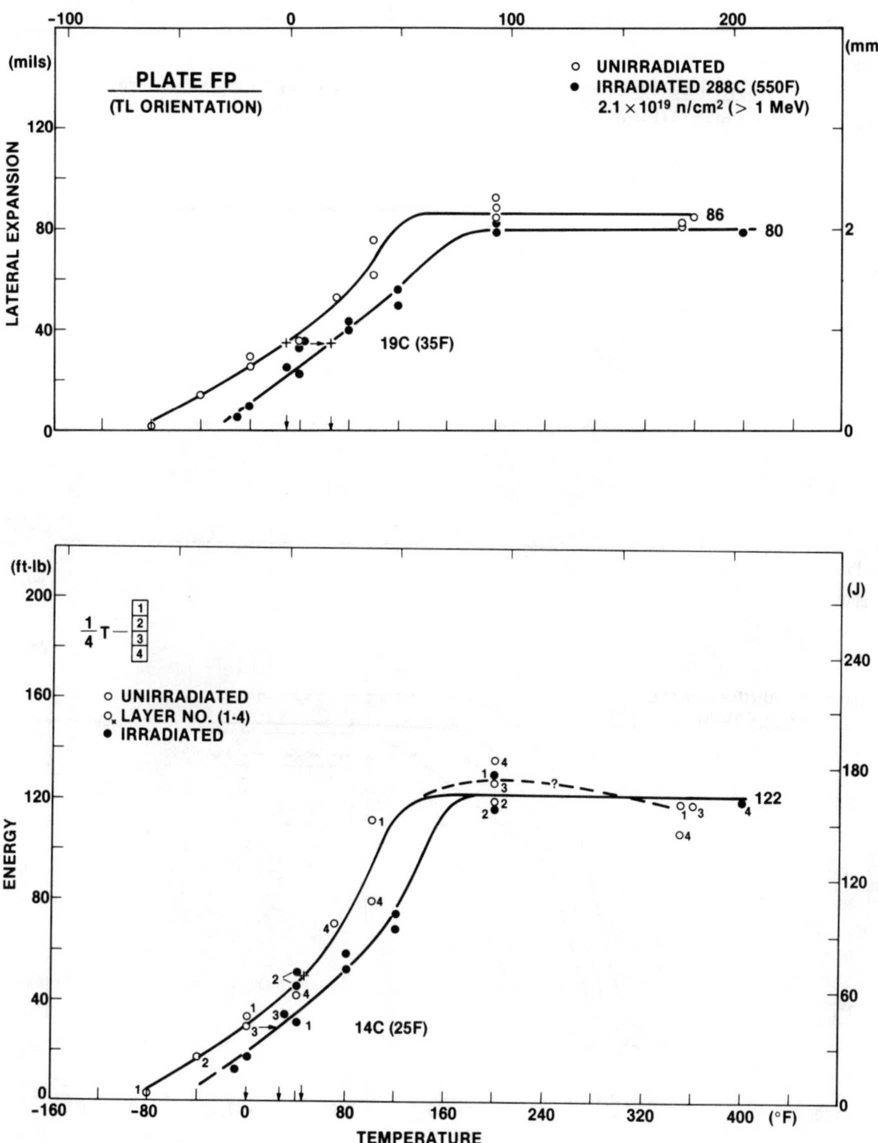

Figure 8 - Charpy-V notch ductility of the A533-B Class 1 steel plate, Code FP, supplied by France.

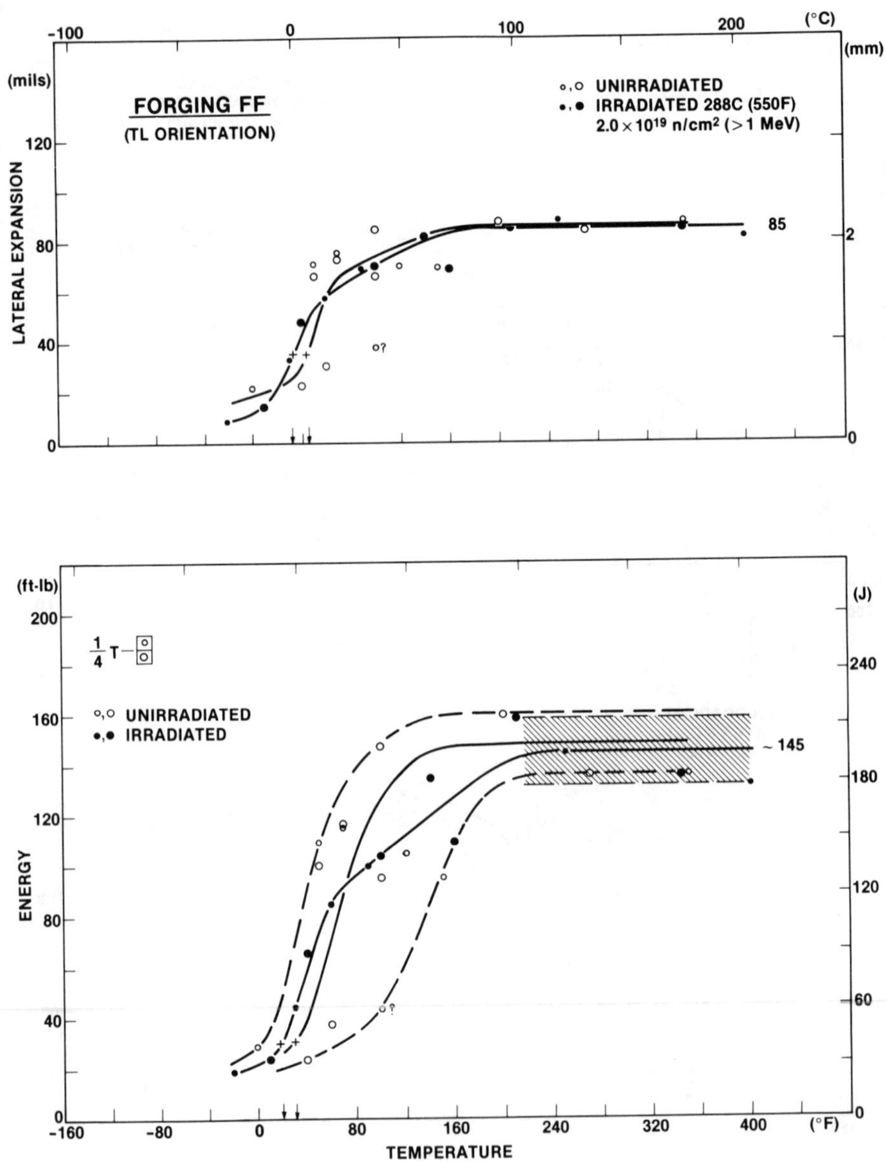

Figure 9 - Charpy-V notch ductility of the A508 Class 3 steel forging, Code FF, supplied by France.

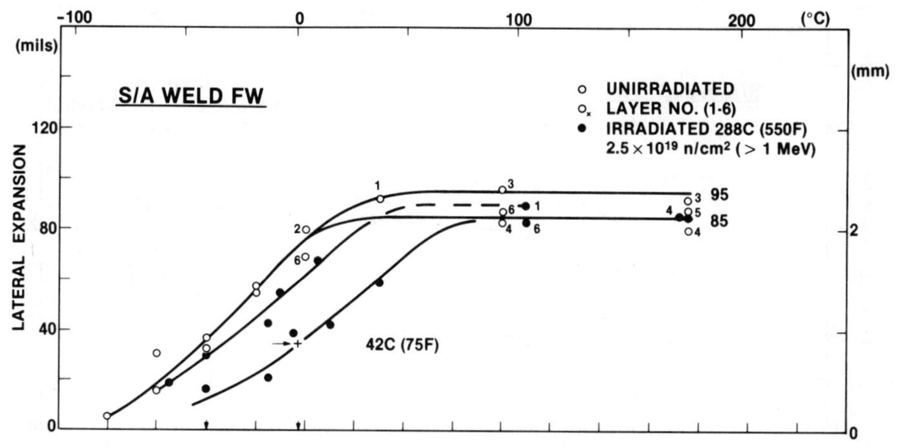

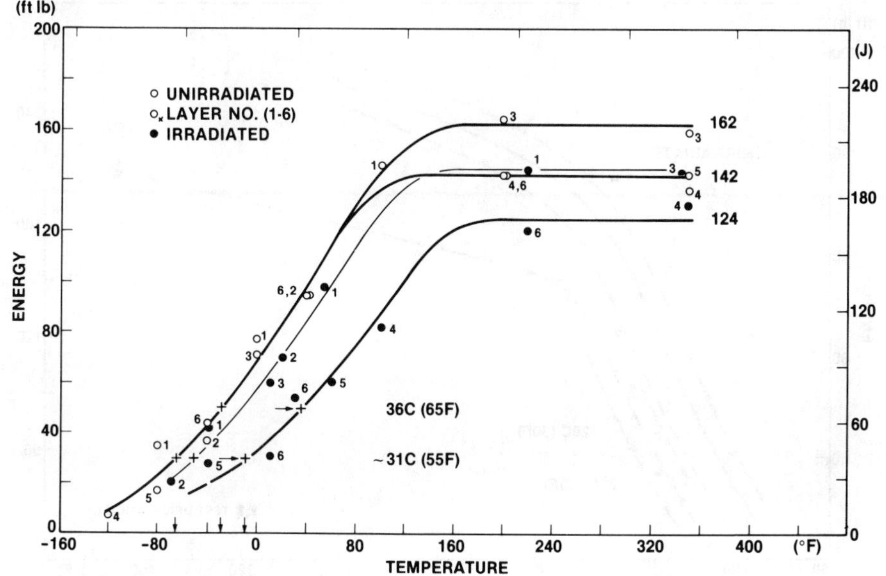

Figure 10 - Charpy-V notch ductility of the submerged arc weld, Code FW, supplied by France. Two levels of radiation embrittlement sensitivity are clearly indicated by the irradiation data, traceable to the original specimen location in the weld deposit thickness (layers 1 to 3 versus layers 4 to 6).

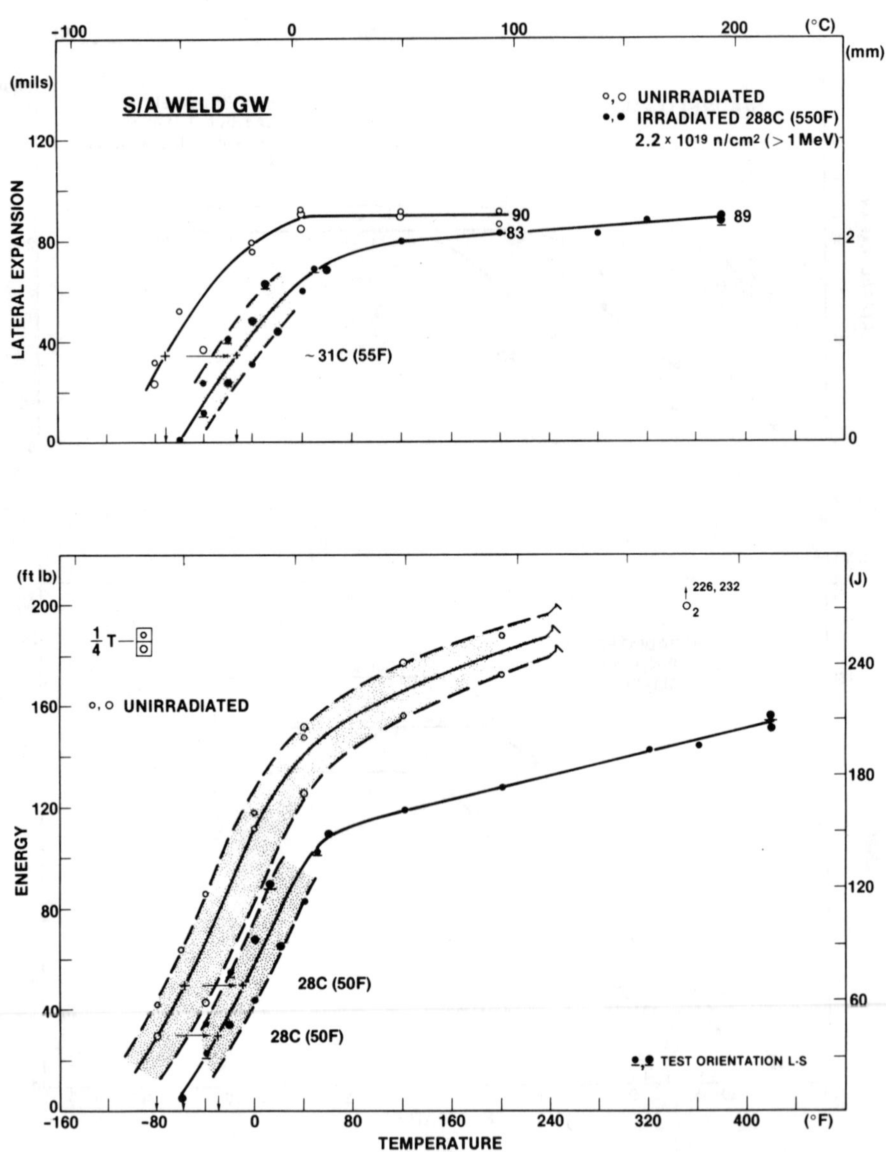

Figure 11 - Charpy-V notch ductility of the submerged arc weld, Code GW, supplied by the Federal Republic of Germany.

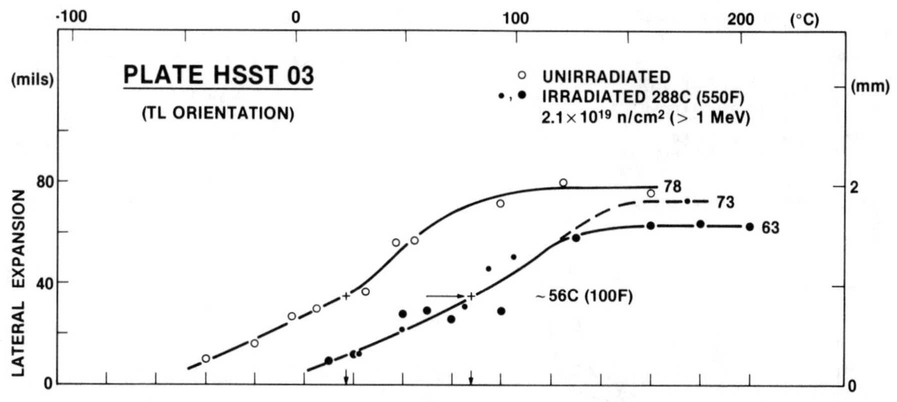

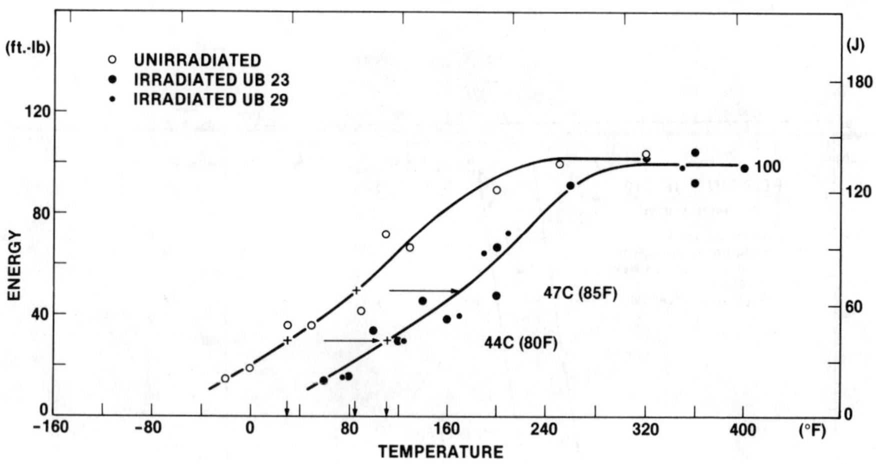

Figure 12 - Charpy-V notch ductility of the A533-B Class reference plate (HSST Program Plate 03), Code 3 MU, supplied by the USA.

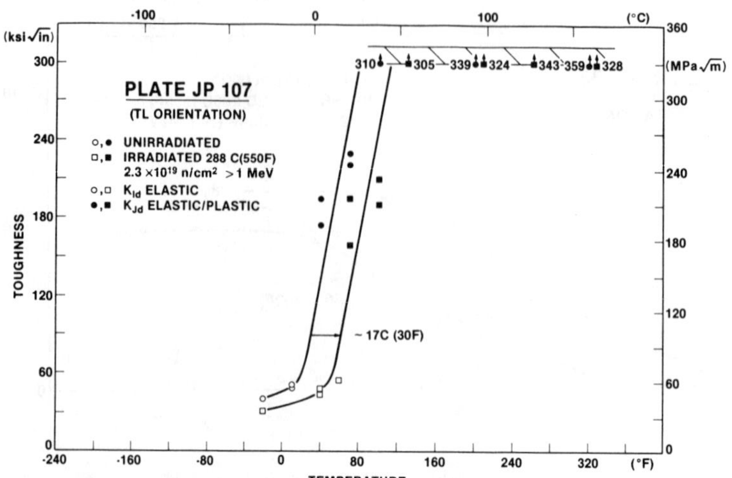

Figure 13 - Fracture toughness of the A533-B Class 1 plate, code JP 107, supplied by Japan (PCC$_V$ Test Method)

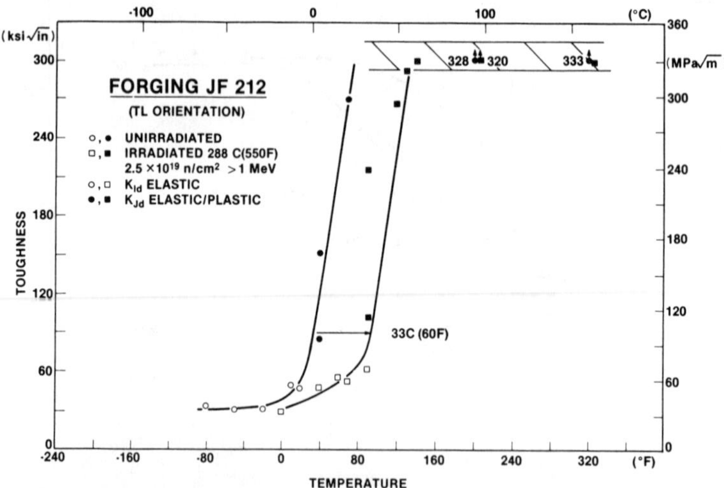

Figure 14 - Fracture toughness of the A508 Class 3 forging, Code JF 212, supplied by Japan (PCC$_V$ Test Method)

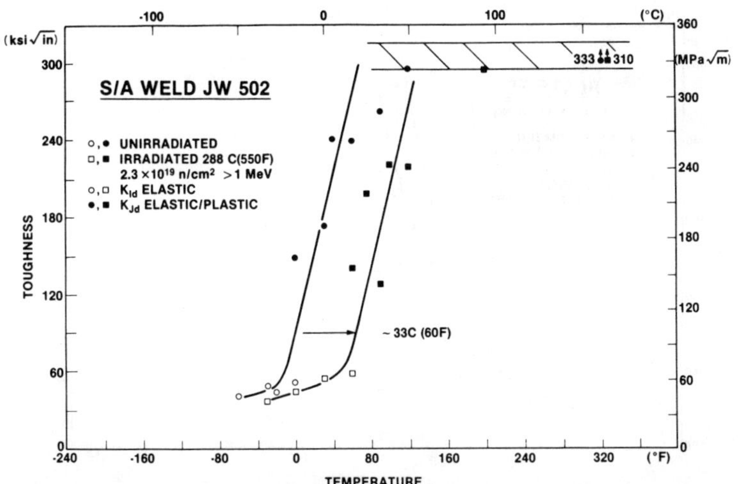

Figure 15 - Fracture toughness of the submerged arc weld, Code JW 502, supplied by Japan (PCC_V Test Method)

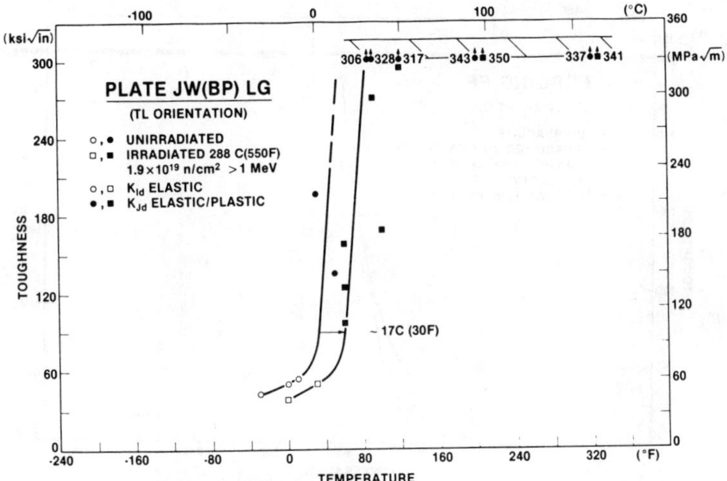

Figure 16 - Fracture toughness of the A533-B base plate, code LG, for the weld Code JW 502 (PCC_V Test Method)

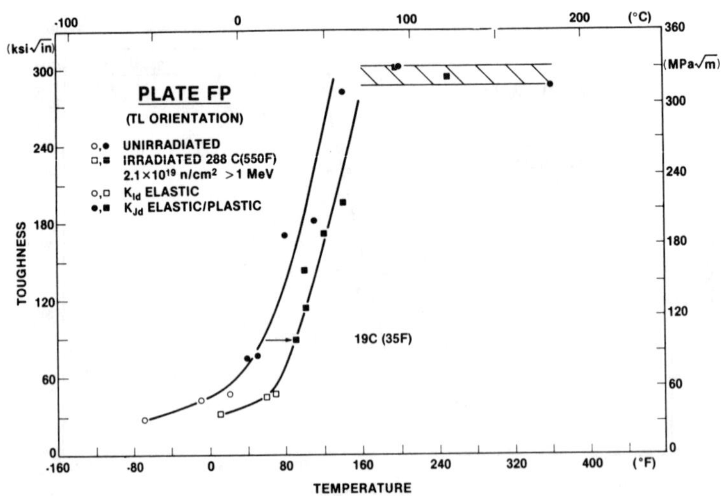

Figure 17 - Fracture toughness of the A533-B Class 1 plate, Code FP, supplied by France (PCC$_V$ Test Method)

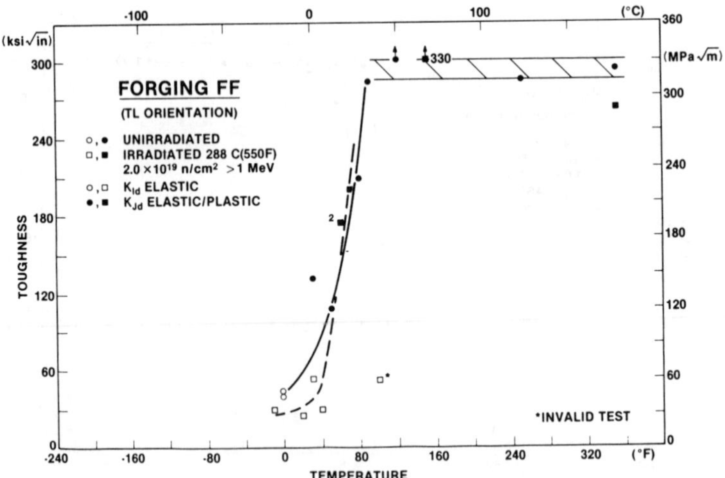

Figure 18 - Fracture toughness of the A508 Class 3 forging, Code FF, supplied by France (PCC$_V$ Test Method)

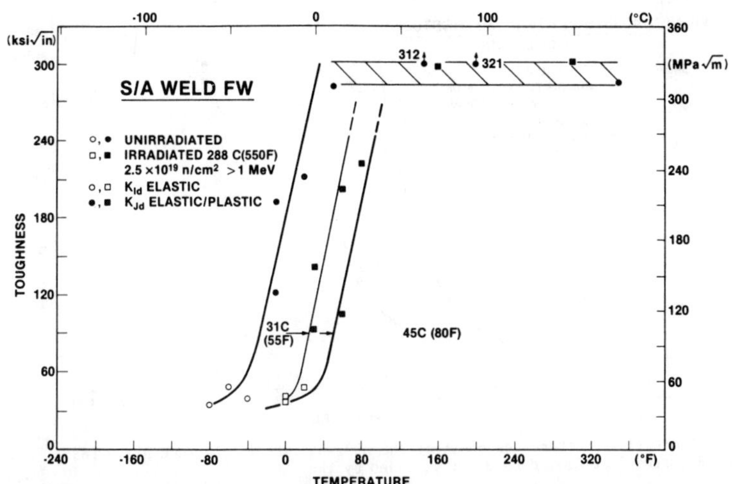

Figure 19 - Fracture toughness of the submerged arc weld, Code FW, supplied by France (PCC$_V$ Test Method)

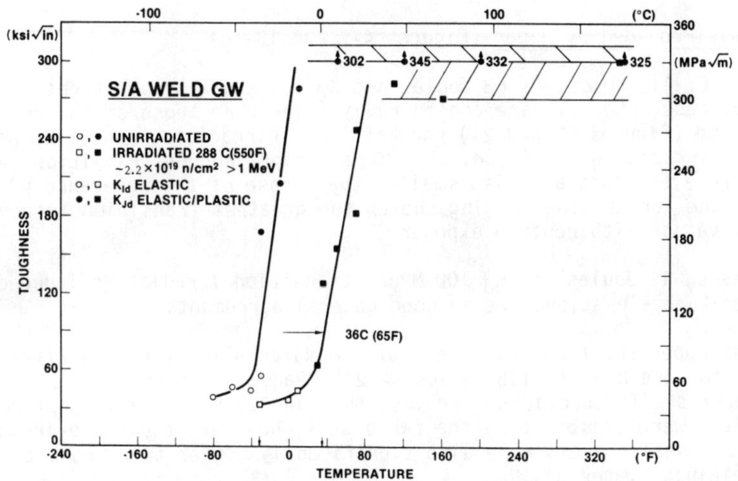

Figure 20 - Fracture toughness of the submerged arc weld, Code GW, supplied by the Federal Republic of Germany (PCC$_V$ Test Method)

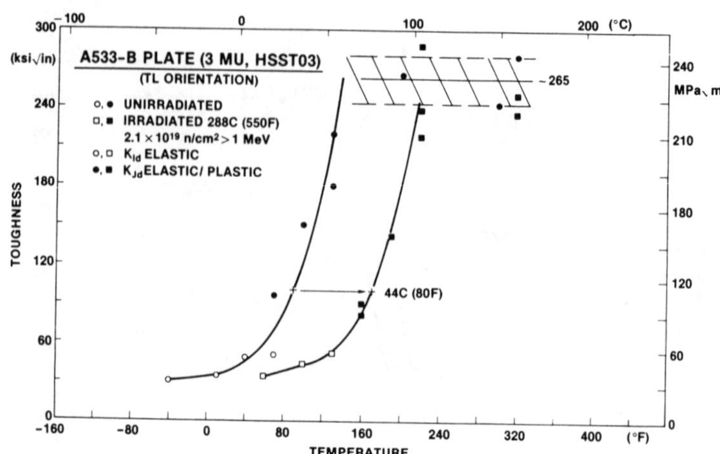

Figure 21 - Fracture toughness of the A533-B Class 1 reference plate (HSST Program Plate 03), code 3 MU, supplied by the USA.

The following observations were made from the data.

Materials Provided by Japan (Figures 4-7 and 13-16)

1. The C_V 41 Joules, C_V 68 Joules and K_J 100 MPa$\sqrt{m}$ temperatures of the Japanese materials are considerably lower than those of the reference plate (Figures 12 and 21) for both the unirradiated condition and for the irradiated condition. The irradiation-induced transition temperature elevations are also smaller than those of the reference plate. Of the former, the forging showed the greatest transition temperature elevation with neutron exposure.

2. The C_V 41 Joules and K_J 100 MPa$\sqrt{m}$ transition irradiation-induced temperature elevations are in good general agreement.

3. The upper shelf energy levels of the materials before and after irradiation are high (> 135 Joules, > 275 MPa$\sqrt{m}$). For the most part, C_V upper shelf reductions were less than 14 Joules. The largest reductions were exhibited by the two plates; however, their pre-irradiation upper shelf levels were also significantly higher than those of the remaining materials.

Materials Provided by France (Figures 8-10 and 17-19)

1. Preirradiation and postirradiation values of C_V 41 Joules, C_V 68 Joules and K_J 100 MPa$\sqrt{m}$ temperatures for French steels are lower than those

of the reference plate with one exception. For the forging, the 41 Joules temperature is about equal to that of the reference plate. In contrast, the 100 MPa$\sqrt{m}$ temperature of the forging is lower than that of the reference plate. Wide data scatter was observed in the C_V tests but not in the PCC_V test of the forging.

2. Transition temperature elevations by irradiation were much smaller than those for the reference plate with the exception of one portion of the weld.

3. Weld code FW data describe two levels of radiation resistance depending on the original specimen location in the weld. Specimens taken from thickness layers 1 to 3 indicate a higher radiation resistance than specimens taken from layers 4 to 6 in terms of 41 Joules temperature elevation and upper shelf level. A less pronounced data separation is also evident in the K_J results. The observations in turn suggest that two, not one, weld wire compositions were used for filling the major groove of the weld from whence the specimens were taken.

4. Each of the materials exhibited a high upper shelf energy level after irradiation; an upper shelf reduction was observed only for the weld material.

<u>Material Provided by the Federal Republic of Germany</u> (Figures 11 and 20)

1. Preirradiation and postirradiation C_V 41 Joules, C_V 68 Joules and K_J 100 MPa$\sqrt{m}$ temperatures for the German steels are significantly lower than those for the reference plate.

2. A relatively large reduction in upper shelf energy was found in postirradiation C_V testing; a small reduction in upper shelf toughness was observed in postirradiation PCC_V tests.

Figure 22 compares the preirradiation and postirradiation transition temperatures and transition temperature elevations of all nine materials. Material-to-material differences are noted; however, the material properties in general appear to be better than those of the USA 305 mm-thick reference plate. The trend can be attributed to processing differences in combination with material thickness differences and to material dissimilarities in copper and phosphorus contents.

Transition Temperature Elevation Vs. Test Method

Figures 23 and 24 compare the measured C_V transition temperature elevations with the K_J 100 MPa$\sqrt{m}$ transition temperatures measured by the PCC_V method. On balance, agreement within 15°C is observed, suggesting a possible correlation which will be of benefit to reactor vessel surveillance programs. An exception to the trend is found in one data set for material code FW (layers 1-3). An explanation for the inconsistency is not available at this time.

A more stringent test of the correlation was permitted recently with the acquisition of additional C_V vs. PCC_V postirradiation test comparisons from other programs at Materials Engineering Associates and the Naval Research Laboratory (12,13). Seventeen of eighteen correlation tests showed agreement within 15°C; transition temperature elevations in these tests ranged as high as 100°C. The sole exception to the trend is that noted in Figure 24 (Code FW, layers 1-3).

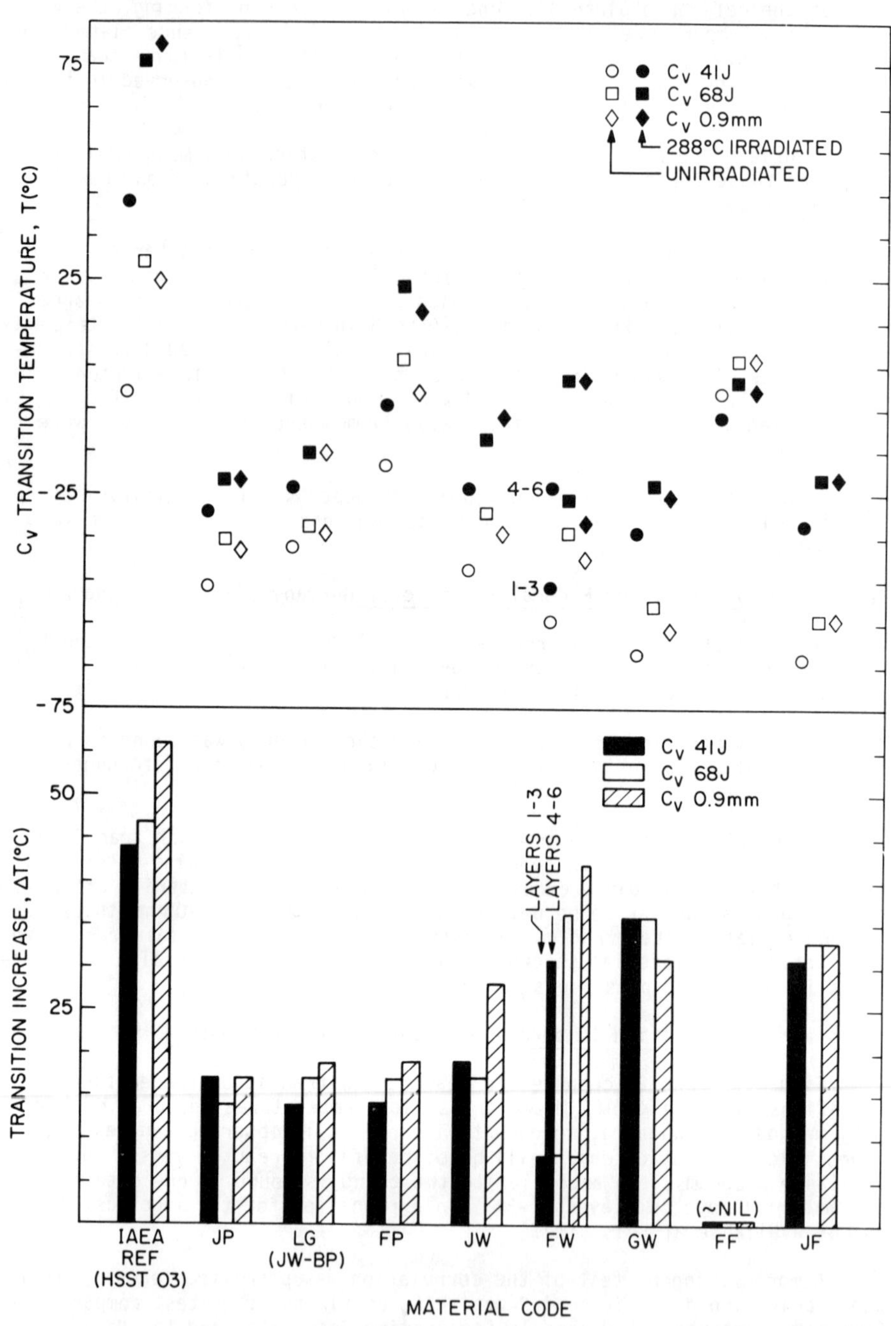

Figure 22 - Summary of C_V 41 Joules, 68 Joules and 0.9 mm transition temperature determinations for unirradiated and irradiated conditions. The upper graph compares transition temperature elevations by irradiation.

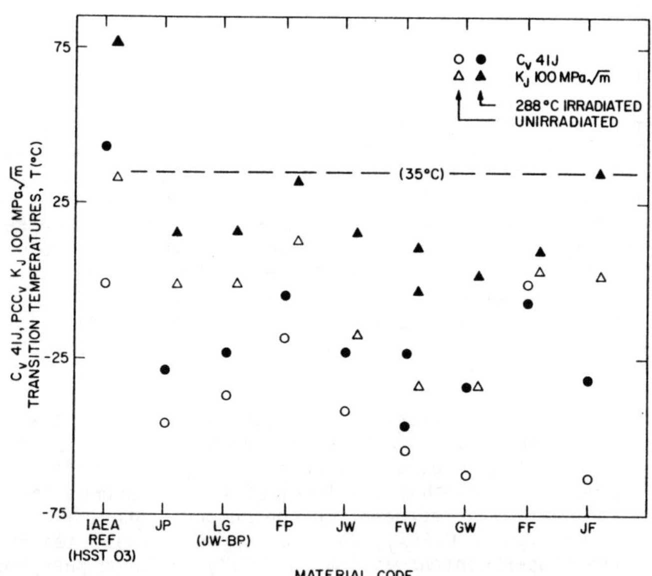

Figure 23 - Summary of 41 Joules and K_J 100 MPa$\sqrt{m}$ transition temperature determinations for unirradiated and irradiated conditions.

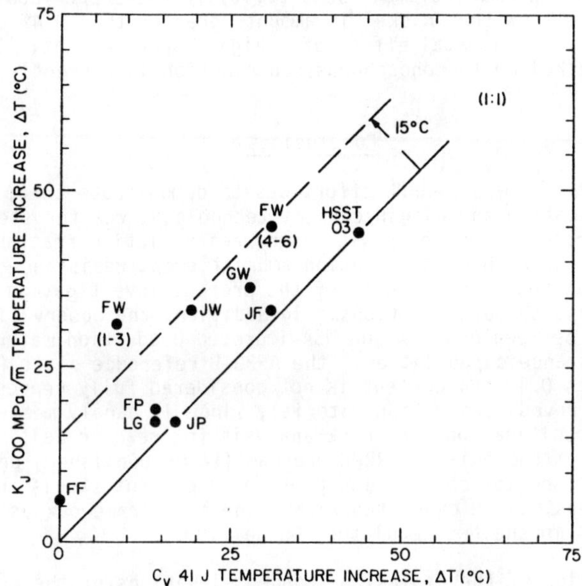

Figure 24 - Comparison of the K_J 100 MPa$\sqrt{m}$ and 41 Joules transition temperature elevations by 288°C irradiation. Agreement within 15°C is observed for all but one irradiation test (weld Code FW, layers 1-3). A small bias toward a greater K_J 100 MPa$\sqrt{m}$ transition temperature elevation is indicated by the data scatter.

Discussion

The postirradiation measurements of the fracture properties of the IWG-RRPC program materials have revealed high radiation resistance qualities and confirm the projected benefits of specifying low copper and low phosphorus impurity contents in reactor vessel steels. These findings compare well with prior determinations for USA-produced steels having similar low impurity levels. The performance of steels from the present study and the trend performance of improved steels and welds representing USA-production (13) are illustrated in Figure 25. The key factor for development of improved radiation resistance at 288°C appears to be the minimization of copper content to suitably low (i.e., $\leq$ 0.10% Cu maximum) levels. Current ASTM and AWS (American Welding Society) specifications for nuclear grade steels specify maximum allowable copper contents of 0.10% and 0.08%, respectively.

Two studies concerning synergisms between elements in radiation sensitivity development are significant to the overall application of the IWG-RRPC program results. One, a very recent study at Materials Engineering Associates, has revealed that the level of the phosphorus contribution to radiation sensitivity is dependent on the copper content (14). Where copper contents are high (> 0.25%), the phosphorus contribution is small. However, when the copper content is low ($\sim$ 0.01%), a large phosphorus contribution is apparent as illustrated in Figure 26. The results of this study help explain some of the inconsistencies found in computer treatments of data bases used to study the phosphorus contribution. The second study confirmed a suspected interaction of nickel content with copper content (15). Whereas nickel contents up to 1% nickel do not appear to influence radiation embrittlement sensitivity if the copper content is low, the results show that nickel in amounts greater than 0.4% can actually enhance the detrimental effect of a high copper content. The influence of nickel on the phosphorus contribution is currently under study.

Conclusions

The intent of the IWG-RRPC effort was to demonstrate that, through careful specification and within current technology, reactor steels and welds can be produced routinely with high preirradiation fracture resistance qualities and with high radiation embrittlement resistance at vessel service temperatures. The results of the present investigation provide a positive response to both questions. In addition, the observations show good agreement between non-USA and USA-improved production materials in radiation resistance capabilities. the A533-B reference plate (HSST 03) by virtue of its 0.12% Cu content is not considered fully representative of current (improved) production material, since it barely meets ASTM supplemental specifications on check analysis for reactor beltline material. Results of the full IWG-RRPC program (to be published) permit a conclusion that the low copper, low phosphorus content steels and welds produced overseas can be evaluated within the same framework as USA-produced steels in the NRC Regulatory Guide 1.99.

A correlation of the radiation-induced elevations of the C_V 41 Joules and K_J 100 MPa$\sqrt{m}$ transition temperatures was also shown.

Acknowledgment

This study was sponsored by the Reactor Safety Research Division, Metallurgy and Materials Branch, U.S. Nuclear Regulatory Commission.

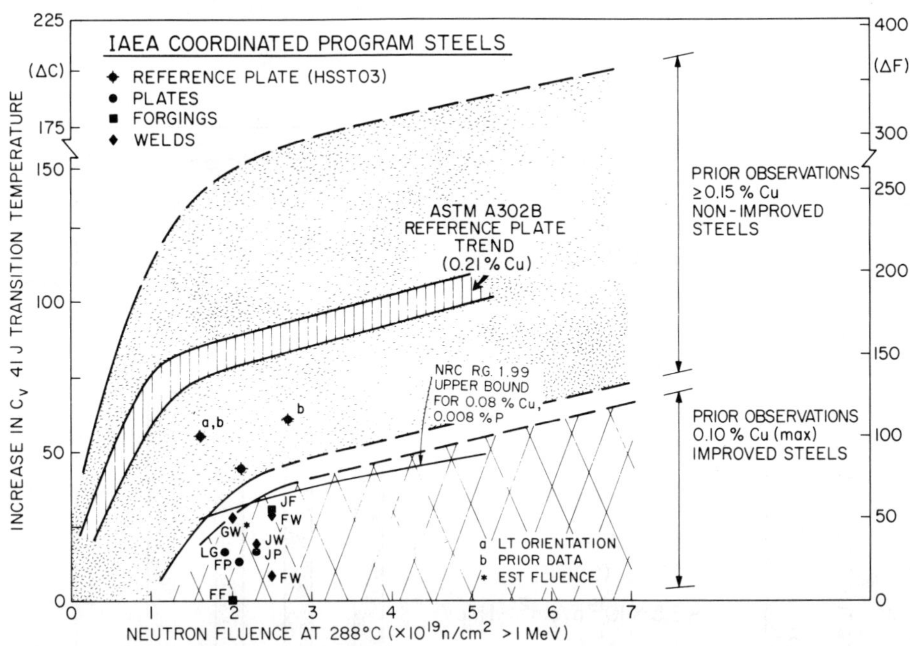

Figure 25 - Comparison of radiation resistance of steels and welds produced by the Federal Republic of Germany, France and Japan (0.01 to 0.07% Cu) with the trend behavior of improved steels produced in the USA. Good agreement is found. Data for the reference plate (0.12% Cu) falls in the lower region of the data trend band for nonimproved (high copper) steels and welds.

References

1. "Coordinated Research Programme on Analysis of the Behavior of Advanced Reactor Pressure Vessel Steels Under Neutron Irradiation," IWG RRPC-78/81, International Atomic Energy Agency, Vienna, Austria, 17-18 Oct. 1977.

2. "Effects of Residual Elements on Predicted Radiation Damage to Reactor Vessel Materials," Regulatory Guide 1.99, Rev. 1, U.S. Nuclear Regulatory Commission, Office of Standards Development, Washington, D.C., April 1977.

3. U. Potapovs and J. R. Hawthorne, "The Effect of Residual Elements on 550°F Irradiation Response of Selected Pressure Vessel Steels and Weldments," Nuclear Applications, 6 (1), Jan. 1969, pp. 27-46.

4. J. R. Hawthorne, "Demonstration of Improved Radiation Embrittlement Resistance of A533-B Steel Through Control of Selected Residual Elements," pp. 96-127 in Irradiation Effects on Structural Alloys for Nuclear Reactor Applications, ASTM STP 484, A. L. Bement, Ed.; American Society for Testing and Materials, Philadelphia, PA, 1971.

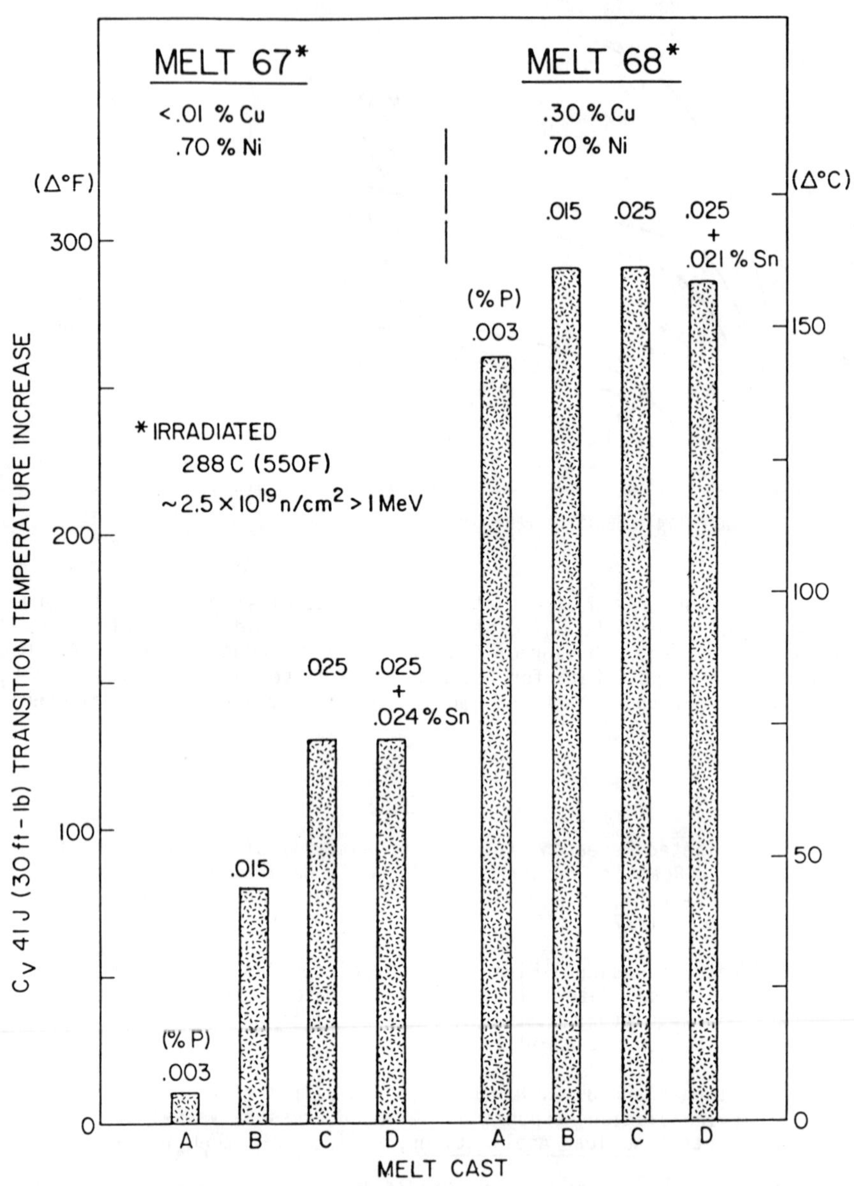

Figure 26. C_V 41 Joules transition temperature increases observed for plates from two 4-way split laboratory melts of A533-B steel. The specimens were irradiated simultaneously in one reactor assembly (14).

5. J. R. Hawthorne, "Radiation Resistant Weld Metal for Fabricating A533-B Nuclear Reactor Vessels," Welding Journal Research Supplement, 51 (7), July 1972, pp. 369s-375s.

6. J. R. Hawthorne, E. Fortner and S. P. Grant, "Radiation Resistant Experimental Weld Metals for Advanced Reactor Vessel Steels," Welding Journal Research Supplement, 49 (10), Oct. 1970, pp. 453s-460s.

7. J. R. Hawthorne, J. J. Koziol, and S. T. Byrne, "Evaluation of Commercial Production A533-B Steel Plates and Weld Deposits with Extra-Low Copper Content for Radiation Resistance," pp. 278-294 in Effects of Radiation on Structural Materials, ASTM STP 683, J. A. Sprague and David Kramer, Eds.; American Society for Testing and Materials, Philadelphia, PA, 1979.

8. "Standard Methods for Notched Bar Impact Testing of Metallic Materials," E23-72, Annual Book of ASTM Standards, Part 10, American Society for Testing and Materials, Philadelphia, PA, 1978, pp. 235-251.

9. D. R. Ireland, W. L. Server and R. A. Wullaert, "Procedures for Testing and Data Analysis, Task A - Topical Report," ETI Technical report 75-43, Effects Technology, Inc., Santa Barbara, CA, Oct. 1975, pp. 5-1 to 5-52.

10. E. P. Lippencott and P. A. Ombrellaro, "Buffalo Light Water Reactor Calculation," Letter Report to J. R. Hawthorne, Nov. 15, 1977.

11. "Fracture Toughness and Surveillance Program Requirements, Appendices G and H. Title 10," Code of Federal Regulations, Part 50 (10 CFR Part 50), U.S. Atomic Energy Commission, Federal Register, Vol. 45 (222), Nov. 14, 1980, pp. 75536-75539.

12. F. J. Loss, ed., "Structural Integrity of Water Reactor Pressure Boundary Components, Quarterly Progress Report, April-June 1980," USNRC Report NUREG/CR-1783, also NRL Memorandum Report 4400, Naval Research Laboratory, Washington, D.C., Feb. 20, 1981.

13. J. R. Hawthorne, ed., "The NRL-EPRI Research Program (RP886-2), Evaluation and Prediction of Neutron Embrittlement in Reactor Pressure Vessel Materials, Final Report," EPRI NP-2782, Electric Power Research Institute, Dec. 1982.

14. J. R. Hawthorne, "Evaluation of Reembrittlement Rate Following Annealing and Related Investigations on RPV Steels," MEA-2032, Materials Engineering Associates, Oct. 1983.

15. J. R. Hawthorne, "Significance of Nickel and Copper Content to Radiation Sensitivity and Postirradiation Heat Treatment Recovery of Reactor Vessel Steels," pp. 375-391 in Effects of Radiation on Materials: Eleventh Conference, ASTM STP 782, H. R. Brager and J. S. Perrin, eds.; American Society for Testing and Materials, 1982.

Ir/PuO_2 COMPATIBILITY: TRANSFER OF IMPURITIES FROM PLUTONIUM DIOXIDE TO IRIDIUM METAL DURING HIGH TEMPERATURE AGING

D.H. Taylor

E. I. du Pont de Nemours & Company
Savannah River Laboratory
Aiken, SC 29808

W. H. Christie

Analytical Chemistry Division
Oak Ridge National Laboratory
Oak Ridge, TN 37830

D. Pavone

Los Alamos National Laboratory
Los Alamos, NM 87545

Summary

Plutonium oxide fuel pellets for powering radioisotopic thermoelectric generators for NASA space vehicles are encapsulated in iridium which has been grain-boundary-stabilized with thorium and aluminum. After aging for six months at 1310°C under vacuum, enhanced grain growth is observed in the near-surface grains of the iridium next to the PuO_2. Examination of the grain boundaries by AES and SIMS shows a depletion of thorium and aluminum. Iron, chromium, and nickel from the fuel were found to diffuse into the iridium along the grain boundaries. Enhanced grain growth appears to result from thorium depletion in the grain boundaries. However, in one instance grain growth was slowed by the formation of thorium oxide resulting from oxygen diffusion along the grain boundaries.

Introduction

Pu-238 oxide is used as a general purpose heat source (GPHS) to power radioisotopic thermoelectric generators for NASA deep space probes, such as the upcoming Galileo mission to Jupiter (launch 1985). After being fabricated into ceramic pellets approximately 2.5 cm in diameter and 2.5 cm long, the PuO_2 is encapsulated in a 700 μm thick iridium shell which is vented on one end to permit escape of the helium decay product. The major purpose of the iridium shell, or clad vent set (CVS), is to provide long term containment and immobilization of the PuO_2 in the event of a mission failure involving a high velocity re-entry impact of GPHS capsules with the ground. An iridium alloy (DOP 26) consisting of 0.3% tungsten, and nominally 50 wt ppm aluminum and 60 wt ppm thorium was chosen as the cladding material because of its high-temperature strength and ductility as well as its chemical inertness, especially with respect to PuO_2. The alloying elements enhance the impact strength and ductility by increasing the coherent strength of the grain boundaries.(1) Perhaps even more importantly these elements, in particular thorium, help maintain ductility during high temperature aging of the iridium by preventing grain growth. (1,2) Thorium is effective as a grain growth inhibitor in part because it exists as discrete $ThIr_5$ particles on the grain boundaries. (1,2)

Although the microstructure and grain boundary chemistry of high-temperature-aged iridium have been well characterized (1,2) the effects of high-temperature aging in the presence of PuO_2 resulted in greater grain growth in the iridium than expected based on aging tests of the iridium alone. In particular, extensive grain growth was observed throughout one-third to one-half of the thickness of the iridium on the side next to the PuO_2. This paper describes analytical work performed to understand and control the cause of this extensive grain growth.

Experimental Details

A photograph of a welded capsule and a diagram of the experimental arrangement of the four capsules (PEF 44, 45, 48 and 49) in a test graphite module are shown in Figure 1. The graphite module and the aging conditions [6 months at 1440°C fuel (∿1310°C iridium) at 1.3×10^{-4} Pa (1.3×10^{-9} atm)] simulate actual service conditions. Samples of the flashing cut from each of the drawn cups used to encapsule the PuO_2 pellets were used as controls to test the effects of aging in the absence of PuO_2 and to obtain a baseline microstructure before aging. After aging, sections were cut from CVS's. Vapor deposited PuO_2 adhering to the inside of the iridium was removed by mechanical abrading and chemical etching in HF/HNO_3. Transverse cross sections of the decontaminated specimens were then mounted in epoxy and metallurgically polished until a smooth mirror-like finish was obtained. The mounted specimens, which showed no smear-removable alpha activity, were then overcoated with 20-30 nm of spectrographic grade carbon, prior to secondary ion mass spectrometry (SIMS), so as the render the entire surface conducting. Fracture surfaces of other samples from each capsule were examined using Auger electron spectrometry (AES).

Key impurities distill out of the plutonium oxide during high-temperature aging. To determine the effects of this distillation, the pellets were doped with different levels of impurities and subjected to one of two heat treatments (Table I).

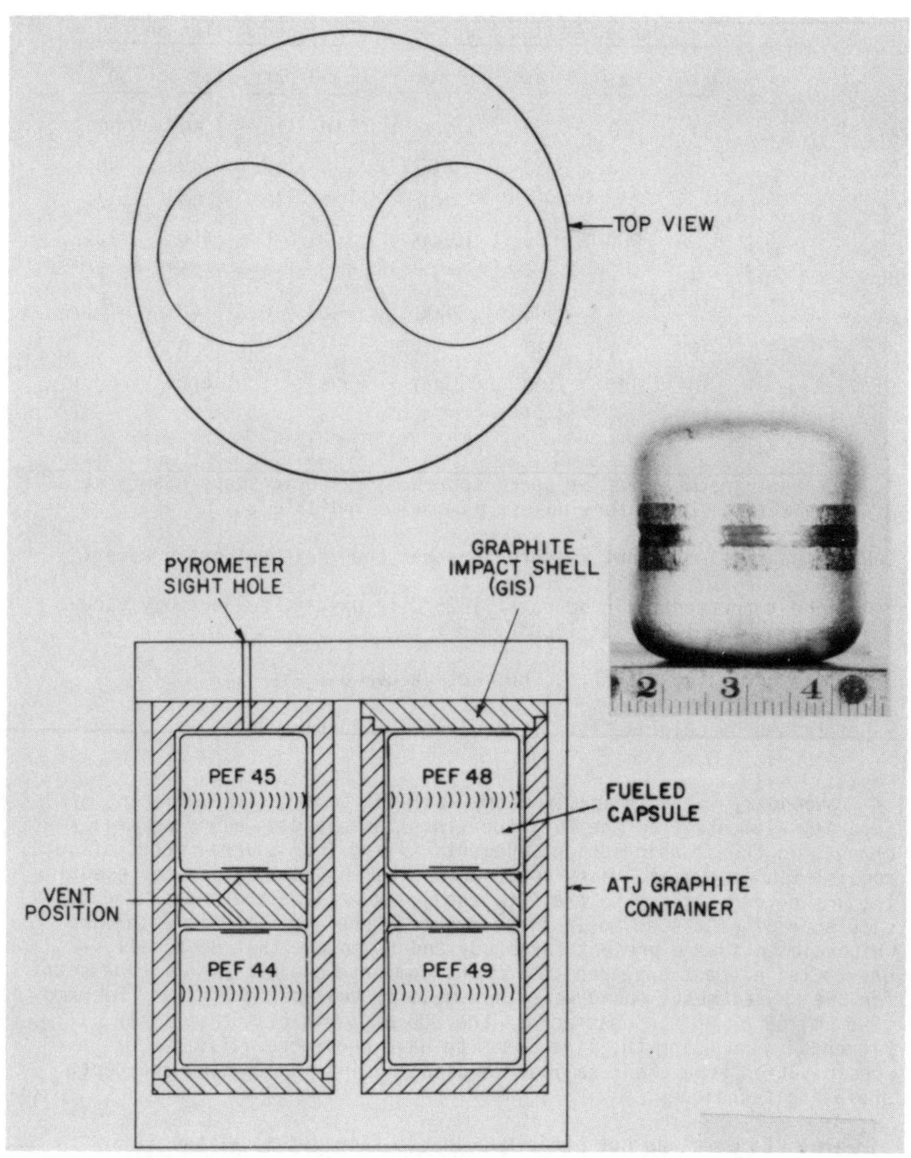

Figure 1. Orientation of compatibility test pellets in the aging furnace. The insert shows a typical fueled capsule. The center band is a girth weld joining two cups.

Table I. Concentrations of Key Impurities in PuO_2 Before and After Aging

Element	Chemical Content (wt. ppm*)							
	Unaged Pellet No.				Aged Pellet No.			
	44[a]	45[b]	48[a]	49[a]	44	45	48	49
Al	30	60	60	900	10	10	60	1000
Ca	2	3	50	600	10	7	40	300
Cr	-[+]	-	1000	2000	20	10	100	7
Fe	900	2000	5000	10000	100	70	400	10
Mg	-	-	2	4	1	5	7	2
Ni	-	-	900	2000	10	0.2	40	0.7
P	4	4	30	40	5[c]	5[c]	20[c]	35[c]
Si	40	100	100	1000	20	7	400	2000
Ti	-	-	<3	5	3	3	20	2

* Concentrations measured by spark source mass spectroscopy (SSMS) at Oak Ridge National Laboratory unless otherwise indicated.

[+] Dash means "element not measured" rather than "element not present."

[a] PuO_2 heat treated for 6 hours at 1525°C in oxygen, followed by vacuum outgassing for 1 hour at 1500°C.

[b] PuO_2 heat treated as in (a), but not vacuum outgassed.

[c] Determined by colorimetry at Los Alamos National Laboratory.

Secondary ion mass spectrometry was used to study the movement of impurity elements from the PuO_2 fuel into the cladding alloy as well as changes in the cladding dopant elements Al and Th. Several modes of data acquisition were used. Mass spectra were taken from regions representing the inside edge (close to the PuO_2 fuel), as well as the center and outside edge of each specimen. These spectra were evaluated to determine which elements were present for study and to ensure that no totally unexpected element appeared that might dominate grain growth. Line scans for the key elements found were then made by traversing a finely focused (2-5 μm) O_2^+ or N_2^+ ion beam across the 700 μm width of the specimen. Elements found using the line scans to have undergone diffusion or precipitation were then examined in detail using ion imaging and depth profiling techniques.

The AES work was not performed in the same detail as the SIMS analyses. Three readings were taken across the cross section of each iridium sample; one at about 100 μm from each surface and one at the center. A 15 μm beam spot size was used.

Results

Iridium aged next to the PuO_2 experienced extensive grain growth in the near-surface regions next to the oxide (Figure 2). In contrast, iridium aged in the absence of PuO_2 showed uniform grain growth with no regions of large grains (Figure 2). A plot of grain size versus thorium content of all samples tested (Figure 3) confirms, at least within the range of thorium concentrations and grain sizes observed in this study, the observations of White and Liu (2) that grain size is controlled by the thorium content on the grain boundaries. This is illustrated graphically by ion micrographs of thorium made using SIMS (Figure 4). Figure 4 shows an ion micrograph of sample 45 with thorium decorating the grain boundaries. A diagram of the grain boundaries visible through the microscope of the SIMS instrument is also shown. Comparing the ion micrograph with the grain boundary diagram reveals that no thorium is visible on the grain boundaries of large grains near the plutonium oxide surface.

Although an apparent correlation existed between grain size and thorium content on grain boundaries, further work was performed to determine what other possible elements might contribute directly to the grain growth or indirectly by leading to removal of thorium.

Broad spectrum scans were performed across the iridium for both light and heavy electropositive elements. Emphasis was placed on those elements known from previous experience to distill out of the fuel and condense in the vent holes of the iridium capsule.

Elements observed in these scans were aluminum, chromium, iron, nickel, thorium and tungsten. Not observed were calcium, magnesium, phosphorus, plutonium, silicon and titanium. All of these elements, both observed and not observed, distill out of the fuel. SIMS has great sensitivity for calcium, plutonium, silicon and titanium. The lack of detection of these elements, even at grain boundaries, suggests they were present at very low levels if at all. Phosphorus was detected on Samples 48 and 49 by AES. However, since the sensitivity of SIMS to electronegative elements using an oxygen ion beam as was done in this study is low, it is not surprising that phosphorus was not detected. A cesium ion beam for analyzing electronegative elements was not available during this study. Ion images were attempted for each of the observed elements. Tungsten forms a solid solution with iridium and is normally homogeneously distributed throughout the iridium. No change in the distribution was observed as a result of aging. Although some tungsten was observed in deposits condensed at the vent hole in the capsule, no loss in tungsten concentration could be measured by SSMS.

Ion micrographs show that thorium is segregated to grain boundaries and also exists as discrete precipitates in the interior of grains and on grain boundaries. The ion micrographs also show that the concentration on the boundaries diminishes in the direction toward the iridium surface that was in contact with the PuO_2 (Figures 4-7). As pointed out earlier, this trend in thorium concentration at the grain boundaries coincides with an increase in grain size. It is evident that the grains tend to grow transverse to the surface, rather than parallel. The data also show that the number of discrete thorium particles within the large near-surface grains is much fewer than in nearly undisturbed grains near the middle of the iridium. It is expected that the thorium particles originally in the near-surface grains have been swept away by grain boundaries in the typical fashion: pinning the grain boundary until the particle "dissolved" after which the boundary continued to move. An exception to this mechanism was found in Pellet 49.

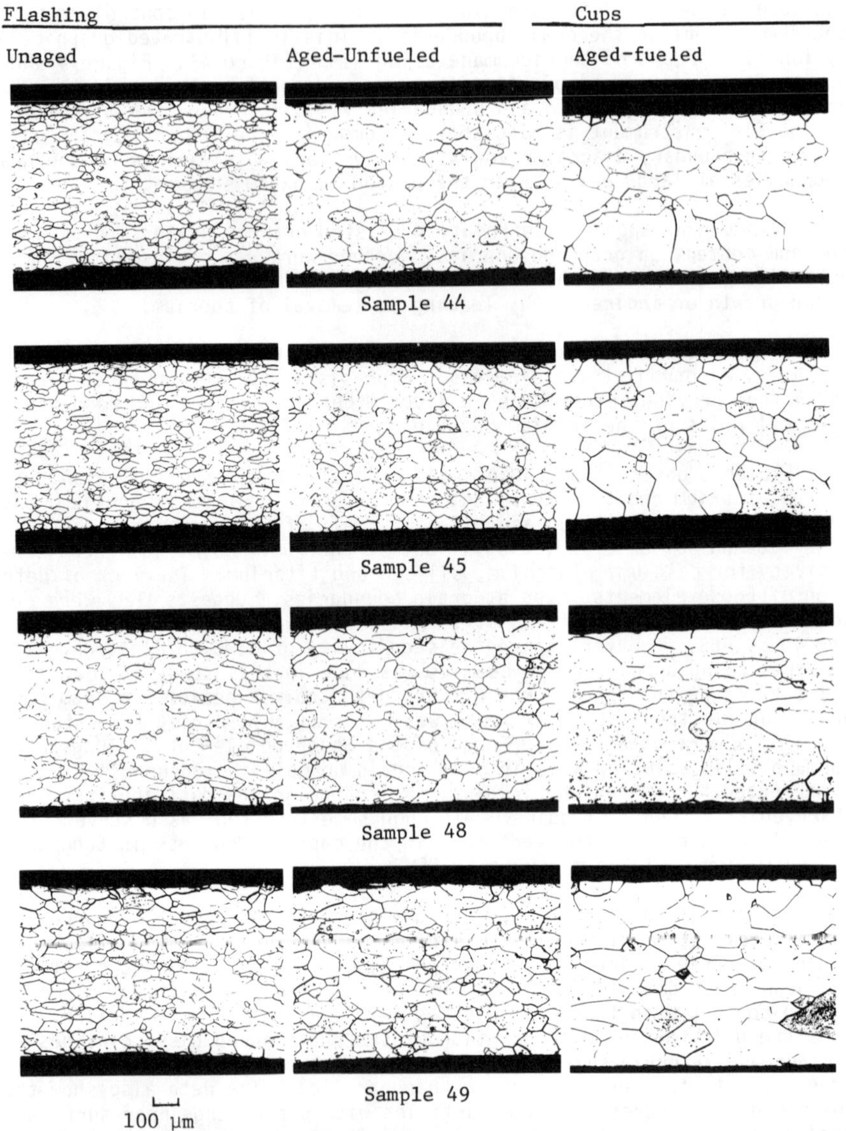

Figure 2 - Comparison of unaged, aged-unfueled and aged-fueled iridium microstructure.

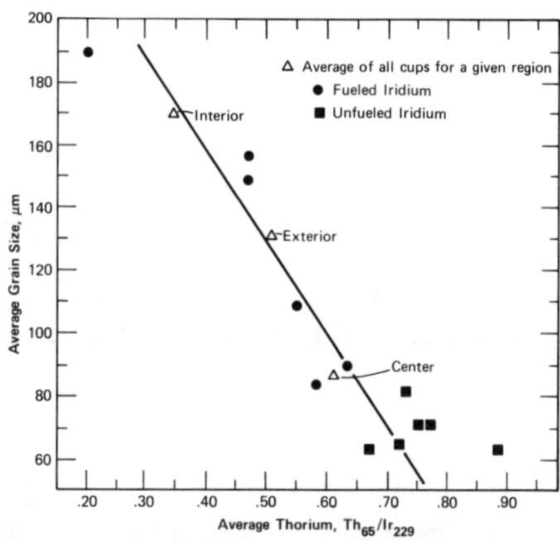

Figure 3 - Average thorium content presented as a ratio of the thorium to iridium signal by Auger Electron Spectroscopy (AES) vs. average grain size.

An ion micrograph of Capsule 49 (Figure 8) shows thorium particles decorating the grain boundaries and two particles located on the inside of the bottom grain in the figure. A corresponding ion micrograph of oxygen reveals that oxygen is associated with thorium on the grain boundaries. Evidently, however, oxygen is not associated with the thorium in the interior of the grains away from the grain boundaries since no oxygen images appear that correspond to the two thorium images located in the interior of the lower grain. Two separate regions on the PuO_2 affected edge of Capsule 49 were studied. Both regions showed ThO_2 precipitates to extend along grain boundaries over 100 μm in depth from the surface exposed to the fuel. These results suggest that if the oxygen partial pressure is high enough (as it must have been in Capsule 49), oxygen can diffuse inward along grain boundaries and oxidize thorium internally.

Evidently the effect of the in-situ oxidation is to stabilize the grain boundaries against further grain growth. Careful examination of Capsules 44, 45, and 48 showed that virtually no ThO_2 precipitates occur on grain boundaries in the PuO_2 region below 10-15 μm in depth, whereas for Capsule 49, precipitates occur along grain boundaries in the same region (cf. Figures 4-7). In the same areas examined, the grain size for Capsule 49 was smaller where boundaries had been pinned by ThO_2 precipitates than in the other capsules where ThO_2 precipitates did not occur.

Not all of the oxygen images observed on the ion micrographs in Figures 4-7 represent oxidized thorium. In most cases the images arise from chemisorbed oxygen on the sample surface. In the case of chemisorbed oxygen, the oxygen signal fades as the N_2^+ beam current is increased and the oxygen layer is sputtered away. In contrast, the oxygen signal from the oxidized thorium precipitates increases as the beam current increases, i.e., as the sputtering rate of the oxide increases (Figure 9). In Figures 4-8, oxidized thorium only appears at grain boundaries in the PuO_2 affected region. These locations have been labeled in Figures 7 and 8. All other oxygen images result from chemisorbed oxygen.

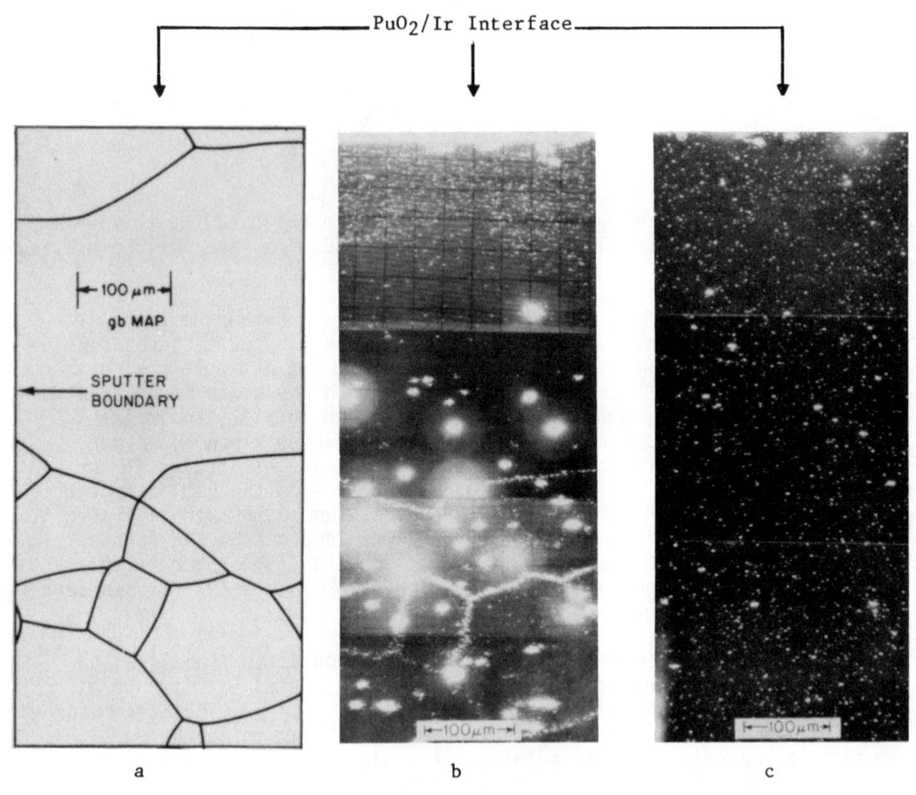

Sample 45

Primary Beam 10 nA N_2^+

Figure 4 - SIMS ion micrographs. (a) Schematic diagram of grain boundaries observed optically after sputtering; (b) Thorium ion micrograph. Thorium can be seen outlining grain boundaries and as particles both on and off grain boundaries. Note the lack of thorium on grain boundaries near the PuO_2/Ir interface; (c) Oxygen ion micrograph. No oxygen is observed on grain boundaries.

PuO_2/Ir Interface

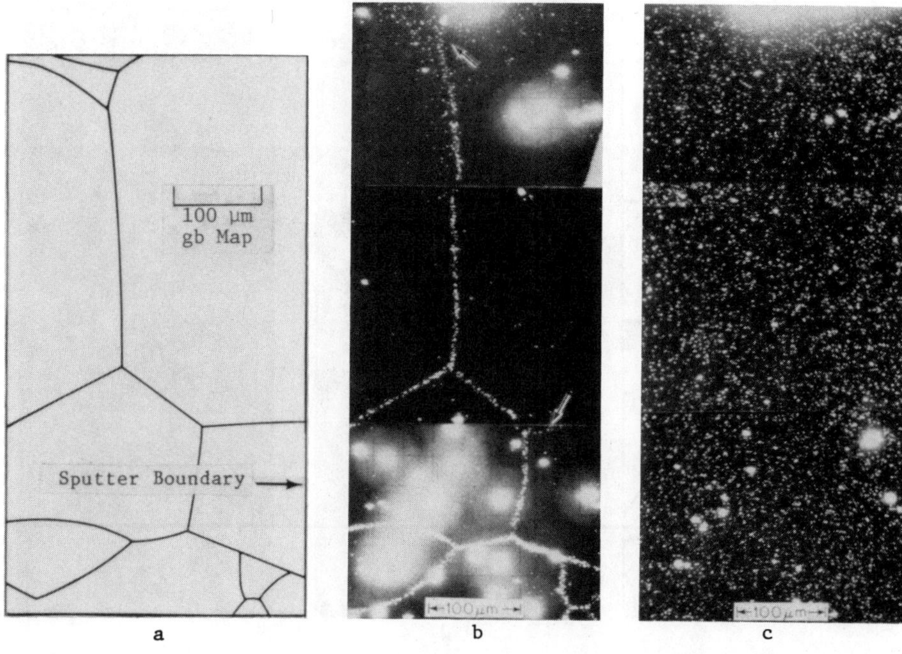

Sample 45
Primary Beam 3 nA N_2^+

Figure 5 - SIMS ion micrographs. (a) Schematic diagram of grain boundaries observed optically after sputtering; (b) Thorium ion micrograph. Thorium depletion is visible on the grain boundary near the PuO_2 interface and on the innermost boundary of the large grain on the right hand side (arrows). (c) Oxygen ion micrograph. The oxygen signal declined with increasing ion beam current, suggesting the presence of only chemisorbed oxygen.

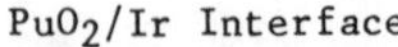

PuO_2/Ir Interface

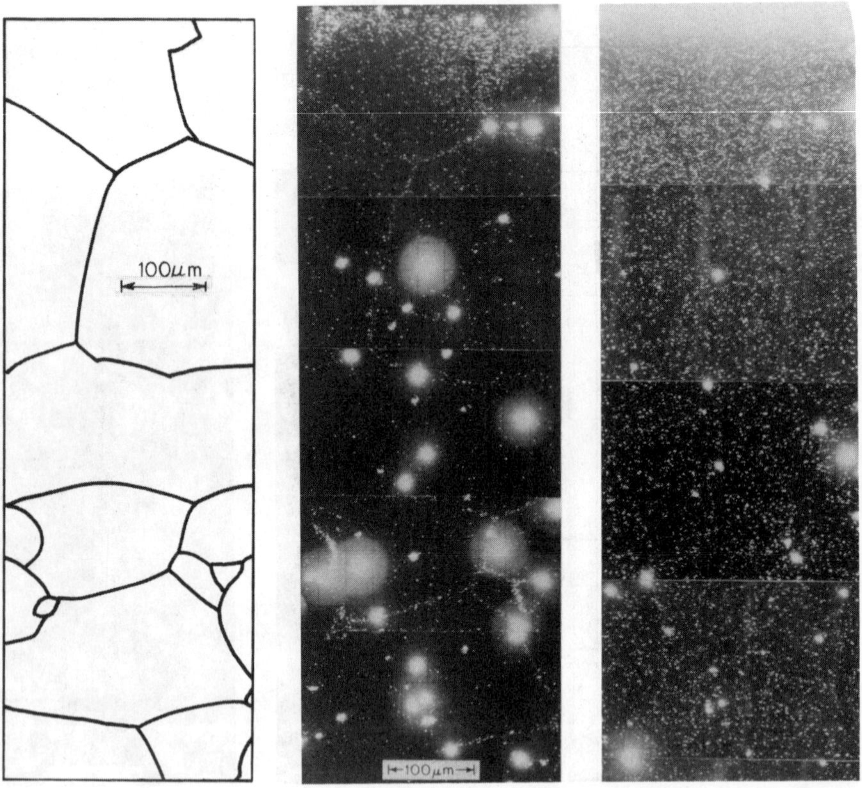

100μm

100μm

a b c

Sample 48

Figure 6 - SIMS ion micrographs. (a) Schematic diagram of grain boundaries observed optically; (b) Thorium; (c) Oxygen images are from chemisorbed oxygen.

PO_2/Ir Interface

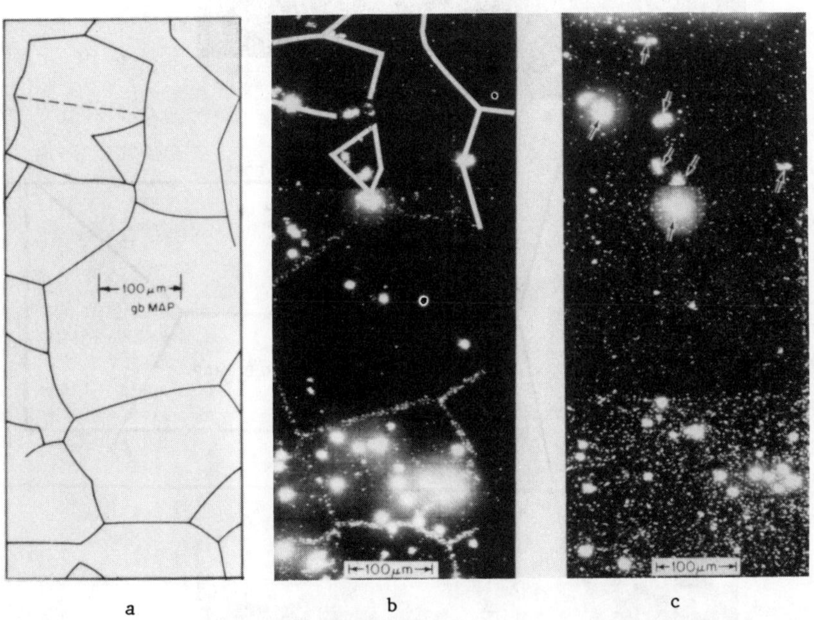

Sample 49
Primary Beam 3 nA N_2^+

Figure 7 - SIMS ion micrographs. (a) Schematic diagram of grain boundaries observed optically; (b) Thorium images lie on grain boundaries (white lines); (c) Corresponding oxygen images (arrows) increased in intensity with increasing beam current suggesting formation of ThO_2. Other images are from chemisorbed oxygen and disappeared with increasing beam current.

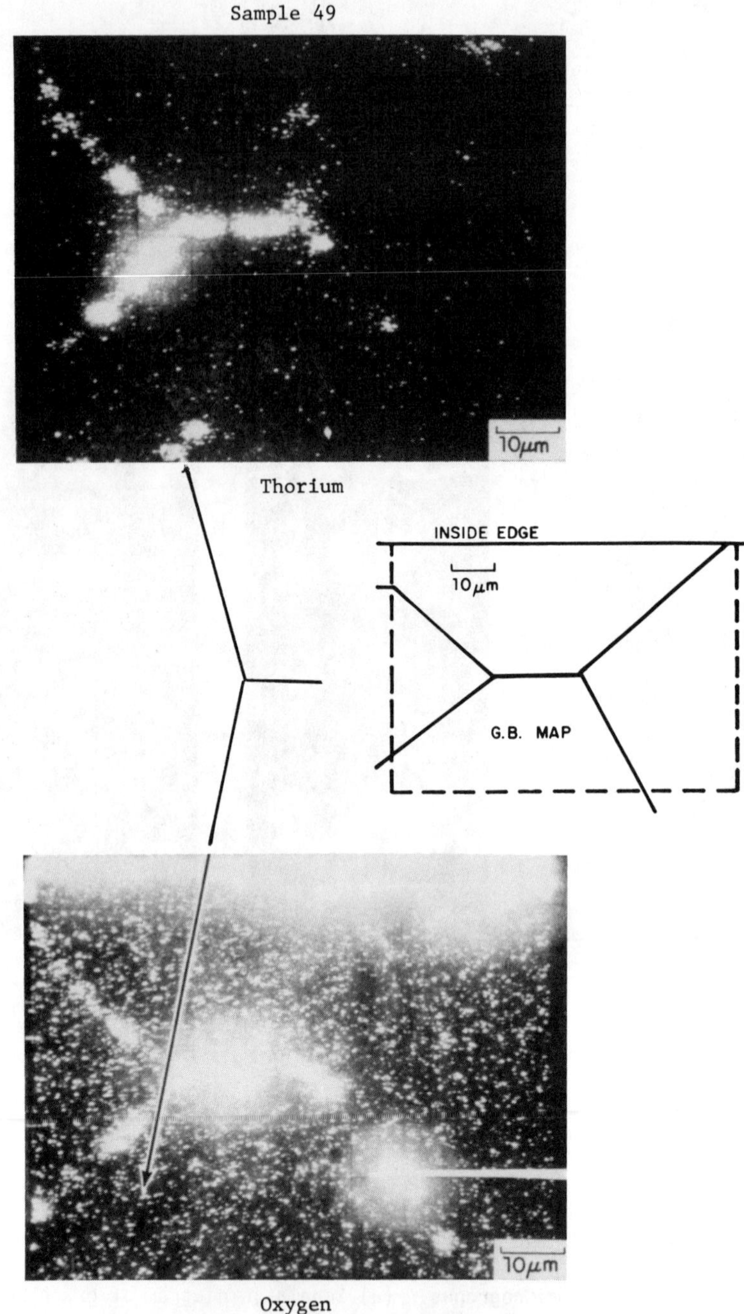

Figure 8 - Ion micrographs showing oxygen combined with thorium on grain boundaries but not with thorium present within grains (arrows).

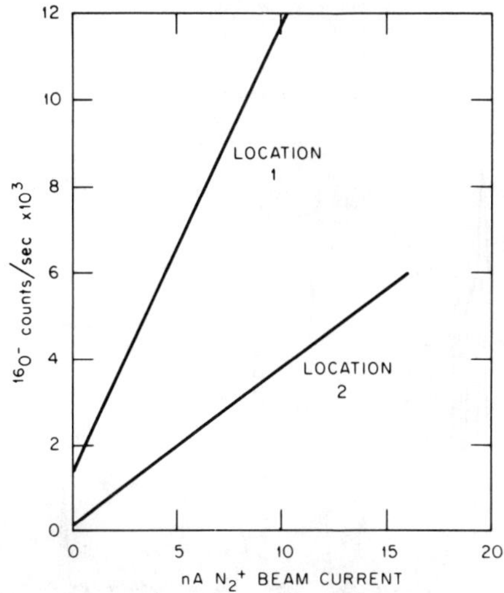

Figure 9 - Oxygen signal vs. ion beam current for oxygen associated with thorium in PuO_2-affected grains of Sample 49.

To confirm that oxygen had not diffused throughout the iridium in detectable quantities, line scans were performed across the thickness of the samples using higher beam currents. The results (Figure 10) confirm that the oxygen is associated with the thorium only in the PuO_2-affected region of enhanced grain size.

Mass spectra measurements showed all four samples of the aged iridium cladding alloy to have significant iron, chromium, and to a lesser extent nickel contamination in the fuel affected region. The deepest penetration (about 220 μm) was observed in Sample 49. Line scans across the specimens demonstrated some variability in the amount and depth of penetration of iron and chromium into the alloy. A typical scan is shown in Figure 11. The amount of iron and chromium observed in any particular line scan and the depth of penetration depended to some extent on how that line scan intersected grain boundaries, but in general it was shown that for Sample 49, the concentration maxima at the inside surface of the iridium was approximately 0.5 to 1.0 atom % and decreased to background (5 to 10 ppm atomic) at a depth of about 220 μm. For Sample 44, the maximum concentration at the surface of iron and chromium was about the same as Sample 49 at the surface, but the depth of penetration was significantly less, being approximately 130 μm. The concentration of iron and chromium at the surface of Samples 45 and 48 was also approximately the same as for Sample 44, and the depth of penetration was midway between that of Samples 44 and 49.

Figure 12 presents data obtained by scanning a 1 nA O_2^+ primary beam across Sample 49 in a direction perpendicular to the inside surface. The scan region chosen was such that no grain boundaries were within 50 μm of the point where the scan started on the inside edge of the sample (see grain boundary map in Figure 12). Presumably, the iron, chromium and nickel that penetrated to a depth of 20 to 30 μm arrived via bulk diffusion. The peaks seen at 70 to 80 μm depth clearly represent diffusion along grain boundaries. SIMS showed that the concentration of iron and chromium was maximum at the iridium grain boundaries and fell off with

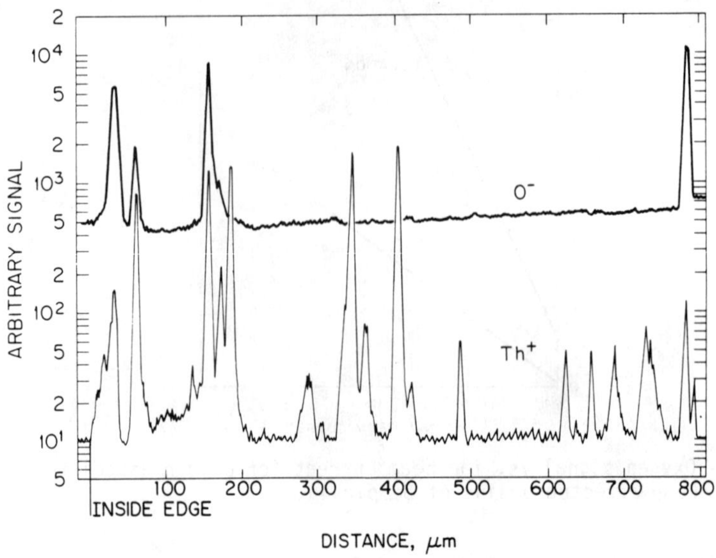

Figure 10 - SIMS line scan for oxygen and thorium across the iridium thickness. Peaks normally occur at grain boundaries or where precipitates are encountered. Maximum oxygen penetration is between 200 and 350 μm.

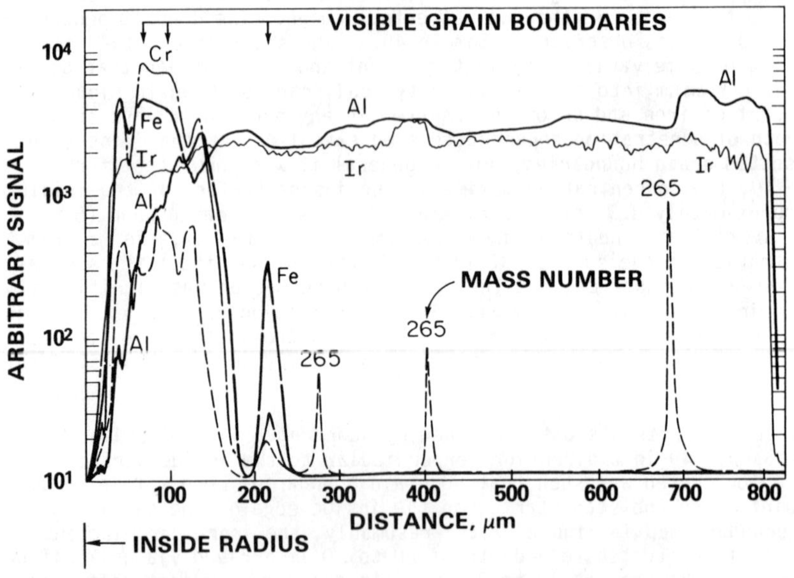

Figure 11 - SIMS line scan showing penetration of Fe and Cr into iridium sample and corresponding depletion of Al (Sample 49).

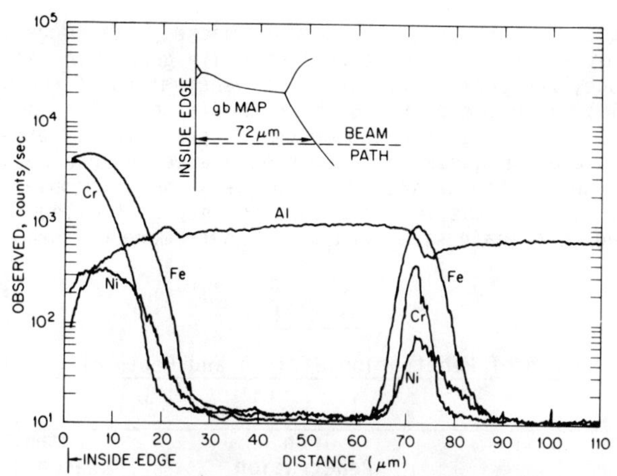

Figure 12 - SIMS line scan showing Fe, Cr and Ni penetration from the fuel into the iridium. Note the depletion of Al at grain boundary at about 75 μm and in bulk iridium near the inside edge.

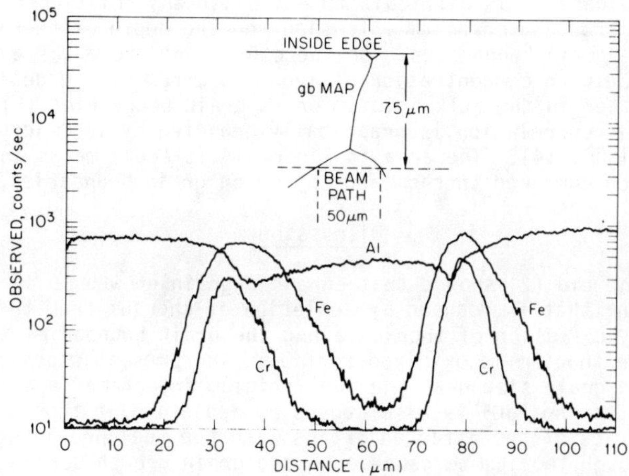

Figure 13 - SIMS line scan showing distribution of Fe and Cr across grain boundary. The greater width of the Fe peaks suggests that Fe diffuses into Ir more rapidly than Cr. Nearer the surface the distribution across grain boundaries was broader. Further away the distribution was narrower.

depth perpendicular to the grain boundaries. For grains near the surface of the iridium sample the concentration fell off very little from the maximum at the grain boundaries. However, for grain boundaries away from the iridium surface, the fall-off became sharper (Figure 13) with depth into the iridium until at the maximum depth of penetration, the iron and chromium were located only at the grain boundaries. Iron appears to diffuse more rapidly along the grain boundaries and into the bulk iridium (e.g. compare the width of the iron and chromium peaks in Figure 13).

Examination of the iron, chromium and nickel data suggests that these elements do not exert a primary effect on grain growth in the iridium. In the first place, the approximate depth of penetration of the iron and chromium do not match the depth to which the large grains have grown (Table II). In addition, iron and chromium appear to be late-comers, diffusing into the boundaries after the boundaries have already been established, whether the grains are relatively small (Sample 49) or very large (Sample 44). If the iron and chromium had controlled grain growth, it is expected that grain sizes would have been more uniform from sample to sample.

TABLE II

Comparison of Depth of Penetration of Iron and Depth of Large Grain Growth

Capsule	Iron Penetration (μm)	Extent of Large Grains (μm)
44	130	210-240
45	160	180-240
49	220	160-300

From the data it is difficult to establish any relationship between the depth of iron and chromium intrusion and the depth of thorium depletion from the grain boundaries. On the other hand there was a strong correlation between concentration of iron and chromium and depletion of aluminum whether in the bulk iridium or at grain boundaries (Figures 12 and 13). This correlation is graphically depicted by ion micrographs of Sample 49 (Figure 14). The area in Figure 14 is the same as that in Figure 8 which compared thorium and oxygen on grain boundaries.

Discussion

White and Liu (2) showed that enhanced grain growth in iridium can be oxygen-driven; that is, caused by depletion of thorium from the grain boundaries by diffusion of thorium along the grain boundaries to the surface where thorium is oxidized to ThO_2. The present study suggests that enhanced grain size near the fuel/iridium interface in aged iridium capsules containing PuO_2 is also caused by depletion of thorium from the grain boundaries of the affected grains with the fuel providing the oxygen. To remove enough thorium to cause enhanced grain growth during six months aging at ∿1330°C, which is similar to this study, White and Liu found an oxygen partial pressure (P_{O_2}) greater than about 10^{-3} Pa (10^{-8} atm) to be necessary. It is evident from the control samples of this study that the oxygen pressure was less than 10^{-3} Pa in the absence of PuO_2 since very little growth of near-surface grains was observed. In fact at 1300°C (estimated graphite temperature) the graphite is expected to maintain an oxygen partial pressure of 10^{-5} Pa (10^{-10} atm). The inside of the capsule is open to the furnance vacuum through a 0.5 mm diameter hole in the iridium and thence through the threads of the graphite module. This gas diffusion path is difficult. Hence, the oxygen partial pressure in the capsule is likely to be greater than in the furnance or module given that the fuel acts as an oxygen source. Work by R. R. Jaeger (Private communication, Mound Laboratory, 1984) has shown that vacuum out-gassing of PuO_2 for 1 hour at 1550°C will reduce the PuO_2 to an O/Pu ratio of only about 1.996. This is likely to be the fuel stoichiometry immediately after fabrication. For this O/Pu ratio, the oxygen partial pressure is calculated to be approximately 10^{-3} Pa (see Appendix). On standing, the decay heat of the oxide is such that it will gradually reoxidize. At an O/Pu ratio of 1.998, the oxygen partial pressure is calculated at 10^{-2} Pa. This is likely to be the fuel stoichiometry when encapsulated. Hence the PuO_2 itself can in principle provide sufficient oxygen to deplete the thorium from the iridium. Moreover, the extent of thorium depletion and therefore enhanced grain growth in the iridium could vary depending on the state of substoichiometry in the fuel. As the capsule ages, oxygen will continue to be pumped out so that the fuel will continue to become more substoichiometric with time. Eventually it will not be able to supply the oxygen to continue thorium depletion.

A potentially stronger oxygen source than the fuel itself exists in the iron impurity found in the fuel (Table I). Electron microprobe analysis of the fuel showed the iron was uniformly distributed throughout the fuel, probably in solid solution. Iron probably exists as Fe_2O_3 because, prior to vacuum outgassing, the fuel is heated for 6 hours in argon/5% oxygen at 1530°C. During aging, oxygen associated with the iron oxide will be generated from the reactions:

$$6Fe_2O_3 \longrightarrow 4Fe_3O_4 + O_2 \qquad (1)$$

and

$$2Fe_3O_4 \longrightarrow 6FeO + O_2 \qquad (2)$$

The equilibrium oxygen partial pressure at 1440°C in reaction (1) is on the order of 10^4 Pa (10^{-1} atm) for pure Fe_3O_4 and, based on ideal solution considerations, 10^2 Pa (10^{-3} atm) when diluted to 10,000 ppm Fe (Sample 49). For reaction (2), the partial pressure of oxygen in equilibrium with the diluted iron oxide in the fuel is of the order of 10^{-3} Pa (10^{-8} atm.). Because this pressure is comparable to that from the PuO_2 initially, it is not clear whether all of the Fe^{+3} is reduced to Fe^{+2}, but certainly oxygen is available for selective removal from the iron oxide impurities and can provide oxygen for thorium depletion. A mass balance was performed to determine available thorium versus available oxygen. Approximately 1.3×10^{-5} moles of thorium are present in the iridium. For 10,000 ppm iron in the fuel, reaction (1) could release 3.7×10^{-5} moles of O_2, which is ample to deplete the thorium from the iridium.

Although the results of this study support the mechanism of thorium diffusing to the iridium surface and subsequent oxidation, the data also suggest that under certain circumstances, such as high oxygen pressure, oxygen may also diffuse into the iridium along grain boundaries to oxidize the thorium in situ. According to this hypothesis, the iron oxide content of Sample 49 is sufficient to supply the required oxygen, whereas the iron oxide content in the other samples is not. Measurements by SIMS on iridium aged in higher oxygen pressures were not available as a means of confirming the consistency of this hypothesis.

The results also suggest that depletion of the thorium and establishment of the large grains occurs relatively early in the aging, with iron, chromium and nickel diffusing into the iridium some time after the grain boundaries have been established. It is expected that thorium depletion and grain growth would occur early in the aging when the oxygen pressure is maximum and that as the oxygen pressure falls below 10^{-3} Pa, further depletion and grain growth effectively stop. It is expected that iron must be reduced to metallic iron before it will diffuse into the iridium. At the early stage of high oxygen partial pressure, reduction of iron oxide to metallic iron would not be thermodynamically favored. The partial pressure of oxygen in equilibrium with Fe and pure FeO at 1440°C is 10^{-4} Pa (10^{-6} Pa in dilute solution in the fuel). This pressure is low enough that the partial pressure of oxygen over the fuel itself (10^{-3} Pa or greater) would prevent formation of metallic iron. Hence diffusion of iron into the iridium would not occur. As aging progresses, iron oxide is vaporized out of the fuel (Table I), so that the partial pressure of oxygen in equilibrium with iron oxide in the fuel would decrease by one to two orders of magnitude from dilution. Also the fuel continues to be reduced. However, it will not be reduced below 10^{-5} Pa (10^{-10} atm) ($PuO_{1.987}$) because it is in equilibrium with the graphite through the vent hole via a CO/CO_2 cyclical transport mechanism.(3) Reduction of the iron oxide in the fuel will still not occur. Hence, metallic iron that diffuses into the iridium is not vaporized directly from the fuel. A more likely source of metallic iron is iron oxide which vaporizes from the fuel and condenses on the cooler iridium cladding. Now an equilibrium can be established with the pure phase. At the temperature of the inside surface of the iridium (estimated to be about 1330 to 1350°C) the equilibrium oxygen partial pressure between Fe and FeO is 10^{-5} Pa. This pressure is about the same as that above the graphite surrounding the capsule, hence, the partial pressure of oxygen above the fuel will also eventually reach that level and permit reduction of the FeO. At this late stage in the aging, metallic iron is formed and diffusion into the iridium occurs. This mechanism would explain why iron appears on grain boundaries after most grain growth has taken place. The progression in depth of penetration among the four samples is consistent with the amount of iron initially available in the fuel (Table I).

Although the above hypothesis offers some insights into the mechanisms at work during aging, it has shortfalls. For example, whereas the oxygen partial pressures are consistent with iron reduction, they are too high for reduction of Cr_2O_3 to Cr metal, yet Cr is also present in the iridium. More work on the vapor species present during aging is needed.

Iron and chromium appear to have no effect on grain growth. This result may be in keeping with previous work by Liu and Inouye (1) who detected no effect of iron on grain size for concentrations of iron in iridium up to 300 ppm. Higher concentrations were not evaluated. It is difficult to compare the concentration of iron on the grain boundaries in the present work with the concentration in the work done by the other workers because the bulk analyses performed in both cases were not sensitive enough to measure concentration more accurately than a factor of two and the distribution of iron in both cases was not determined (as by SIMS).

Although no effect on grain growth has been clearly observed, the presence of iron and chromium on the grain boundaries will necessarily affect boundary mobility because of the change in grain boundary free energy they cause. In the present case, however, this effect will be small because presumably the grain boundaries had already developed prior to the arrival of the iron and chromium until little driving force for growth remained.

An indirect effect of iron and chromium on grain boundaries may arise from their impact on the presence of aluminum. In this study, aluminum was found by SIMS to decrease as the concentration of iron and chromium increased in both the bulk (Figure 12) and at grain boundaries (Figures 12-14). At concentrations up to about 100 ppm, Liu and Inouye found that aluminum appeared to improve impact ductility (no strong influence on grain growth was detected). The present study suggests that the impact ductility of the iridium may be reduced by the action of the iron and chromium on the aluminum. However, loss of aluminum does not appear to contribute to the enhanced grain growth observed since it does not extend into the iridium as deeply as the large grains have grown.

Calcium, aluminum and silicon appeared in Sample 49 as a glass-like second phase. Although no second phase was seen in the other samples, these elements probably exist combined in them as well. These elements were observed to vaporize out of the fuel and condense in the iridium vent hole and at other locations around the inside of the clad vent sets. However, neither calcium or silicon was found by SIMS to diffuse into the iridium.

Considerable amounts of plutonium oxide vapor condensed on the inside of the iridium. After mechanically removing all visible PuO_2 and acid pickling to dissolve any residual oxide, the iridium sample still radiated approximately 10^4 d/m (disintegrations/minute) alpha from incorporated plutonium even though no transferrable contamination from the surface could be measured. However, no plutonium was found in the iridium by SIMS even though SIMS has a great sensitivity for plutonium.

Conclusions

Aging iridium next to PuO_2 leads to enhanced grain growth in the near-surface regions next to the PuO_2. The cause of the enhanced grain growth appears to be depletion of thorium from the grain boundaries by a mechanism of diffusion down the grain boundaries to the iridium/PuO_2 interface where thorium is oxidized. The oxygen partial pressure required for this may come from the fuel but more likely arises from iron impurities in the fuel. If the oxygen partial pressure is high enough oxygen can diffuse along the iridium grain boundaries to oxidize the thorium in situ. ThO_2 on the grain boundaries appears to be stable and to inhibit further grain growth. Iron and chromium in the fuel are depleted from the fuel during aging, with significant amounts of both being transferred to the iridium. Although both elements diffuse readily along grain boundaries as well as into the bulk, their presence does not appear to contribute directly to the observed enhanced grain growth. Other impurities in the fuel, especially calcium and silicon, are evaporated out of the fuel but do not diffuse into the iridium. Although substanital amounts of plutonium vapor are formed, plutonium does not diffuse into the iridium grain boundaries in quantities detectable by SIMS.

Acknowledgments

The authors appreciate helpful discussions with D. F. Bickford and P. K. Smith of Savannah River Laboratory; D. Petersen and R. N. R. Mulford of Los Alamos National Laboratory; and T. M. Besmann of Oak Ridge National Laboratory. The SIMS work of R. E. Eby, Oak Ridge National Laboratory, is gratefully acknowledged.

The information contained in this article was developed during the course of work under Contract no. DE-AC09-76SR00001 with the U.S. Department of Energy.

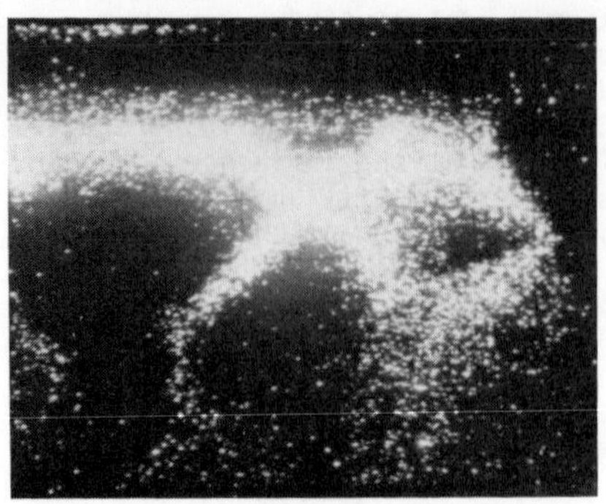

Iron

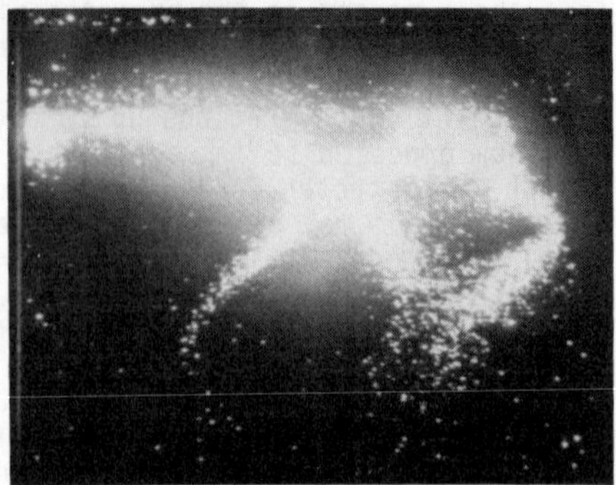

Chromium

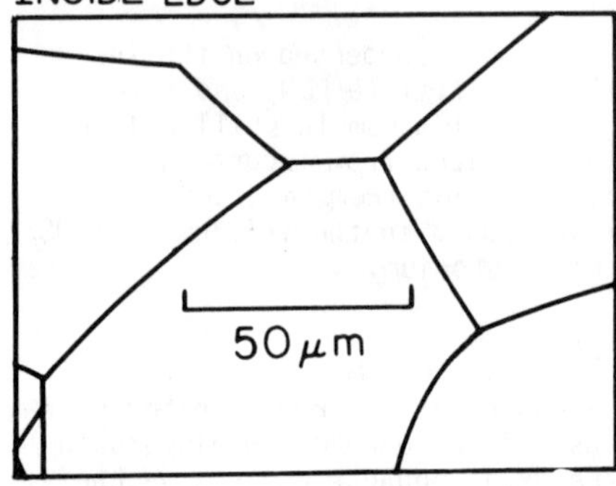

Aluminum

Figure 14 - Ion micrographs showing penetration of the Fe and Cr and depletion of aluminum (Sample 49).

References

1. C. T. Liu and H. Inouye, "Development and Characterization of an Improved Ir-0.3% W Alloy for Space Radioisotopic Heat Sources," Oak Ridge National Laboratory Report ORNL-5290 (1977), available from NTIS, U.S. Department of Commerce, Springfield VA 22161.

2. C. L. White and C. T. Liu, "Outward Diffusion and External Oxidation of Thorium in Iridium Alloys," Acta Metallurgica, 29 (2) (1981) pp. 301-310.

3. D. E. Petersen and D. Pavone, "A Study of Carbon Monoxide Production from a Multi-Hundred Watt Fuel Sphere Assembly," Los Alamos National Laboratory Report LA-8414-MS, available from DOE-TIC, P.O. Box 62, Oak Ridge, TN 37830, (1980).

4. T. L. Markin and M. H. Rand, "Thermodynamic Data for Plutonium Oxides," p. 145 in Thermodynamics, Vol. I, IAEA, Vienna (1966).

5. L. M. Atlas and G. J. Schlehman, "Defect Equilibria of PuO_{2-x}," p. 407 in Thermodynamics, Vol. II, IAEA, Vienna (1966).

6. R. A. Kent and R. W. Zocher, "Reduction of PuO_2 by CO and Equilibrium Partial Pressures Above Plutonia," Los Alamos Report LA-6534 (November 1976), Available from NTIS, U.S. Department of Commerce, Springfield, VA 22161.

Appendix

Oxygen Pressure Over PuO_2

The oxygen pressure (P_{O_2}) was calculated using an empirical expression derived by Mulford* who performed a least squares fit of three independent sets of data which relate equilibrium oxygen pressure to fuel composition and temperature (4-6). The following expression is assumed to represent the P-C-T relation for PuO_{2-x}:

$$\text{Log}_{10}\, P(\text{atm}) = a \log x + b\,(10^4/T) + C \qquad (1)$$

where a = - 3.7007
b = - 4.7339
c = 10.6329

x is the deviation from stoichiometry in 2-x.
T is the temperature in Kelvin.

The data cover range up to $PuO_{1.998}$ (x = 0.002). The fitted equation is assumed to apply for all x less than 0.1. The pressure x = 0 is not defined; the pressure isotherm approaches the exact stoichiometry (PuO_2) ordinate asymptotically.

*Private communication, R. N. R. Mulford, Los Alamos National Laboratory, March 1980.

Theoretical support studies

CALCULATED ALLOYING EFFECTS ON IRRADIATION HARDENING OF FERRITIC STEELS

E. P. Simonen

Pacific Northwest Laboratory
Richland, Washington 99352
USA

Summary

Both copper precipitates and depleted zones contribute to calculated irradiation hardening of ferritic steels. In addition, predicted magnitudes of radiation induced solute segregation indicate that segregation can be significant and may influence the irradiation sensitivity of steel at low fluence. Radiation enhanced diffusion accelerates the predicted growth of precipitates but the magnitude of acceleration was less than expected from measured hardening kinetics. The discrepancy between predicted and measured kinetics may be attributed to limitations in the precipitate nucleation and hardening assumptions. The dependencies of precipitate growth and hardening on copper content, precipitate number density and copper diffusion kinetics were predicted. Assumed depleted zone formation resulted in predicted rates of hardening that were consistent with experimental measurements. Simultaneously, growth of precipitates and formation of depleted zones were evaluated. Significant influences of radiation on solute segregation were also calculated. The hardening and segregation influences occurred at low fluence and were insensitive to temperature.

Introduction

Irradiation hardening of pressure vessel steels can result from the production of irradiation defect aggregates (1) and the enhancement of copper precipitation (2). Both interstitials and vacancies are mobile at Light Water Reactor (LWR) operating temperatures, i.e., 560 K. Aggregates of interstitials and vacancies can form and radiation induced segregation can occur at low fluences. Furthermore, irradiation induced vacancy concentrations are significant and can accelerate diffusion kinetics. The accelerated diffusion kinetics may result in enhanced copper precipitation and hardening in steels containing copper. At temperatures below 500 K, the irradiation sensitivity of steel occurs even in the absence of copper (1). At temperatures above 600 K, hardening can be induced by aging without irradiation (2). The hardening induced by aging is clearly caused by copper precipitation. The two processes, irradiation damage and aging, are competitive processes at LWR operating temperatures. The competition between the irradiation damage process and the precipitation process is the subject of this paper as well as radiation induced solute segregation.

The effects of irradiation fluence and steel composition on radiation hardening and embrittlement have been determined from analyses of test reactor and surveillance specimen irradiations (1-6). The fluence dependence of embrittlement has been shown to have a form, A $(\text{fluence})^n$, where n ranges from about 0.2 to 0.3. End of life fluences for pressure vessel steels are less than 0.1 displacements per atom (dpa) which is equivalent to 6×10^{23} n/m^2 (E >1 MeV). Copper content has been known to influence the irradiation sensitivity and nickel has been recently shown to enhance the sensitivity of embrittlement caused by copper. Nickel in the absence of copper is not detrimental. Copper may directly contribute to hardening by precipitating during the irradiation exposure. In that case, the influence of irradiation is to enhance precipitation kinetics and hence hardening kinetics. Alternatively, copper precipitates may indirectly affect hardening caused by point defect aggregates. The development of defect aggregates is influenced by sink competition from precipitates.

The dimensional scale of irradiation induced defect aggregates in steel is below the resolution of transmission electron microscopy. Detailed characteristics of irradiation damage in pressure vessel steels have not been determined. Limited characterization has come from Field Ion Microscopy (FIM) examinations. A high density ($\sim 8 \times 10^{23}$ m^{-3}) of 0.6 nm defect clusters in Fe - 0.34 w/o Cu alloy irradiated to 3×10^{23} n/m^2 (E > 1 MeV) at 560 K was observed (7). It was not possible to identify if the clusters were composed of copper atoms or vacancies. In a later FIM study (8), copper rich regions were identified; however, they were larger and at a lesser density ($<10^{22}$ m^{-3}) than clusters observed in the first study.

A radiation damage model was developed to predict the influence of low fluence development of microstructures on hardening (9). Stabilized depleted zones were found to produce more hardening than unstable depleted zones and interstitial loops. When depleted zones were stabilized as microvoids, their lifetime was long and their contribution to hardening was significant. However, when depleted zones were assumed to be collapsed into vacancy loops their lifetimes were short and their contribution to hardening was insignificant. The lifetimes of vacancy loops were short because they possessed an interstitial bias. The interstitial bias caused more interstitial annihilation than vacancy annihilation at the depleted zones when zones were in the form of vacancy loops compared to microvoids. The result was predicted rapid zone dissolution and low predicted densities for depleted zones in the form of vacancy loops. The simultaneous influence of hardening from depleted zones and interstitial loops was also calculated. The presence of interstitial loops caused an influence on the fluence coefficient, A, for hardening while having a minimal influence on the fluence exponent, n.

The purpose of the present paper is to report calculations of hardening caused by irradiation induced defect aggregates and irradiation enhanced copper precipitation. The bases for the calculations are described and the influences of irradiation and material parameters on copper precipitation, depleted zone formation and solute segregation are shown.

Theory

The rates of growth of copper precipitates and the production of depleted zones are dependent on defect production, migration and annihilation rates. Growth of copper precipitates requires long range defect diffusion and hence can be calculated from predicted vacancy and interstitial concentrations. Depleted zone formation, however, occurs directly in cascades and does not require long range diffusion of point defects. The depleted zone lifetimes are dependent upon annihilation rates of interstitials and vacancies and on thermal emission rates of vacancies.

The rates of change of defect concentrations, depleted zone number density and precipitate radius were calculated from the numerical solutions of the following four equations (9).

$$\frac{dC_i}{dt} = K_i - \alpha\, C_i C_v - Z_i^d\, D_i C_i \rho_d - Z_i^z\, D_i C_i\, 4\pi\, r_z\, \rho_z - D_i C_i\, 4\pi\, r_p\, \rho_p \tag{1}$$

$$\frac{dC_v}{dt} = K_v - \alpha\, C_i C_v - Z_v^d\, D_v\, (C_v - \bar{C}_v^e)\, \rho_d - Z_v^z\, D_v\, (C_v - \bar{C}_v^z)\, 4\pi\, r_z\, \rho_z - D_v\, C_v\, 4\pi\, r_p\, \rho_p \tag{2}$$

$$\frac{d\rho_z}{dt} = K_z - \frac{3\rho_z}{r_z^2}\, (Z_i^z\, D_i C_i - Z_v^z\, D_v\, C_v + Z_v^z\, D_v \bar{C}_v^z) \tag{3}$$

$$\frac{dr_p}{dt} = \frac{3\, D_v^{cu}\, C_v\, C_{cu}}{4\pi\, r_p} - D_v^{cu} C_v\, r_p^2\, \rho_p \tag{4}$$

Precipitate growth (Equation 4) was calculated according to the method of Nelson et al. (10). The subscripts refer to interstitial (i), vacancy (v), zone (z) and precipitate (p) parameters. The defect production rates, K, biases, Z, and diffusivities, D, are defined in Table I. Other defect parameters that influence hardening include the recombination parameter, α, the dislocation density, ρ_d, the Burgers vector, b_v, and the equilibrium vacancy concentration, C_v^e. The initial matrix copper concentration is C_{cu}.

Table I. Parameter Values for Steel Used in the Present Calculations

Parameter	Value
α (s^{-1})	5.3×10^{20} ($D_i + D_v$)
ρ_d (m^{-2})	10^{14}
r_p (nm)	0.3
ρ_p (m^{-3})	0, 10^{23} or 10^{24}
b_v (nm)	0.248
D_i (m^2/s)	10^{-4} exp (-0.3/kT)
D_v (m^2/s)	5×10^{-5} exp (-1.3/kT)
D_v^{cu} (m^2/sec)	5×10^{-5} exp ($-E_v^{cu}$/kT)
E_v^{cu}	0.9, 1.1, or 1.3 eV
$\bar{C}_v^e$	exp (1.5) exp (-1.7/kT)
$\bar{C}_{v,z}$	$\bar{C}_{v,z} = \bar{C}_v^e \exp\left(\frac{0.138\ \mathrm{nm}}{kT\ r_{v,z}}\right)$
Z_i^d	1.1
Z_v^d	1.0
Z_i^z, Z_v^z	1.0
K_i (dpa/s)	5×10^{-10}
K_v (dpa/s)	4.55×10^{-10}
K_z (zones/s)	5×10^{-12}
E_{vs}^b (eV)	0
E_{isa}^b (eV)	0.1 or 0.2
E_{isb}^b (eV)	3/5 E_{isa}^b
E_{isa}^m (eV)	0.3 + 3/5 E_{isa}^b

The increase in yield stress, $\Delta\sigma_y$, was estimated using an Orowan hardening model where

$$\Delta\sigma_y = \alpha_h \mu b_v \sqrt{2 r_i \rho_i} . \quad (5)$$

The particle radius is r_i and the particle number density is ρ_i. The assumed shear modulous, μ, was 8.27 x 10^4 MPa and the hardening parameter was 0.5.

Radiation induced solute segregation was calculated using the computer code developed at Argonne National Laboratory (11). The material parameters were assumed to be the same as those given by Johnson and Lam (11) unless alternative values are shown in Table I. Solute-defect interactions assumed in the model include the vacancy-solute binding energy, E^b_{vs}, the type-a interstitial-solute binding energy, E^b_{isa}, the type-b interstitial-solute binding energy, E^b_{isb}, and the type-a interstitial-solute migration energy. The solution for the spherical geometry was used assuming an inner radius of 1 nm and an outer radius of 20 nm. A microvoid outer radius of influence of 20 nm was consistent with a microvoid spacing of 40 nm and a microvoid density of about 10^{23} m^{-3}. The size of radius element used in the theory varied from 0.1 nm near the inner radius to 3.8 nm near the outer radius. The assumed initial-solute concentration in the matrix was 10^{-4} atom fraction.

Results

Fluence dependencies for copper precipitate radius, depleted zone number density and radiation induced solute segregation were calculated. Precipitate growth and depleted zone formation were found to be interdependent processes. The precipitate radius and depleted zone number density were related to the fluence dependence of increase in yield stress. Specifically, depleted zone hardening exhibited fluence exponents greater than fluence exponents for precipitate hardening. Radiation induced solute segregation was predicted to be significant at low fluences and for small interstitial-solute binding energies.

Precipitate Growth

Calculated growth of precipitates was accelerated by irradiation but not to the extent expected from measured rates of irradiation hardening and embrittlement (1-6). The enhanced growth was predicted as a function of fluence, copper content and precipitate number density. In addition, the increase in yield strength caused by the precipitates was calculated.

The fluence dependencies of precipitate radius calculated from the solutions to equations (1-4) are shown in Figure 1 for three choices of copper/vacancy migration energy, i.e., 0.9 eV, 1.1 eV, and 1.3 eV. For 1.3 eV, the predicted growth was significant only at high fluences. For 1.1 eV, the predicted growth initiated at low fluences and continued to high fluences. Lastly for 0.9 eV, the predicted growth was rapid at low fluences and saturated at intermediate fluences. Depletion of copper content in the matrix caused the predicted saturation. The choice of 1.3 eV implied that copper diffusion rates were equal to iron self diffusion rates. The choice of 0.9 eV and 1.1 eV implied that the activation energy

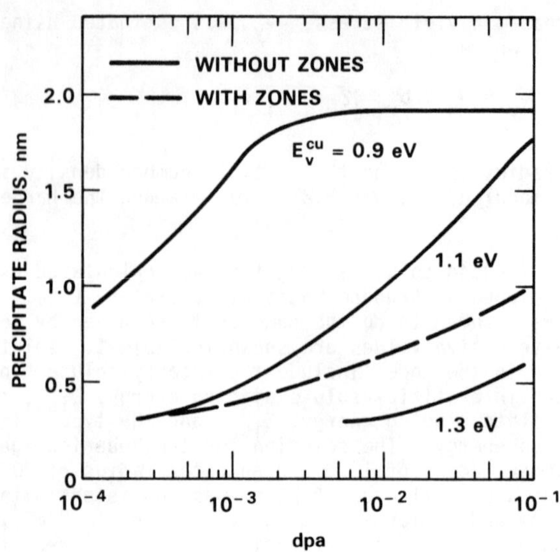

Figure 1 - Precipitate radius as a function of fluence is shown for three choices of copper/vacancy migration energies. For 1.1 eV, the predictions are shown for conditions without depleted zones and with depleted zones.

for copper diffusion was 0.4 eV and 0.2 eV less than for iron diffusion, respectively. Calculations reported below were made for an assumed copper/vacancy migration energy of 1.1 eV.

Predicted fluence dependencies of radiation induced hardening caused by copper precipitates are shown in Figure 2. The fluence dependence was calculated for copper concentrations of 0.2% and 0.3% and for precipitate number densities of 10^{23} and 10^{24} m^{-3}. Initial hardening at low fluence resulted from the assumption that the precipitates had an initial radius of 0.3 nm. Therefore, the large magnitude of predicted hardening at low fluences is an artifact of this assumption. To show only the irradiation enhanced contribution to hardening, the initial increase in predicted yield stress at zero fluence, $\Delta\sigma_y$ (0), was subtracted from the predicted values, $\Delta\sigma_y$, shown in Figure 2. The difference between these two values is shown in Figure 3. The fluence exponent of this relative increase in yield stress was about 0.3 (Figure 3) whereas the fluence exponent of the absolute increase in yield stress was about 0.15 (Figure 2).

At a constant fluence (0.05 dpa), increasing the copper matrix concentration increased the predicted increase in yield stress as shown in Figure 4. The dependence of predicted hardening on copper concentration was greater at low concentrations than at high concentrations. As the copper concentration in the matrix approached zero, the increase in yield stress approached the hardening caused by the assumed initial precipitate radius of 0.3 nm.

Calculated enhanced diffusion kinetics were insensitive to assumed temperatures near 560 K, i.e., LWR temperatures. Steady state defect fluxes were predicted for the assumption of no depleted zones and no precipitates and are shown as a function of temperature in Figure 5. The

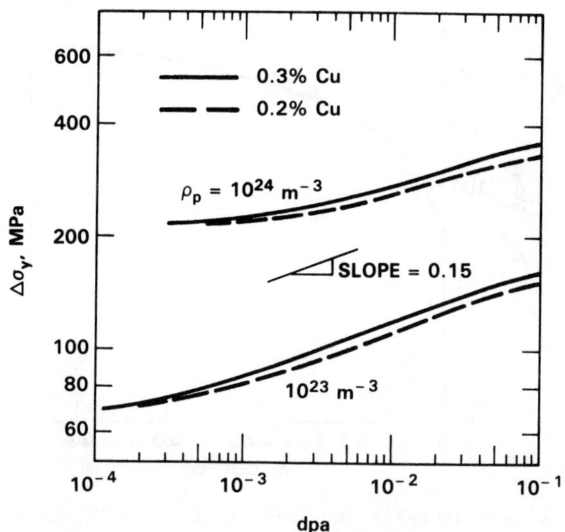

Figure 2 - Calculated increase in yield stress caused by copper precipitates is shown as a function of dpa for two choices of copper concentration and precipitate number density. The slope of the curves indicates the fluence exponent for hardening.

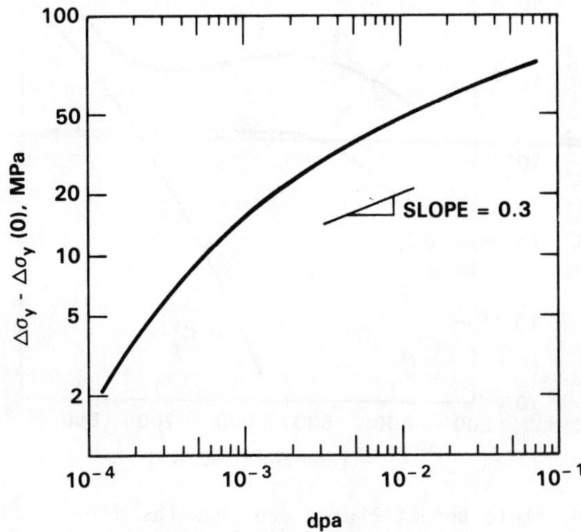

Figure 3 - The predicted difference between the increase in yield stress after irradiation and the assumed initial unirradiated increase in yield stress is shown as a function of dpa. The slope of the curve indicates the fluence exponent for hardening.

flux of vacancies from thermal emission, $D_v\bar{C}_v^e$, was trivial at 560 K. The interstitial flux, D_iC_i, and the vacancy flux, D_vC_v, were comparable below

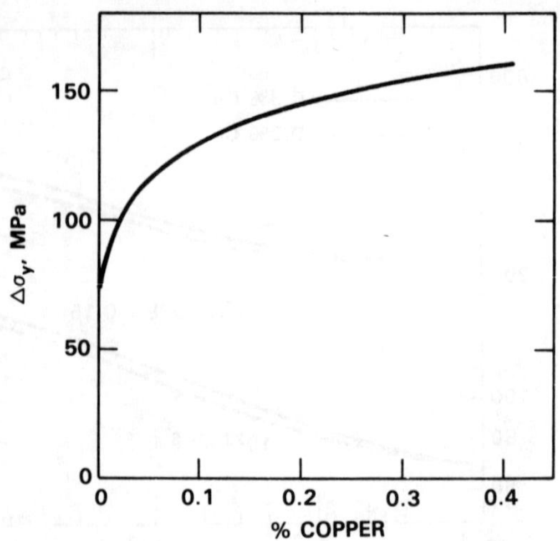

Figure 4 - Calculated increase in yield stress at 0.05 dpa is shown as a function of atom percent copper.

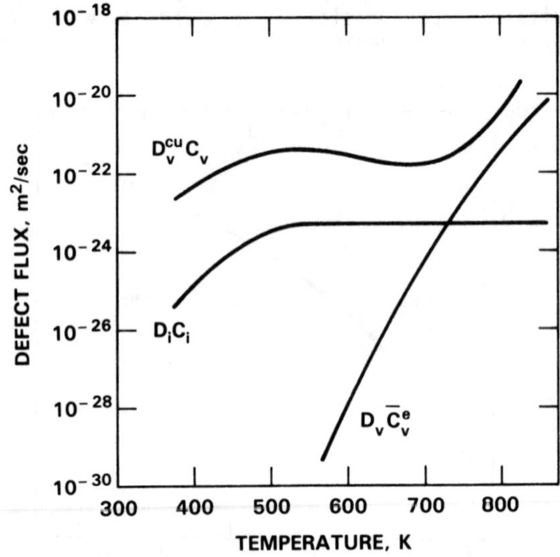

Figure 5 - Steady state defect fluxes are shown as a function of temperature. Copper/vacancy flux, interstitial flux and equilibrium vacancy flux are plotted.

700 K and differ only slightly because of the slight difference in bias factors for interstitials and vacancies at dislocations. Enhanced copper diffusivity, $D_v^{cu}\ C_v$, was calculated for use in Equation 4. The copper flux shown in Figure 5 was greater than the interstitial flux (and hence

vacancy flux) because the activation energy for a copper/vacancy jump was assumed to be 1.1 eV, whereas the activation energy for an iron/vacancy jump was assumed to be 1.3 eV.

Depleted Zone Influences

Depleted zones affected the predicted hardening directly as dislocation glide barriers and indirectly as sinks that influenced the growth kinetics of precipitates. The predicted increase in yield stress caused by depleted zone formation is shown in Figure 6 for three precipitate number densities, 0, 10^{23}, and 10^{24} m^{-3}. The fluence exponent dependence of yield stress is near to 0.25 as determined from the slope of the curves in Figure 6. Increasing the assumed precipitate number density resulted in an increase in the hardening caused by depleted zones but the fluence exponent of hardening was insensitive to the assumed precipitate number density.

The hardening increase caused by precipitate growth was predicted to be much less than the hardening increase caused by depleted zones as shown in Figure 6 for zones and Figure 2 for precipitates. Also the assumed presence of 10^{23} m^{-3} zones reduced the predicted precipitate growth compared to the assumed absence of zones as is evident in Figure 1.

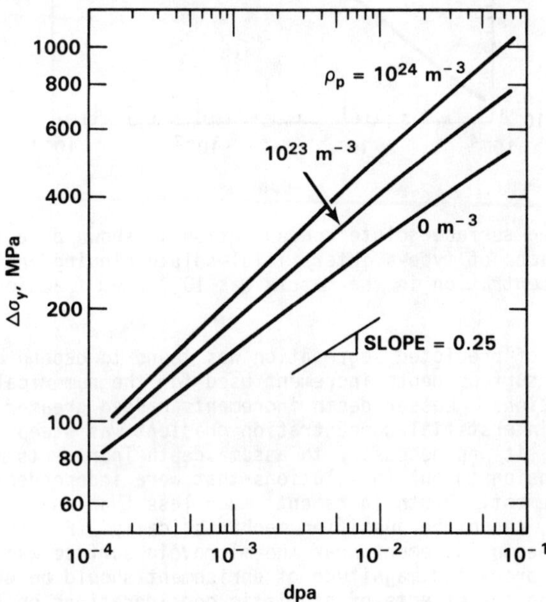

Figure 6 - Calculated increase in yield stress caused by depleted zones is shown as a function of dpa for three choices of copper precipitate number density. The slope of the curves indicates the fluence exponent for hardening.

Radiation Induced Segregation

Radiation induced solute segregation was predicted to cause large increases in solute concentrations at microvoid surfaces. The calculated fluence dependence of solute concentration at the surface was strong but the temperature dependence was weak. Surface concentration as a function of fluence and interstitial-solute binding energy, E_{isa}^{b}, is shown in Figure 7. Large increases in surface concentration were predicted even at 10^{-4} dpa and the surface concentration saturated near 10^{-2} dpa. The initial matrix concentration was assumed to be 10^{-4} atom fraction of solute. Assumed small binding energies of 0.1 and 0.2 eV resulted in substantial surface enrichment. The enrichment was, however, dependent on numerical approximations in the theory. None the less, the results shown in Figure 7 give a qualitative estimate of the segregation kinetics.

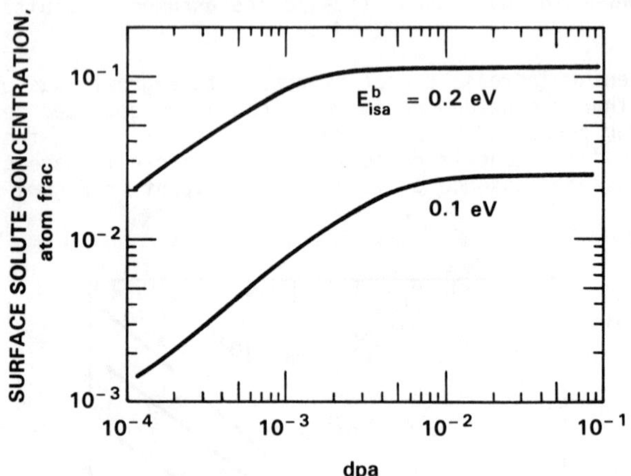

Figure 7 - Predicted surface solute concentration is shown as a function of dpa for two choices of type-a interstitial-solute binding energy. The initial solute concentration in the matrix was 10^{-4} atom fraction.

The magnitude of predicted segregation was found to depend on the choice of the near surface depth increment used for the numerical solution to the defect equations. Lesser depth increments caused greater predicted segregation. The interstitial concentration gradient was steep near the surface; therefore, it was necessary to assume depth increments much less than one atom dimension to obtain solutions that were independent of the assumed depth increment. Depth increments much less than one atomic dimension, however, cannot be justified mechanistically. For the present calculations, the depth increment near the microvoid surface was assumed to be 0.1 nm. The predicted magnitude of enrichment should be evaluated further to determine the effects of atomistic considerations on the continuum assumptions in the segregation model.

Solute segregation was predicted to be insensitive to assumed temperatures near LWR temperatures. Calculated dependencies of interstitial flux on temperature, e.g., see Figure 5, indicated that radiation enhanced solute segregation should not be a strong function of temperature.

Discussion

Calculated precipitate growth kinetics were too slow to account for the measured kinetics of hardening when the increase in hardness was calculated from the Orowan hardening mechanism. On the other hand, the calculated increase in hardening caused by depleted zone formation and Orowan hardening was in agreement with the measured kinetics. Precipitate growth and depleted zone formation were interrelated because of point defect sink interactions. Increased numbers of depleted zones retarded the predicted precipitate growth; but increased numbers of precipitates enhanced depleted zone numbers. Predicted rates of radiation induced solute segregation were rapid because of the assumed low fluence rate, 5 x 10^{-10} dpa/sec, and temperature, 560 K, for LWR pressure vessels. The microstructural and segregation responses were found to be insensitive to temperature and fluence rate for LWR conditions because neither mutual recombination of point defects nor thermal emission of vacancies were dominant defect processes.

Precipitate Growth

Disagreement between predicted and measured hardening kinetics suggests that either precipitates do not cause hardening or that certain model assumptions may not be appropriate. The fluence exponent for the predicted increase in hardness was about 0.15 when Orowan hardening was assumed, whereas an exponent of about 0.25 was expected based on reported measured dependencies. The concentration of precipitates was assumed to be independent of fluence but in reality may not be. If the concentration was assumed to increase with increasing fluence, then the fluence exponent for predicted hardening would be larger and closer to measured fluence exponents. Furthermore, the Orowan hardening model may be an improper estimate of hardening caused by soft copper precipitates (2). Alternative forms of hardening as a function of precipitate radius could result in precipitate hardening kinetics greater than expected from Orowan hardening.

When the predicted initial (unirradiated) precipitation hardening was subtracted from the predicted irradiation hardening, the fluence exponent of hardening was in accord with measured fluence exponents. The stronger fluence dependence for this relative increase in hardness, $\Delta\sigma_y - \Delta\sigma_y(o)$, compared to the absolute increase in hardness, $\Delta\sigma_y$, suggested that the assumed initial precipitate conditions are responsible for much of the predicted low sensitivity of hardening to irradiation fluence.

Calculated defect fluxes were insensitive to temperature near LWR temperatures. The temperature insensitivity also suggested that hardening should be insensitive to fluence rate because of coupled temperature effects and fluence rate effects. Below 500 K, mutual recombination of vacancies and interstitials caused a temperature dependence in the defect fluxes, whereas above 700 K, thermal emission of vacancies caused a significant temperature dependence in the vacancy flux. At intermediate temperatures the vacancy, interstitial, and copper diffusivities were predicted to be insensitive to temperature.

Depleted Zone Influences

High depleted zone densities or precipitate densities suppressed predicted defect concentrations which reduced rate processes that were controlled by defect diffusion. The assumed presence of high number densities of depleted zones caused a reduction in the predicted vacancy concentration and hence in the predicted growth of precipitates. Similarly, the assumed presence of high number densities of precipitates caused a reduction in the predicted interstitial concentration and hence the annihilation kinetics of depleted zones. The lower rate of interstitial annihilation stabilized the zones and increased the rate of depleted zone accumulation and hardening with increasing fluence.

The dependence of predicted depleted zone hardening on copper precipitates was similar to the measured dependence of hardening on copper content. In both cases, the increase in yield stress as a function of fluence has the form A $(\text{fluence})^n$. The predicted copper precipitate effect and the measured copper content effect were observed to affect the constant factor, A, and not the fluence exponent, n.

Radiation Induced Segregation

The segregation calculations indicated that solute elements that bind with interstitials, even weakly, will strongly segregate to defect sinks during pressure vessel irradiations. The subsequent solute redistribution could significantly affect the bias of irradiation clusters and hence the development of microstructures that control irradiation hardening.

The low fluence sensitivity of predicted radiation induced segregation was expected based on other calculations of segregation (11). The steep predicted interstitial concentration gradients near microvoids was caused by the low fluence rate, 5×10^{-10} dpa/sec, that was assumed and by the spherical geometry of the microvoid having a small assumed radius, 1 nm. The low damage rate and the assumed high density of microvoids assured that all interstitials that were produced would annihilate at the microvoid surface. Therefore, an optimum segregation efficiency existed.

Conclusion

Independent and synergistic effects of copper precipitate growth and depleted zone formation on irradiation hardening of pressure vessel steels were calculated. The fluence dependence and the copper precipitate influence on the predicted increase in yield stress caused by depleted zones agreed with measured effects reported in the literature. The calculated increase in yield stress caused by the growth of copper precipitates for the stated model assumptions did not agree with measured effects. Fluence dependent nucleation and precipitate radius dependencies of hardening are suggested modifications to the theory that would indicate how copper precipitates can account for the observed hardening kinetics.

Segregation effects were predicted to be strong even for weak interstitial-solute binding energies. Minor impurity elements can be readily enriched at sinks which may cause significant alteration in the microstructural development and hence hardening kinetics.

Acknowledgment

This work was supported by the Office of Basic Energy Sciences, Division of Materials Sciences, U.S. Department of Energy under contract DE-AC06-76RLO 1830.

References

1. S. H. Bush, "Structural Materials for Nuclear Power Plants," Journal of Testing and Evaluation, JTEVA, 2 (6) (1974) pp. 435-462.

2. G. R. Odette, "On the Dominant Mechanism of Irradiation Embrittlement of Reactor Pressure Vessel Steels," Scripta Metallurgica, 17 (10) (1983) pp. 1183-1188.

3. L. M. Davies, T. Ingham, and R. L. Squires, "Evaluation of Advanced Reactor Pressure Vessel Steels Under Irradiation," pp. 433-463 in Effects of Radiation on Materials: Eleventh Conference, ASTM STP 782, H. R. Brager and J. S. Perrin, eds.; ASTM, Philadelphia, PA., 1982.

4. J. R. Hawthorne, "Status and Knowledge of Radiation Embrittlement in USA Reactor Pressure Vessel Steels," pp. 100-115 in Radiation Embrittlement and Surveillance of Nuclear Reactor Pressure Vessels: An International Study, ASTM STP 819, L. E. Steele, ed.; ASTM, Philadelphia, PA., 1983.

5. G. L. Guthrie, "Pressure Vessel Steel Irradiation Embrittlement Formulas Derived from PWR Surveillance Data," Transactions of American Nuclear Society, 44 (1983) pp. 222-223.

6. J. D. Varsik, "An Empirical Evaluation of a Transition Temperature Shift in Light Water Reactor Pressure Vessel Steels," Transactions of American Nuclear Society, 44 (1983) pp. 223-224.

7. S. S. Brenner, R. Wagner, and J. Spitznagel, "Field Ion Microscope Detection of Ultrafine Defects in Neutron-Irradiated Fe-.34 Pt Cu Alloy," Metallurgical Transactions, 9A (1978) pp. 1761-1765.

8. R. G. Lott, "A Microstructural Basis for the Prediction of Embrittlement in Pressure Vessel Steels," pp. 37-44, in Topical Conference on Ferritic Alloys for Use in Nuclear Energy Technologies, J. W. Davis and D. J. Michel, eds.; AIME, New York, N.Y., 1984.

9. E. P. Simonen, "Theory of Irradiation Microstructure and Hardening of Pressure Vessel Steels," pp. 647-653, in Topical Conference on Ferritic Alloys for Use in Nuclear Energy Technologies, J. W. Davis and D. J. Michel, eds.; AIME, New York, N.Y., 1984.

10. R. S. Nelson, J. A. Hudson and D. J. Mazey, "The Stability of Precipitates in an Irradiation Environment," Journal of Nuclear Materials, 44 (1972) pp. 318-330.

11. R. A. Johnson and N. Q. Lam, "Solute Segregation in Metals Under Irradiation," Physical Review B, 13 (10) (1976) pp. 4364-4375.

NUMERICAL SIMULATION OF VOID NUCLEATION AND GROWTH

UNDER CYCLIC FUSION IRRADIATION CONDITIONS

S. J. Primeau
Department of Nuclear Engineering

and

K. C. Russell
Department of Materials Science and Engineering
Department of Nuclear Engineering
Massachusetts Institute of Technology
Cambridge, Massachusetts 02139

Summary

A computer simulation of time dependent void nucleation and growth in a tokamak first wall was performed using a theory developed for the homogeneous nucleation of cavities in irradiated metal. A plasma burn period of 3000 s was used with a displacement rate of 5.77 x 10^{-7} dpa/s and a temperature of 500°C, followed by a cooldown period of 1000 sec with the temperature varied from 300-500°C. The material properties of the metal were typical of austenitic stainless steel. The void/matrix surface energy was varied from 0.5 - 1.0 J/m^2. Total void number densities and size distributions were calculated during fifty consecutive cycles of operation. The effects of transmutation-produced inert gas were not considered.

The total void number density was very strongly dependent on surface energy, varying by as much as twenty orders of magnitude after fifty cycles. There was little void dissolution during the cooldown periods; only at the higher cooldown temperatures did some of the voids shrink by thermal vacancy emission. Most of the voids which became supercritical during the cooldown cycle dissolved due to the burst of interstitials at the start of the succeeding burn.

With several exceptions, the average void radius was relatively unaffected by changes in surface energy and cooldown temperature. The average radius after fifty cycles was approximately 40 Å; the maximum radius ranged from about 60 to 70 Å.

Introduction

The phenomenon of bubble nucleation in irradiated metals and alloys was first discovered in the late 1950's (1-3). Bubbles were formed by the transmutation of certain elements in the metal (especially nickel, boron, or lithium) into helium by (n,α) reactions. The helium atoms, being an inert gas and very insoluble in metal, co-precipitated with vacancies as equilibrium bubbles in the metal matrix, resulting in dimensional swelling of up to several percent as well as varying degrees of embrittlement.

In 1967, Cawthorne and Fulton (4) revealed that, under fast breeder reactor irradiation conditions, cavities (also called voids) could form in metal without the aid of helium pressurization to equilibrium bubbles. They showed that although some helium was, in fact, present, the cavities were certainly not equilibrium bubbles. The structural components of fast reactors and fusion devices were expected to have serious void swelling, dimensional stability, and embrittlement problems due to this irradiation-induced void formation.

The first theoretical treatment of the void swelling problem was performed by Harkness and Li (5); they approached their analysis by ignoring the effect of excess self-interstitials in the matrix, which proved to be an over-simplification. The first theories to consider the effect of interstitials appeared in 1971 (6,7) and were based on extensions of classical nucleation theory. These early theories have been modified since then to account for inert gases, soluble impurities, and heterogeneous nucleation sites which often occur in irradiated materials (8-13).

Much theoretical work has also been devoted to the understanding of void growth (14-21), assuming either a pre-voided metal matrix or a constant void nucleation rate. Some of this recent work has been reviewed by Mansur (15), Mansur and Yoo (22), and Ghoniem and Kulcinski (23).

In 1978 Russell (24) presented a theory of void nucleation in metals which took into account all the relevant physics involved in both homogeneous and heterogeneous (helium-assisted) nucleation mechanisms. Several numerical studies of this theory have already been performed (25-29).

Early modeling studies of void nucleation and growth assumed the essentially steady state conditions of such fission reactors as the LMFBR. By contrast, both magnetically and inertially confined fusion reactors are expected to operate in a pulsed mode, with cyclic variations in temperature, atomic displacement rates, and point defect concentrations. Accordingly, a time dependent analysis is needed to describe void nucleation and growth in these devices.

A number of analyses have been performed of void growth under pulsed conditions (16-21,23). By contrast, relatively few nucleation studies have been performed.

Ghoniem and Cho (30) modeled void nucleation during a single pulse. Odette and Meyers (31) and Choi, et al. (32) modeled void nucleation over a number of pulses, but represented the defect concentrations by square waves. This assumption is removed in the present analysis.

This paper uses a version of Russell's theory (7,29) to predict void nucleation behavior under these first wall conditions. Direct Frenkel pair recombination is assumed to occur only at dislocation lines, and the effects of transmutation-induced helium are ignored. As a result of these approximations, calculations will be representative of materials under relatively low temperature irradiations, where void nucleation is expected to be homogeneous with little aid from inert gas (25), high dislocation densities, where direct vacancy:interstitial recombination is relatively unimportant, and to early times in the wall lifetime when voids are not yet significant defect sinks.

It is hoped that calculations such as these will help provide an understanding of the behavior of metals under irradiation, and may ultimately be of use in screening candidate alloys for fusion first wall materials.

Point Defect Concentrations

Void formation in irradiated metals and alloys takes place in the presence of point defect supersaturations caused by high-energy neutrons or charged particles which knock atoms out of their equilibrium lattice sites, creating vacancy and self-interstitial pairs (Frenkel defects).

The interstitials have higher strain fields associated with them than do vacancies; hence, they tend to annihilate preferentially at dislocations, where the strain fields can be accommodated. The rate at which point defects annihilate at such sinks is determined by the sink strength, which is defined as the inverse of the square of the mean free path of a point defect in the matrix between creation and annihilation.

Because of their higher mobilities and sink preference, the interstitials annihilate rapidly; consequently, the arrival rate of vacancies at voids is typically a few percent greater than that of interstitials. Thus, vacancies may agglomerate into voids.

Analyses of point defect kinetics and void growth in irradiated alloys cover a wide range in sophistication (14-23). The more sophisticated treatments consider such factors as:

- A variety of point defect sinks.
- The effects of solute segregation on defect absorption at sinks.
- The image forces which exist as a point defect approaches a void surface.
- Variations in defect concentration with time.

The calculations presented herein are for comparative purposes, to show qualitatively the effects of surface energy, cooldown temperature, and number of cycles on void number density and size. As such, the simple, mathematically transparent approach of Brailsford and Bullough (14) was used.

Brailsford and Bullough (14) have derived conservation equations for point defect concentrations in irradiated metals:

$$\frac{dC_i}{dt} = K - D_i k_i^2 C_i - \alpha C_i C_v \quad , \tag{1}$$

$$\frac{dC_v}{dt} = K - D_v k_v^2 (C_v - C_v^e) - \alpha C_i C_v \quad , \tag{2}$$

where C_v^e is the thermal vacancy concentration, D_v and D_i are the diffusion coefficients for vacancies and interstitials respectively, and C_v and C_i are actual vacancy and interstitial concentrations. K is the atomic displacement rate, α is the Frenkel pair direct recombination coefficient, and k_v^2 and k_I^2 are the sink strengths for vacancies and interstitials, respectively. The first term in each equation is the production rate due to irradiation, the second term is the loss due to annihilation at dislocations, voids and other fixed sinks, and the third term is the loss rate due to direct recombination of interstitials and vacancies. In the present analysis the effect of recombination is ignored in order to make the numerical analysis tractable; the third term is therefore omitted. As noted earlier, the sink strength of voids is ignored for computational simplicity. Then, $k_v^2 = Z_v\rho_d$ and $k_i^2 = Z_i\rho_d$, where ρ_d is the dislocation density, and Z_v and Z_i are the sink bias factors for vacancies and interstitials, respectively. Both Z_v and Z_i are only slightly greater than unity.

We consider a tokamak fusion burn cycle, consisting of a plasma burn period with a constant displacement rate in the first wall, followed by an instantaneous shutdown and a cooldown period in which the "ash" is cleared out of the vacuum chamber before the cycle begins again. Each burn and cooldown period may last from ten to a thousand seconds or longer. Therefore, the steady-state point defect concentrations may not be assumed, as is customary, and a transient treatment is needed.

Integration of the simplified forms of Eqs. (1,2) from time zero to time t within a burn or cooldown period gives the following expressions:

$$C_i = \frac{(K - D_i k_i^2 C_i^o)\exp(-D_i k_i^2 t) - K}{-D_i k_i^2} , \tag{3}$$

$$C_v = \frac{[K - D_v k_v^2 (C_v^o - C_v^e)]\exp(-D_v k_v^2 t) - K - D_v k_v^2 C_v^e}{-D_v k_v^2} , \tag{4}$$

where C_i^o and C_v^o are the concentrations at $t = 0$. With these explicit expressions for point defect concentrations during the burn and cooldown periods, the void nucleation behavior at any time in the fusion cycle can be determined.

Nucleation Theory

The vacancy supersaturation in the irradiated metal, $S_v = C_v/C_v^e$, is the driving force for void nucleation; however, this driving force is countered by the void/matrix surface energy, γ , and interstitial arrivals which tend to shrink the void. The nucleation barrier in the bulk of the metal as a function of n, the number of vacancies in a void, is shown schematically in Figure 1 and given by the following:

$$\frac{\Delta G'_n}{kT} = \sum_{j=1}^{n} \ln\left\{\frac{\beta_i^o}{\beta_v^o} + \exp\left[-\ln S_v + \frac{1}{kT}(4\pi)^{1/3}(3\Omega)^{2/3}\gamma\left(j^{2/3} - (j-1)^{2/3}\right)\right]\right\} , \tag{5}$$

where Ω is the atomic volume, and β_i^o and β_v^o are interstitial and vacancy arrival rates at a hypothetical void of one vacancy. Void nucleation may occur only when $\beta_i^o/\beta_v^o < 1$, so that vacancies must arrive at a greater rate than do interstitials.

It should be emphasized that $\Delta G'_n$ is the activation barrier to forming a void at a <u>particular</u> lattice site. Although a positive $\Delta G'_n$ makes void formation improbable on a local basis, statistically voids will form on some of the sites in a macroscopic volume of metal. It is the task of nucleation theory to calculate this average rate of formation.

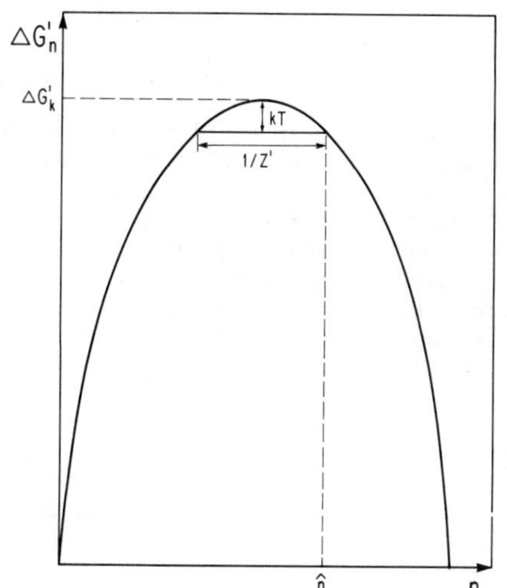

Figure 1 - Schematic plot of nucleation barrier versus void size.

As a void increases in size by accumulating vacancies, it may pass the maximum in the nucleation barrier; at that point, it may grow steadily and is considered to be nucleated. The steady state void nucleation rate, J_s, is the rate at which voids pass the maximum in the $\Delta G'_n$ curve, and is given (7) by the following:

$$J_s = \frac{Z'\beta_k}{\Omega} \exp(-\Delta G'_k/kT) \quad , \tag{6}$$

where $\beta_k = \beta_v^o \hat{n}^{1/3}$,

$\hat{n}$ = critical void size for growth (see Figure 1),

$\beta_v^o = \dfrac{4\pi D_v C_v}{a^2}$,

a = lattice jump distance.

Z' is the Zeldovich factor, the inverse of the width of the $\Delta G'_n$ curve kT units below the maximum, as shown in Figure 1.

Void nucleation does not begin immediately upon the onset of irradiation. An incubation time must pass, during which vacancies can agglomerate into clusters and grow to the critical void size. Furthermore, once a void has reached the maximum of the $\Delta G'_n$ curve, it can still fluctuate in size due to its own thermal energy, possibly decaying down the subcritical side of the curve.

The point at which a void is considered safe from such random fluctuations is to the right of the $\Delta G'_n$ peak at a point kT units below the maximum. This is the critical void size for growth, $\hat{n}$, and is shown on Figure 1. The incubation time, τ, defined as the time required for a cluster to random-walk across the distance 1/Z' at the top of the $\Delta G'_n$ curve and reach the point $\hat{n}$, is given by Russell (7) as:

$$\tau = (2Z'^2\beta_k)^{-1} \quad . \tag{7}$$

It should be noted that any sudden change in irradiation conditions may require a resetting of the incubation time. Thus, when beginning each burn period or cooldown period, a new incubation time may need to pass before any nucleation can occur. The one exception to this requirement is when a change in experimental conditions gives a decrease in $\hat{n}$.

As the vacancies accumulate into voids and grow, an equilibrium distribution of subcritical clusters develops along the subcritical side of the $\Delta G'_n$ curve (from $n = 0$ to $n = \hat{n} - 1/Z'$). Should the value of $\hat{n}$ drop suddenly to a value $\hat{n}' < \hat{n}$, due to a sudden change in irradiation, then all the voids between $\hat{n}'$ and $\hat{n}$ which were previously subcritical will now be supercritical and are by definition nucleated. This is the only case in which voids can become nucleated without an incubation time. Conversely, a sudden increase in $\hat{n}$ will cause nucleated voids to become subcritical. Our calculations of void sizes and number densities account for both circumstances.

Growth

Once a void of radius r has been nucleated, it can then grow according to the following equation:

$$\frac{dr}{dt} = \frac{D_v C_v}{r} \left\{ 1 - \frac{C_v^e}{C_v} \exp\left(\frac{2\sigma\Omega}{rkT}\right) - \frac{\beta_i^o}{\beta_v^o} \right\} \quad . \tag{8}$$

The first term in the brackets represents the total capture rate of vacancies by the void, the second term represents shrinkage due to void size dependent thermal vacancy emission, and the third term represents shrinkage due to arrival of interstitials. The relationship between void size n and void radius r is given by:

$$\frac{4}{3}\pi r^3 = n\,\Omega \ .$$

As noted earlier, a void, once nucleated, may later become subcritical due to a change in irradiation conditions or point defect concentrations. The growth law accounts for this; if a void of radius r becomes subcritical, dr/dt will become zero or negative and the void will shrink until it reaches the divacancy radius and then annihilates.

It should also be noted that if a nucleated void becomes subcritical later on, it will still be counted in the void number density totals computed in the program. Once a void is nucleated, it is counted into the void density and followed by the growth law until it shrinks and annihilates or the calculation ends. Only those subcritical clusters which have never been nucleated are not counted into the void density totals.

Calculations

A Fortran program was used to calculate point defect concentrations, critical void sizes, nucleation rates, and incubation times at various points in time within the tokamak fusion cycle. At specified intervals, a histogram of void number density versus void radius was printed out so that the progress of void swelling in the first wall could be monitored. Computational details are given elsewhere (33).

We assume a burn period of 3000 s at a displacement rate of 5.77×10^{-7} dpa/s, and a temperature, T_b, of 500°C. The cooldown period is 1000 s with a zero displacement rate and a constant temperature, T_c, of either 300, 350,

400, 450, or 500 °C. The displacement rates and temperatures are thus step functions of time within the cycle, as can be seen in Figure 2. The calculations were performed for fifty repeated cycles.

The values of displacement rate and temperature in the first wall were assumed as typical for tokamak operation at a power loading of approximately 1 MW/m^2 (32,34).

All material parameters were fixed at values appropriate for type 316 stainless steel (28) which is the basis for several candidate alloys for the first wall of developmental tokamak devices.

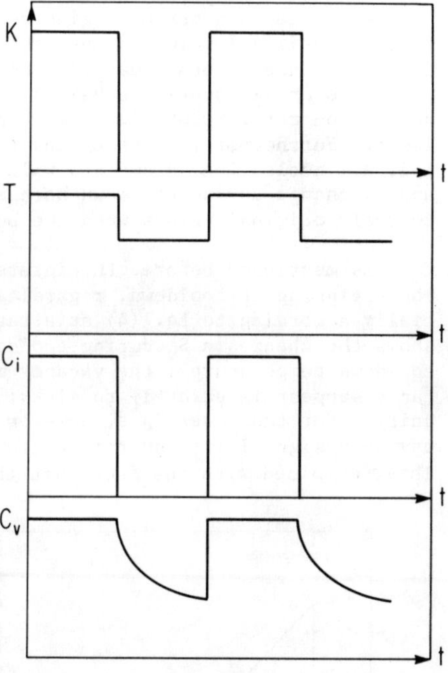

Figure 2 - Displacement rate, temperature, and interstitial and vacancy concentrations as functions of time for two cycles (not to scale).

The only disposable variable is the void/matrix surface energy, γ, which was set at either 0.5, 0.75, or 1.0 J/m^2. This parameter was varied because of uncertainty in the actual value of the surface energy in the bulk of the metal, as opposed to the free surface. These three surface energies, together with the five different cooldown temperatures, produce an experimental matrix of fifteen different conditions.

At various intervals within each of fifty cycles, the program printed out the following: time within the period, total number of periods elapsed, and instantaneous values of displacement rate, temperature, point defect concentrations, nucleation rate, incubation time, critical void size, Zeldovich factor, normalized nucleation barrier ($\Delta G'_k/kT$), and vacancy supersaturation. In addition, a histogram of void number density versus radius was printed periodically. An average void radius, being the arithmetic mean computed over the histogram range, was printed along with the histogram values.

Results and Discussion

Point Defect Behavior

Figure 2 shows schematically the general behavior of C_i and C_v during the fusion cycle. Note that the concentrations reach steady state almost instantly and remain constant during the burn cycle. At the beginning of cooldown, however, the interstitials disappear to sinks within one second, while the vacancy concentration decays more slowly. This difference is due to the much higher mobility of interstitials in the matrix.

When the burn period begins again, C_i and C_v almost immediately return to their original values; thus, the point defect behavior is completely cyclic. Since no other material conditions are changing (such as sink strengths or dislocation density), the pertinent nucleation values such as nucleation rate, incubation time, and critical size are all completely periodic. Furthermore, since C_i and C_v are both constant during the burn period, all nucleation parameters will also be constant. These values (J_s, τ, and $\hat{n}$) change during cooldown because C_v is decreasing but instantly return to their original values when the burn restarts.

As mentioned before, the interstitials all disappear immediately after the beginning of cooldown, regardless of T_c. The vacancies decay exponentially according to Eq. (4) at a rate dependent on the value of T_c. Figure 3 shows the change in S_v during cooldown as a function of T_c. For the higher cooldown temperatures, the vacancy mobility is high enough that the vacancies can disappear immediately to sinks; consequently, S_v goes very quickly to unity. For the lower T_c's, however, vacancy motion is slow enough that there may be a significant supersaturation immediately after the burn period. This, combined with the fact that the interstitial arrival rate, β_i^o, is zero

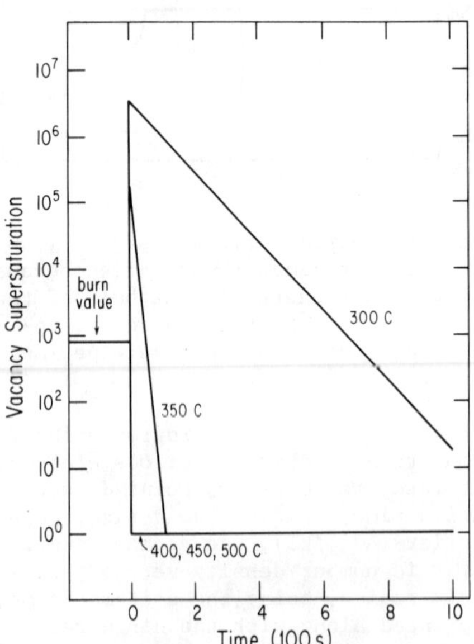

Figure 3 - Vacancy supersaturation versus cooldown time.

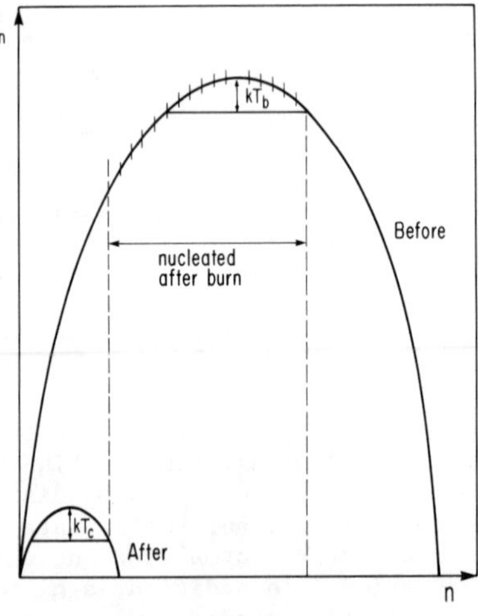

Figure 4 - Nucleation barrier versus void size before and after the end of the burn period.

for the entire cooldown period, means that the interval directly after the end of the burn period may be favorable for nucleation.

Figure 4 shows schematic graphs of $\Delta G'_n$ versus n directly before and after the end of the burn cycle. One second after the irradiation stops, the interstitials have all annihilated, while there is still a relatively large concentration of vacancies. Consequently, the nucleation barrier shrinks significantly; exactly how much it shrinks is determined by the values of T_c and γ. This shrinkage means that all the vacancy clusters between the old and new values of $\hat{n}$--the dashed part of the $\Delta G'_n$ curve--are now supercritical and have nucleated, as was discussed earlier.

As the vacancies decay exponentially, the driving force for nucleation decreases and the nucleation barrier increases. As a result, both $\Delta G'_k$ and $\hat{n}$ increase, after the initial shrinkage. It is probable, then, that $\hat{n}$ will catch up with at least some of those voids that have already nucleated and render them subcritical. At that point, the growth rate (dr/dt) would become negative and these voids would shrink. Again, the amount by which the nucleation barrier shrinks and the speed with which it expands again are functions of cooldown temperature and surface energy.

Nucleation Parameters

Figure 5 shows the change in critical void size during the cooldown period. For the lower cooldown temperatures, the decrease and subsequent increase in $\hat{n}$ is clearly evident. At the higher temperatures, the vacancies diffuse to sinks so rapidly that S_v goes very quickly to unity and the driving force for nucleation disappears (see Figure 3). The increase in $\hat{n}$ is thus precipitous and almost instantaneous. The differences in $\hat{n}$ during the

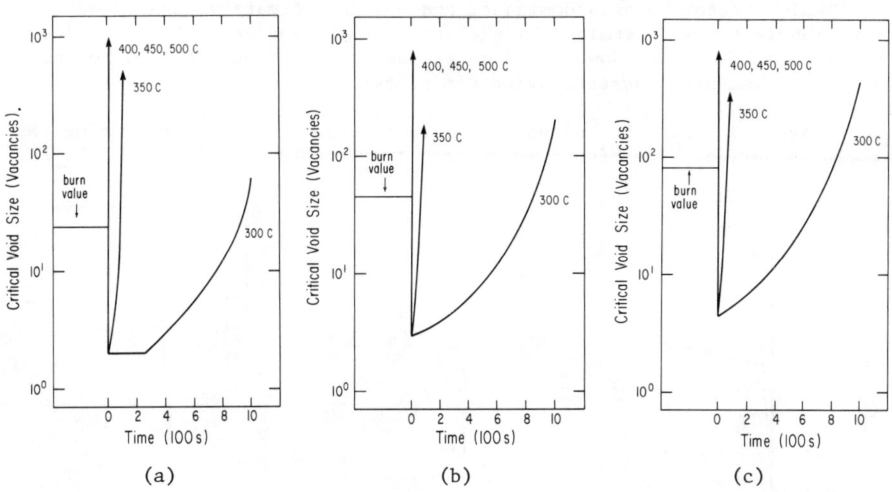

Figure 5 - Critical void size versus cooldown time. (a) $\gamma = 0.5$ J/m^2. (b) $\gamma = 0.75$ J/m^2. (c) $\gamma = 1.0$ J/m^2.

burn period are due to lower surface energies, giving less resistance to nucleation, and a lower critical size.

Figure 6 shows the nucleation rate during the burn and cooldown periods. The very large differences in J_s during the burn period show the strong dependence of nucleation rate on surface energy.

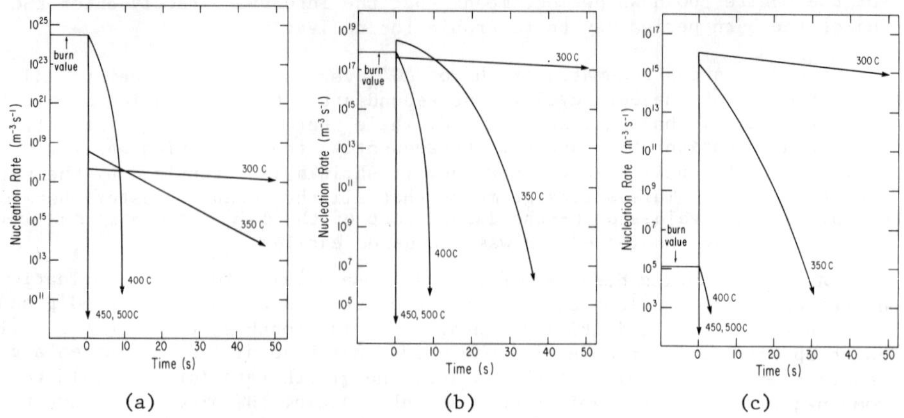

Figure 6 - Nucleation rate versus initial cooldown time. (a) $\gamma = 0.5$ J/m^2. (b) $\gamma = 0.75$ J/m^2. (c) $\gamma = 1.0$ J/m^2.

Several competing factors influence the behavior of J_s during the burn/cooldown transient. At the end of irradiation, before C_v starts to drop, the drop in temperature increases S_v, which by itself would decrease $\Delta G'_k$ and thus increase J_s. However, $\Delta G'_k/kT$ is approximately proportional to $1/T^3$; therefore, a drop in T alone would increase $\Delta G'_k/kT$, which in turn would decrease J_s. A drop in T also decreases the vacancy diffusivity, which decreases J_s proportionally.

Which factor becomes dominant, and how J_s ultimately behaves during the transient, is determined by the value of the surface energy, as demonstrated in Figure 6. However, J decreases later during cooldown for each T_c as the vacancy supersaturation diminishes.

Figure 7 shows the change in incubation time prior to and during the cooldown period. The figure shows that τ increases sharply with time for

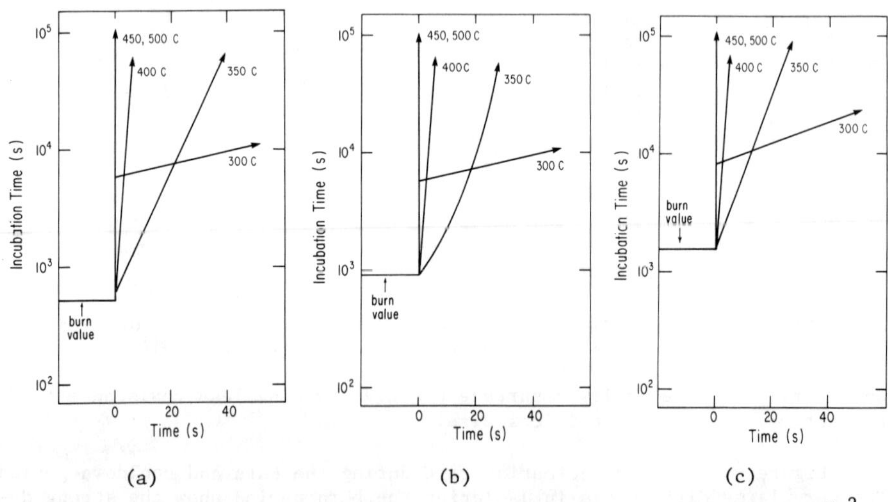

Figure 7 - Incubation time versus initial cooldown. (a) $\gamma = 0.5$ J/m^2. (b) $\gamma = 0.75$ J/m^2. (c) $\gamma = 1.0$ J/m^2.

each cooldown temperature, regardless of surface energy. At the higher temperatures, the vacancies disappear very rapidly. At the lower temperatures, though the decrease in β_i^o aids the nucleation process, the decrease in vacancy diffusivity dominates and τ increases.

Under all conditions of cooldown, τ passes 1000 s (the length of the cooldown period) within at most a few seconds. Because of this sharp increase in τ, there can be no void nucleation after the earliest part of the cooldown period. Nucleation may next take place in the succeeding burn period, after a new incubation time has passed.

A lower surface energy yields a lower incubation time during the burn period, as might be expected. Values for the various nucleation parameters during irradiation are summarized in Table I.

Table I. Nucleation Values During Irradiation

	γ (J/m^2)		
	0.5	0.75	1.0
$\frac{\Delta G'_k}{kT}$	6.3	21	51
$1/Z'$	22	36	54
$\hat{n}$	24	46	82
J_s (m^{-3}s^{-1})	3×10^{24}	8×10^{17}	1×10^{5}
τ (s)	521	918	1540

Void Swelling Parameters

Figure 8 shows the void number densities during the first five cycles of operation. For the lower cooldown temperatures, there is a significant increase in void density at the beginning of each cooldown period, due to the sudden decrease in $\hat{n}$ and subsequent nucleation of subcritical clusters.

The relative increase in void density is much larger for the higher surface energies because a relatively larger number of subcritical clusters "piles up" on the subcritical side of the $\Delta G'_n$ curve during the burn period, unable to pass over the large nucleation barrier. When the irradiation stops and $\hat{n}$ drops, these voids become nucleated and join the void density total.

For the higher T_c's, the nucleation barrier does not shrink, or shrinks very little, during cooldown and as a result, few or no subcritical clusters are nucleated.

The void densities remain constant throughout the cooldown period; there is no significant void loss during this time. However, when the burn period begins again after the lower T_c's, many of the small clusters that were nucleated in the burn/cooldown transient are annihilated by the burst of irradiation-induced interstitials. The densities at the higher T_c's do not drop at the start of the burn because no small clusters were nucleated at the start of the cooldown. Void densities during each burn period do not increase until their respective incubation times have passed.

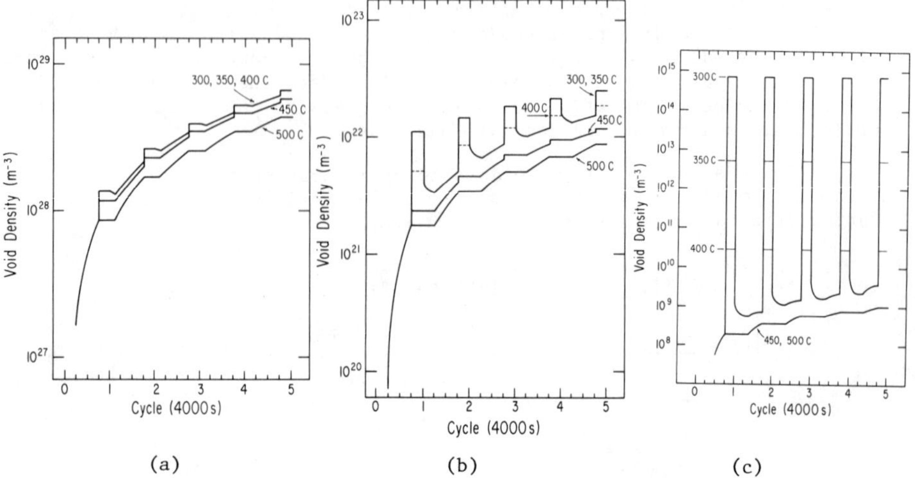

Figure 8 - Void density versus time during first five cycles. (a) γ = 0.5 J/m^2. (b) γ = 0.75 J/m^2. (c) γ = 1.0 J/m^2.

Figure 9 shows average void radius over five cycles for a low and high T_c. Figure 9(a) shows that the average radius takes a sharp drop at the beginning of the cooldown because of the nucleation of large numbers of small

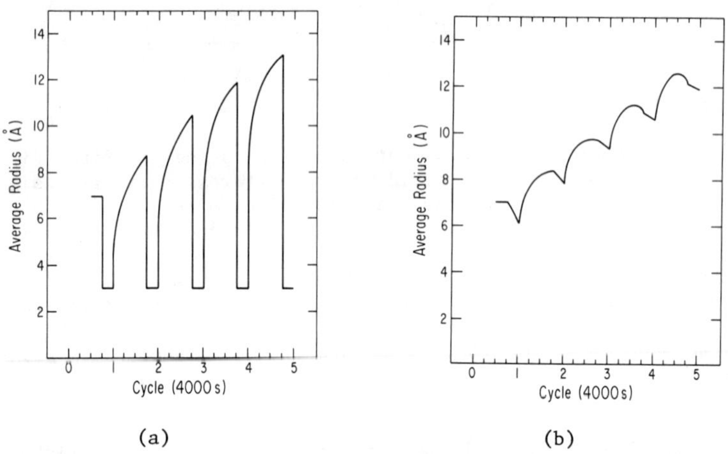

Figure 9 - Average radius versus time during first five cycles. (a) γ = 1.0 J/m^2, T_c = 300°C. (b) γ = 1.0 J/m^2, T_c = 500°C.

clusters which drives down the mean radius. There is no shrinkage or annihilation of voids during cooldown because D_v is too low. When the burn period begins again, many small clusters are destroyed because of the increase in β_i^o; this drives the average radius up.

In Figure 9(b) the radius does not drop sharply at the start of cooldown because no small voids are nucleated. However, the average radius decreases slightly during cooldown due to shrinkage by thermal vacancy emission.

Figure 10 shows total void number densities after fifty continuous cycles of operation. The points on these graphs were taken at the end of each cooldown period. Note first of all the very large differences in void density between the three surface energies--as many as twenty orders of magnitude after fifty cycles. A high surface energy is thus very effective in suppressing void nucleation.

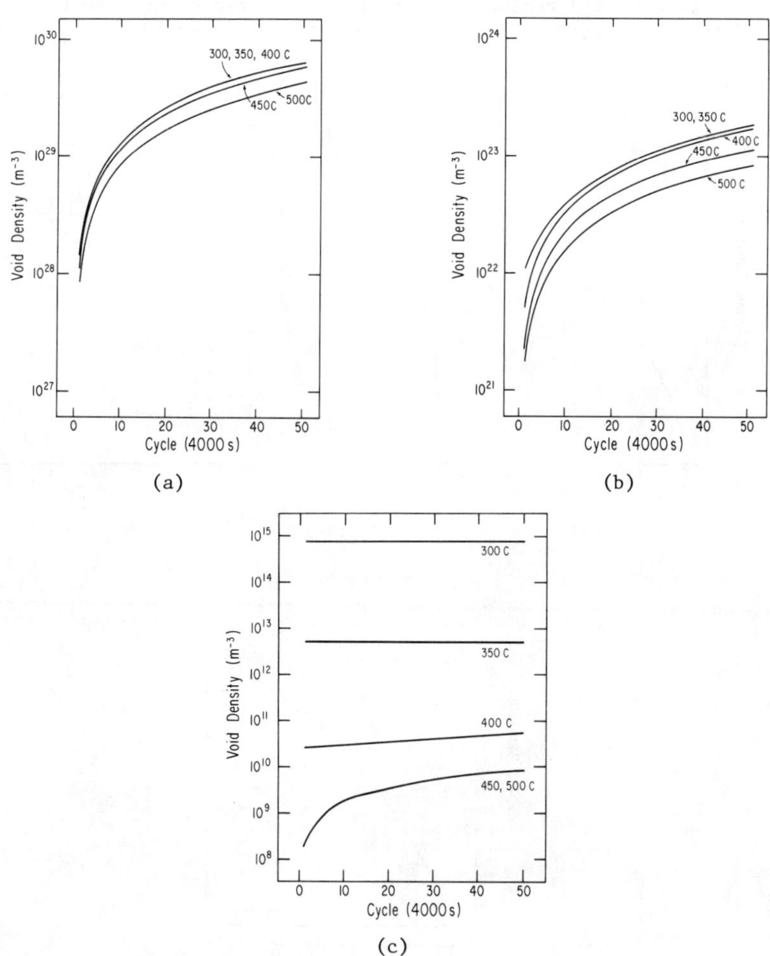

Figure 10 - Void number density versus time for fifty cycles. (a) $\gamma = 0.5$ J/m^2. (b) $\gamma = 0.75$ J/m^2. (c) $\gamma = 1.0$ J/m^2.

The number densities are higher for the lower T_c's because of the addition of clusters nucleated during the burn/cooldown transient. The reason for the large difference in densities between the low and high T_c's for $\gamma = 1.0$ J/m^2 is the addition of very large numbers of clusters nucleated at the beginning of cooldown. The extremely high void densities given for $\gamma = 0.5$ J/m^2 represent a non-physical situation.

Figure 11 shows the average void radius during fifty cycles, again at the end of each cooldown period. Except for $\gamma = 1.0$ J/m^2, there is very little difference in the average radius among either the different T_c's or surface energies. At lower temperatures and $\gamma = 1.0$ J/m^2 the very large number of clusters nucleated at the start of each cooldown period keeps the average radius lower than for the other cases.

The maximum void radius after fifty cycles ranges from 60 to 70 Å for most of the calculated runs; in no case does it exceed 70 Å.

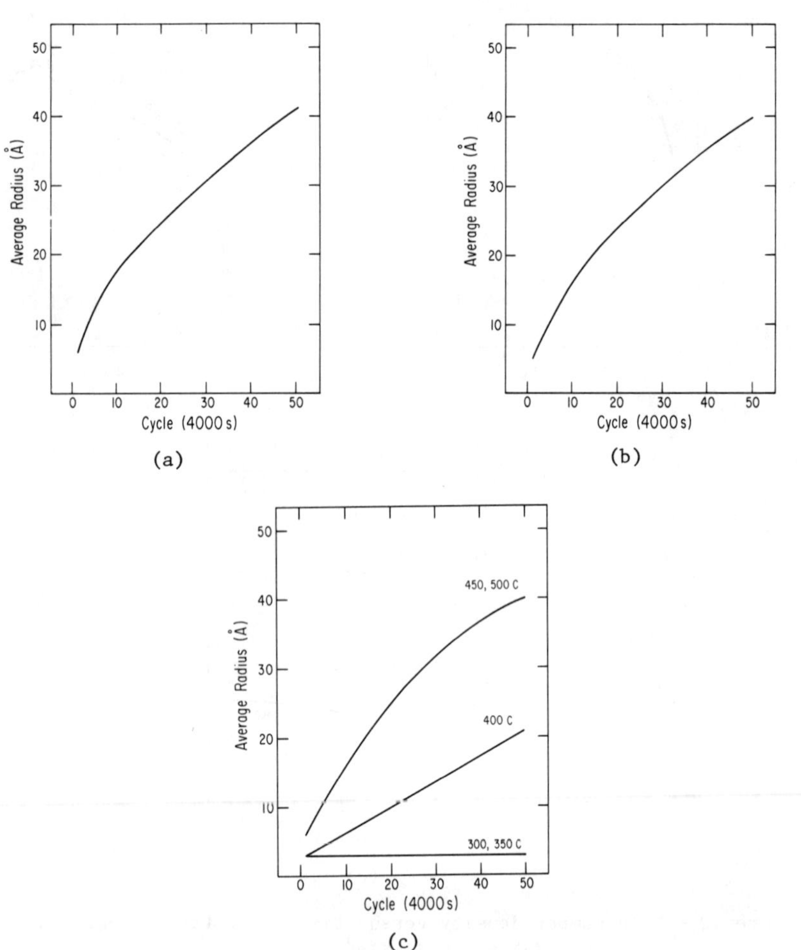

Figure 11 - Average radius versus time for fifty cycles. (a) $\gamma = 0.5$ J/m^2. (b) $\gamma = 0.75$ J/m^2. (c) $\gamma = 1.0$ J/m^2.

Conclusions

The following conclusions are drawn for the conditions studied.

1. In the absence of inert gas the void/matrix surface energy has a very strong effect on void nucleation rate, with a high surface energy depressing nucleation by many orders of magnitude.
2. Little void dissolution occurs in the cooldown period; only at the higher cooldown temperatures do some of the voids shrink by thermal vacancy emission. No nucleation occurs during cooldown after the initial burn/cooldown transient.
3. The rapid disappearance of self interstitials during the cooldown period gives a much smaller critical nucleus size than during the burn so that a large number of subcritical-sized voids may become nucleated. Most of these voids are destroyed by the burst of self interstitials at the start of the next burn cycle.
4. The average void radius is relatively unaffected by variations in cooldown temperature or surface energy except at the highest surface energy and lowest temperatures. The average radius after fifty continuous cycles of operation is approximately 40 Å, with a maximum radius ranging from 60 to 70 Å.
5. The final void number density and size distribution for cyclic irradiation are not greatly different than would have been obtained after a single burn cycle equal to the sum of the individual burn times.

Acknowledgments

Financial support of the National Science Foundation's Division of Materials Research, under Grant number DMR-80-21244, is gratefully acknowledged. Numerical computations were performed at the M.I.T. Information Processing Services. We are grateful to Dr. R. A. Johnson of the University of Virginia for explaining possible ambiguities in the meaning of the activation barrier for void nucleation.

References

1. R. S. Barnes and G. W. Greenwood, "The Effect of Gases Produced in Reactor Materials by Irradiation," in Second United Nations International Conference on Peaceful Uses of Atomic Energy, Vol. 5, United Nations, Geneva, 1958, pp. 481-485.
2. R. S. Barnes, "Gases in Reactor Materials," Nuclear Metallurgy, 6 (1959) pp. 21-26.
3. D. W. Lillie, "Effects of Radiation-Generated Helium and Tritium on the Properties of Aluminum-Lithium Alloys," Transactions of the AIME, 218 (1960) pp. 270-276.
4. C. Cawthorne and E. J. Fulton, "Voids in Irradiated Stainless Steel," Nature, 216 (1967) pp. 575-576.
5. S. D. Harkness and C. Y. Li, "A Model for Void Formation in Metals Irradiated in a Fast-Neutron Environment," pp. 189-213 in Radiation Damage in Reactor Materials, Vol. II, Int'l Atomic Energy Agency, Vienna, 1969.
6. J. L. Katz and H. Wiedersich, "Nucleation of Voids in Materials Supersaturated with Vacancies and Interstitials," Journal of Chemical Physics, 55 (1971) pp. 1414-1425.

7. K. C. Russell, "Nucleation of Voids in Irradiated Metals," Acta Metallurgica, 19 (1971) pp. 753-758.

8. K. C. Russell, "Nucleation of Voids in Irradiated Metals. II.The General Case," Scripta Metallurgica, 6 (1972) pp. 209-214.

9. K. C. Russell, "Thermodynamics of Gas-Containing Voids in Metals," Acta Metallurgica, 20 (1972) pp. 899-907.

10. J. L. Katz and H. Wiedersich, "Effect of Insoluble Gas Molecules on Voids in Materials Supersaturated with Both Vacancies and Interstitials," Journal of Nuclear Materials, 46 (1973) pp. 41-45.

11. K. C. Russell, "Nucleation of Voids in Irradiated Metals. III.Impurity Effects," Scripta Metallurgica , 7 (1973) pp. 755-760.

12. H. Wiedersich, J.J. Burton and J.L. Katz, "Effect of Mobile Helium on Void Nucleation in Materials During Irradiation," Journal of Nuclear Materials, 51 (1974) pp. 287-301.

13. S. A. Seyyedi, M. Hadji-Mrizai and K. C. Russell, "Void Nucleation at Heterogeneities," pp. 101-106 in Proceedings of the International Conference on Irradiation Behavior of Metallic Materials for Fast Reactor Core Components, Gif-sur-Yvette, France, 1979.

14. A. D. Brailsford and R. Bullough, "The Rate Theory of Swelling Due to Void Growth in Irradiated Metals," Journal of Nuclear Materials, 44 (1972) pp. 121-135.

15. L. K. Mansur, "Void Swelling in Metals and Alloys Under Irradiation: An Assessment of the Theory," Nuclear Technology, 40 (1978) pp. 5-34.

16. N. M. Ghoniem and G. L. Kulcinski, "A Rate Theory Approach to Time Dependent Microstructural Development During Irradiation," Radiation Effects, 39 (1978) pp. 47-56.

17. N. M. Ghoniem and G. L. Kulcinski, "The Effect of Damage Rate on Void Growth in Metals," Journal of Nuclear Materials, 82 (1979) pp.392-402.

18. L. K. Mansur, W.A. Coghlan and A.D. Brailsford, "Swelling with Inhomogeneous Point Defect Production: A Cascade Diffusion Theory," Journal of Nuclear Materials, 85 & 86 (1979) pp.591-595.

19. N. M. Ghoniem and G. L. Kulcinski, "The Use of the Fully Dynamic Rate Theory to Predict Void Growth in Metals," Radiation Effects, 41 (1979) pp. 81-89.

20. N. M. Ghoniem and G. L. Kulcinski, "The Effect of Pulsed Irradiation on the Swelling of 316 Stainless Steel in Fusion Reactors," Nuclear Engineering Design, 52 (1979) pp. 111-125.

21. L. N. Kmetyk, J. Weertman, W. D. Green and W. F. Sommer, "Void Growth and Swelling for Cyclic Pulsed Radiation," Journal of Nuclear Materials, 98 (1981) pp.190-205.

22. L. K. Mansur and M. H. Yoo, "Advances in the Theory of Swelling in Irradiated Metals and Alloys," Journal of Nuclear Materials, 85 & 86 (1979) pp.523-532.

23. N. M. Ghoniem and G. L. Kulcinski, "A Critical Assessment of the Effects of Pulsed Irradiation on the Microstructure, Swelling, and Creep of Materials," Nuclear Technology/Fusion, 2 (1982) pp. 165-198.

24. K. C. Russell, "The Theory of Void Nucleation in Metals," Acta Metallurgica, 26 (1978) pp. 1615-1630.

25. C. A. Parker and K.C. Russell, "Cavity Nucleation Calculations for Irradiated Metals," Journal of Nuclear Materials, 119 (1983) pp. 82-94.

26. S. I. Maydet, Numerical Analysis of a Theory of Void Nucleation in Metals Under Irradiation, S.M. Thesis, M.I.T. (1977).

27. S. I. Maydet and K. C. Russell, "Numerical Simulation of Void Nucleation in Irradiated Metals," Journal of Nuclear Materials, 82 (1979) pp. 271-285.

28. C. A. Parker, Numerical Simulation of Void Nucleation Under Irradiation with Continuous Helium Generation, S.M. Thesis, M.I.T. (1981).

29. R. W. Powell and K. C. Russell, "Computer Evaluation of Nucleation of Voids in Irradiated Metals," Radiation Effects, 12 (1972) pp. 127-131.

30. G. R. Odette and R. Meyer, CTR Quarterly Progress Report, HEDL TME/7590, Hanford Engineering Development Laboratory (April-June 1975) pp. 2-19.

31. N. H. Ghoniem and D. D. Cho, "The Simultaneous Clustering of Point Defects During Irradiation," Physica Status Solidi a, 54 (1979) pp.171-178.

32. Y. H. Choi. A. L. Bement and K. C. Russell, "The Effect of Fusion Burn Cycle on First Wall Swelling," pp. II.1-II.17 in International Conference on Radiation Effects and Tritium Technology for Fusion Reactors, eds. J. S. Watson and F. W. Wiffen, U.S. Dept. of Commerce, Springfield, VA, 1976.

33. S. J. Primeau, Numerical Simulation of Void Nucleation in Metals Under Cyclic Irradiation, S.M. Thesis, M.I.T. (1983).

34. K. C. Russell, "Helium Induced Swelling in Reactor Cladding," pp. 41-45 in Fifth Symposium on Engineering Problems of Fusion Research, IEEE Nuclear and Plasma Sciences Society, New York, 1974.

IRRADIATION-INDUCED CAVITY NUCLEATION ON DISLOCATION LINES

B. Bendriem
Department of Materials Science and Engineering

and

K. C. Russell
Department of Materials Science and Engineering
Department of Nuclear Engineering
Massachusetts Institute of Technology
Cambridge, Massachusetts 02139

Summary

The theory of cavity (or void) nucleation is adapted to describe strain energy assisted nucleation on dislocation lines in irradiated metals. The resulting equations are evaluated for type 316 stainless steel at a fast reactor displacement rate of 10^{-6} dpa/s, $T = 0.5\,T_m$, and an arrival rate of transmutation induced atoms which increases linearly with time. The critical size and the activation barrier for cavity nucleation are reduced by formation on a dislocation line so that nucleation is facilitated to a significant degree. The nucleation site density reaches a steady state value after a characteristic relaxation time of about $1/\sqrt{\lambda} = 10^5$ sec where λt equals the helium arrival rate at the unit void. The incubation time for steady state nucleation conditions is nearly the same for nucleation on dislocation lines as for nucleation in the matrix if the same gas arrival rates are assumed at each.

Introduction

Changes in mechanical and physical properties occur in structural materials subjected to high fluences of energetic neutrons. In particular, volumetric swelling and embrittlement due to the nucleation and growth of cavities (also called voids) in metals are significant problems in the development of fast breeder and fusion reactors.

The fundamentals of cavity nucleation are developed in a general theory of void nucleation in metals (1). A cavity is formed when a sufficient number of irradiation-induced vacancies and transmutation-produced helium atoms agglomerate at a specific location in the metal. The internal helium pressure and the vacancy supersaturation counteract the surface tension and cause the cavity to grow.

Dislocation lines play several important roles in the nucleation process. Interstitials and vacancies move quickly to the dislocation lines which are excellent point defect sinks. In addition, interstitials have a stronger elastic interaction with dislocations than do vacancies, which results in a slight excess of vacancies arriving at cavities and defect sinks other than dislocations. These effects result in a supersaturation of vacancies (number of vacancies much greater than thermal equilibrium number of vacancies) which are then able to gather and form a cavity.

In addition, it has been observed that dislocation lines may act as preferential cavity nucleation sites. Dislocation lines are helium traps (1) which should facilitate cavity nucleation, and furthermore, the energetics of cavity formation may be assisted by the release of the high strain energy densities at the dislocation core (2,3). However, dislocations which can climb will absorb vacancies and interstitials and thereby reduce the local point defect supersaturation to unity. Accordingly, only interfacial dislocations and other partial dislocations which cannot climb and absorb vacancies are expected to be the effective cavity nucleation sites.

Although the dislocation stress field geometry first implies the development of a cylindrically shaped cavity, calculations have shown that a spherical cavity, by minimizing surface energy, is energetically favored as the cavity nucleus.

Calculations are performed for the $\dot{n}$ nodal lines for both homogeneous nucleation and nucleation on dislocations. Then, the concentration of embryos for cavity nucleation and the time for the concentration of these embryos to reach steady state are calculated. Finally, the activation barrier for nucleation and the resulting nucleation rates are calculated.

The ability of certain interfacial partial dislocations to trap helium without absorbing vacancies and self interstitials may be used in materials optimization. Introduction of a high number density of fine precipitates which are semi-coherent with the matrix would promote nucleation of large numbers of very small high pressure bubbles. A large amount of helium could be trapped economically in this way and prevented from promoting grain boundary embrittlement and volumetric cavity-induced swelling.

Theory

Cavity or void nucleation can be described in terms of movement of a point in a (n,x) plane where n is the number of vacancies and x is the number of gas atoms in the cavity. The velocities $\dot{n}$ and $\dot{x}$ (net number of vacancies and gas atoms exchanged with matrix per unit time) have been previously derived (1) as:

$$\dot{n} = \beta_v^o n^{1/3} [1 - \frac{\beta_i^o}{\beta_v^o} - \exp(\frac{1}{kT} \frac{\partial \Delta G^o(n,x)}{\partial n})] \quad , \tag{1}$$

$$\dot{x} = \beta_x^o n^{1/3} [1 - \frac{x K_x^c}{\beta_x^o n^{1/3}} - \exp(\frac{1}{kT} \frac{\partial \Delta G^o(n,x)}{\partial x})] \quad , \tag{2}$$

where β_v^o, β_i^o, β_x^o are the arrival rates (defects/sec) of vacancies, interstitials and gas atoms at a hypothetical cavity of one vacancy, and the factor of $n^{1/3}$ accounts for the size dependence of the capture rate. K_x^c is the rate at which gas atoms are removed from the cavity by irradiation displacement events (resolution/atom•sec), kT is Boltzmann's constant times absolute temperature, and $\Delta G^o(n,x)$ is the Gibbs free energy of forming a cavity of n vacancies and x gas atoms in the absence of self interstitials.

The different nucleation processes are visualized on nodal line plots, which are the loci of points (n,x) where $\dot{n}=0$ or $\dot{x}=0$. An embryo beneath the $\dot{n}$ nodal line loses vacancies and above the $\dot{n}$ nodal line gains vacancies. It gains gas atoms below the $\dot{x}$ nodal line and loses gas atoms above the $\dot{x}$ nodal line.

In the case shown in Figure 1 the $\dot{n}$ and $\dot{x}$ nodal lines do not cross and three different nucleation mechanisms are possible:

(1) Homogeneous nucleation may occur without the assistance of gas by a fluctuation along the abscissa, from the origin to a point to the right of the $\dot{n}$ nodal line. This mechanism would tend to be favored at high vacancy supersaturations and low helium concentrations.

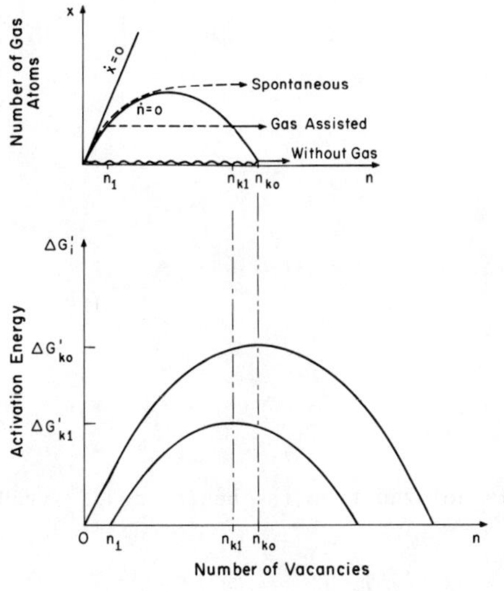

Figure 1 - Nodal lines for nucleation showing the path for nucleation without gas, nucleation on gas-containing embryos, and spontaneous nucleation, each shown in relation to a schematic plot of the activation barriers for homogeneous nucleation and for nucleation on a gas-containing embryo.

n_{ki} = critical nucleus size.

$\Delta G'_{ki}$ = activation barrier.

(2) Nucleation may occur on a cluster or embryo of 1, 2, 3 or more gas atoms, as shown by the horizontal dashed line in Figure 1.

(3) Cavities may grow spontaneously after acquiring enough helium to reach the maximum in the $\dot{n}$ nodal line. This mechanism is favored by high helium concentrations which would give a high helium arrival rate at the cavity. It has been calculated (4) that under most irradiation conditions $\beta_x^o << \beta_v^o$ so that cavities tend to creep along the $\dot{n} = 0$ nodal line.

In a general way, the nucleation of a cavity can be described by a fluctuation from the point (n_i, x_i) to the point (n_{ki}, x_i), the latter designated as the critical nucleus size (Figure 1). Beyond this point it is both energetically and kinetically favorable for a cavity to grow without further gas capture. This fluctuation from n_i to n_{ki} is characterized by an "activation barrier" whose expression is given as (1):

$$\Delta G'_{ki} = kT \sum_{n=n_i}^{n=n_{ki}} \ell n \left[\frac{\beta_i^o}{\beta_v^o} + \exp\left(\frac{1}{kT} \frac{\partial \Delta G^o(n, x_i)}{\partial n} \right) \right] . \quad (3)$$

If $n_i=0$ then $x_i=0$ and the nucleation is that described in case (1). If $\Delta G'_{ki}$ is less than kT, as occurs for clusters located near the top of the $\dot{n}$ nodal line, nucleation is effectively spontaneous since only a trivial thermal fluctuation is necessary to bring the cavity to the area to the right of the $\dot{n}$ nodal line loop where it can grow [case (3)]. Intermediate situations are in case (2).

The nucleation rate ($cm^{-3} sec^{-1}$) is given by (1) as:

$$J_i = Z_i \beta_v^o n_{ki}^{1/3} N_i \exp\left[- \frac{\Delta G'_{ki}}{kT} \right] , \quad (4)$$

where N_i is the density of nucleation sites, and the Zeldovich factor, Z_i, is given by (1) as:

$$Z_i = \left(\frac{-1}{2\pi kT} \frac{\partial^2 \Delta G'_i(n, x_i)}{\partial^2 n} \right)_{n_{ki}}^{1/2} , \quad (5)$$

or, equivalently, as:

$$Z_i = \frac{1}{n_{i2} - n_{i1}} , \quad (6)$$

where n_{i2} and n_{i1} are defined by:

$$\Delta G'_i(n_{i2(1)}, x_i) = \Delta G'_{ki} - kT . \quad (7)$$

The onset of steady state nucleation is delayed from the beginning of irradiation by an incubation time (1):

$$\tau_i = (2 Z_i \beta_v^o n_{ki}^{1/3})^{-1} . \quad (8)$$

Effects of Dislocations

The Gibbs free energy of forming a (n,x)-mer in the matrix is (1,4):

$$\Delta G^o(n,x) = -nkT\ln S_v + (36\pi\Omega^2)^{1/3}\gamma n^{2/3} - P_e\bar{V}_e x + xkT\ln\frac{P}{P_e} \tag{9}$$

where S_v is the supersaturation (ratio of actual and equilibrium vacancy concentrations), Ω is the metal atomic volume, γ is the cavity:matrix interfacial energy (J/m^2), V_e is the specific volume of the gas at the pressure P_e, P_e is the equilibrium gas pressure, and P is the actual gas pressure in the cavity. Gas pressures are typically so high that the ideal gas law is invalid and the van der Waals formulation is required. The reduced form of the van der Waals gas law is expressed as:

$$P\left[\frac{n\Omega}{x} - b\right] = kT \quad , \tag{10}$$

where b is a constant (m^3/atom). More complex gas laws give at most a trivial increase in accuracy (4).

Formation of a cavity centered on a dislocation line is assumed to release the strain energy formerly resident in the volume now occupied by the cavity. The release of the strain energy when a spherical cavity of radius r is created is (2):

$$\Delta G_r = -A\left[\ln\left(\frac{2\alpha r}{a_o}\right) - 1\right]2\Omega \quad . \tag{11}$$

α characterizes the dislocation core strain energy ($\alpha \approx 2$), a_o is the Burgers vector (approximately one lattice distance), and A depends on the Burgers vector and habit plane of the dislocation and on the materials parameters of the matrix.

The cavity radius is related to n by:

$$r = \left(\frac{3\Omega}{4\pi}\right)^{1/3} n^{1/3} \quad , \tag{12}$$

so that

$$\Delta G_r = -\left(\frac{6\Omega}{\pi}\right)^{1/3}\frac{A}{3}\, n^{1/3}\ln\left[\frac{6\alpha^3\Omega n}{a_o^3 e^3\pi}\right] \quad , \tag{13}$$

where e is the natural logarithm base. The Gibbs free energy for cavity formation on a dislocation line is then:

$$\Delta G^o(n,x) = -A'n^{1/3}\ln(C'n) + B'n^{2/3} - nkT\ln S_v - P_e\bar{V}_e x + xkT\ln\frac{P}{P_e} \tag{14}$$

where

$$A' = \frac{1}{3}\left(\frac{6\Omega}{\pi}\right)^{1/3} A \quad ,$$

$$B' = (36\pi\Omega^2)^{1/3}\gamma \quad ,$$

$$C' = \frac{\alpha^3}{a_o^3 e^3} \frac{6\Omega}{\pi} \quad .$$

It is now possible to calculate the partial derivatives $\partial\Delta G^o/\partial x$ and $\partial\Delta G^o/\partial n$ and then to obtain the equations for the nodal lines. We obtain for $\dot{n}=0$:

(1) using the ideal gas law: $x = n[f(n) - \ln S_e)$ (15)

(2) using the van der Waals gas law: $x = \frac{n}{\frac{b}{\Omega}} \frac{[f(n) - \ln S_e]}{[f(n) - \ln S_e] + 1}$ (16)

where

$$f(n) = -\frac{A'}{kT}[1 + \frac{1}{3}\ln(C'n)]n^{-2/3} + \frac{2}{3}\frac{B'}{kT}n^{-1/3} \quad , \tag{17}$$

and S_e is the effective supersaturation:

$$S_e = S_v(1 - \frac{\beta_i^o}{\beta_v^o}) \quad .$$

Since the release of strain energy is assumed to be independent of the presence of gas atoms, the $\dot{x}$ nodal line is still given in n,x space (4) as:

$$x = \text{minimum of } [\frac{\beta_x^o}{K_x^c} n^{1/3} \; , \; \frac{n\Omega}{b}] \quad . \tag{18}$$

Dislocation lines are known to be good helium traps. The arrival rate of gas at cavities on dislocations is thus expected to be much higher than at cavities in the matrix, so that a relatively small gas concentration will elevate the $\dot{x}$ nodal line above the $\dot{n}$ nodal line, as shown in Figure 1. This nodal line arrangement will be assumed to be the case in the calculations to follow.

Evaluation of Nucleation Site Density

We may now calculate the densities of the combined vacancy and gas atom clusters on dislocations which will serve as heterogeneous nucleation sites for cavities. It is assumed that β_x^o is linearly increasing with time due to (n,α) reactions. During all the time of irradiation β_v^o is assumed to be greater than β_x^o so that the formation rate of embryos for nucleation sites is controlled by the arrival of gas atoms. Since the sites are located on the ascending portion of the $\dot{n}$ nodal line, their state is completely determined by their content of gas atoms.

We denote by ρ_i the density of sites of size (n_i, i). The change in the number density with time is given by (5):

$$\frac{\Delta\rho_i}{\Delta t} = \rho_{i-1}\, n_{i-1}^{1/3}\, \beta_x^o + \rho_{i+1}\,(i+1)\, K_x^c - \rho_i\,[\,K^* + n_i^{1/3}\,\beta_x^o + iK_x^c\,] \quad . \tag{19}$$

The loss of sites due to nucleus formation is neglected. K^* is the destruction rate of cavities by displacement cascades and K_x^c is the cavity resolution rate of helium. ρ_o is taken equal to ρ_d/a, the number of atomic sites on dislocation lines, and ρ_i is taken to be zero before irradiation. The arrival rate of helium at the unit void in the matrix has been calculated by (1) from resolution kinetics to be:

$$\beta_x^o = \frac{C_x^o\, K_x^T}{a_o^2\, k_x^2} \quad , \tag{20}$$

where C_x^o is the total gas concentration, K_x^T is the gas resolution rate from traps other than cavities, k_x^2 is the total sink strength for gas, and a_o is the lattice parameter. C_x^o is equal to $K_x^o t$, where K_x^o is the gas production rate due to transmutation and t is the elapsed time of irradiation. Therefore:

$$\beta_x^o = \lambda t \quad , \tag{21}$$

where

$$\lambda = \frac{K_x^o\, K_x^T}{a_o^2\, k_x^2} \quad . \tag{22}$$

We assume this same equation for β_x^o on a dislocation line, for lack of a better value. It is probably an underestimate since the high concentration of gas atoms trapped on dislocation lines should give a higher β_x.

At the beginning of irradiation we can make the assumption that $\rho_{i-1} >> \rho_i$. (This assumption will be justified later.) Therefore:

$$\frac{\Delta\rho_i}{\Delta t} = \rho_{i-1}\, n_{i-1}^{1/3}\, \beta_x^o \quad . \tag{23}$$

Then:

$$\frac{\Delta\rho_1}{\Delta t} = \rho_o\, \lambda\, t \qquad \text{and} \qquad \rho_1 = \frac{\rho_o \lambda t^2}{2} \quad ,$$

$$\frac{\Delta\rho_2}{\Delta t} = \frac{n_1^{1/3}\rho_o\lambda^3 t^3}{2} \qquad \text{and} \qquad \rho_2 = \frac{n_1^{1/3}\rho_o\lambda^2 t^4}{2\cdot 4} \quad ,$$

and so on, to:

$$\rho_p = \frac{(n_1\cdot n_2\cdot\ldots n_{p-1})^{1/3}\,\rho_o\,\lambda^p\,t^{2p}}{2^p\cdot p\;!} \quad . \tag{24}$$

Indeed, the result in Eq. (24) can also be obtained by assuming a solution for Eq. (19) of the form:

$$\rho_i = \sum_{j=0}^{\infty} a_{ij} t^j \quad .$$

From Eq. (19) we then obtain a relation between the coefficients a_{ij}:

$$ja_{ij} = n_{i-1}^{1/3} \lambda a_{i-1,j-2} + (i+1) K_x^c a_{i+1,j-1}$$
$$- (K^* + i K_x^c) a_{i,j-1} - \lambda n_i^{1/3} a_{i,j-2} \quad . \tag{25}$$

The boundary conditions can be expressed as:

$$a_{oo} = \rho_o \quad \text{and for any p,} \quad a_{po} = 0 \quad .$$

It can be shown by mathematical induction that for a given p the first non-zero coefficient is $a_{p,2p}$ and its value is the one found for p-1.

Moreover, the different site densities tend to equalize after a time of approximately $1/\sqrt{\lambda}$. After this time β_x^o becomes the important parameter in Eq. (19). We can then say:

$$\frac{\Delta\rho_i}{\Delta t} = n_{i-1}^{1/3} \rho_{i-1} \beta_x^o - n_i^{1/3} \rho_i \beta_x^o \quad . \tag{26}$$

We find a steady state value for ρ_i of:

$$n_{i-1}^{1/3} \rho_{i-1} = n_i^{1/3} \rho_i \quad .$$

Then:

$$\rho_1 = \frac{\rho_o}{n_1^{1/3}} \quad \text{and} \quad \rho_p = \frac{\rho_o}{n_p^{1/3}} \quad . \tag{27}$$

Results and Discussion

Calculations were performed for cavity nucleation on screw dislocations in type 316 stainless steel. The displacement rate assumed is 10^{-6} dpa/s, characteristic of the environments experienced by fast reactor cladding or fusion reactor first wall materials. The vacancy and interstitial concentrations are obtained by the method of Brailsford and Bullough (6), the surface energy is 1 J/m^2, and the temperature is half the melting temperature. The cavity sink strength is zero, assumed for simplification. We assume that approximately 1% of the 4 x 10^9 dislocations/cm^3 are unable to absorb vacancies, hence are potential cavity nucleation sites.

An important parameter which is not well understood is the ratio of the interstitial bias for dislocations to the vacancy bias for dislocations. This ratio is equal to (1) β_v^o/β_i^o; it must be greater than one, otherwise void nucleation is impossible. β_v^o/β_i^o is expected to be less for nucleation on dislocations than for nucleation in the matrix. Accordingly, a value of 1.02 was chosen for β_v^o/β_i^o. Dislocations provide both traps and high diffusivity paths for inert gas atoms. High gas concentrations and high diffusion coefficients will lead to a high gas arrival rate at embryonic cavities. Since the details of gas trapping and migration in dislocations are not known, the gas arrival rate calculated by Parker and Russell (4) for homogeneous nucleation in the matrix is used as an underestimate. For type 316 stainless steel under conditions of fast reactor irradiation and continuous helium generation, β_x^o is obtained from Eqs. (20) and (21):

$$K_x^o = 3 \times 10^{-6} \text{ appm/s} \quad ,$$

$$K_x^T = 100 \text{ K} \quad , \qquad \text{where K = atomic displacement rate,}$$

$$k_x^2 = 4 \times 10^9 \text{ cm}^{-2} \quad \text{(approximate dislocation density).}$$

Then, $\beta_x^o = \lambda t$, where $\lambda = 10^{-10}$ sec^{-2}.

Figure 2 shows the $\dot{n}$ nodal line calculated on the basis of the ideal gas law and the van der Waals gas law. The $\dot{n}$ nodal line for matrix nucleation is shown for comparison. The ideal gas law is clearly unrealistic since cavities of only a few vacancies are predicted to contain many gas atoms. Figure 2 demonstrates the influence of the release of dislocation strain energy. The nodal line loop shrinks so that an embryo needs fewer gas atoms to grow, and for a given embryo the critical cavity size is smaller and nucleation is easier.

Figure 3 presents calculated nucleation site densities versus irradiation time for embryos of 1, 2, and 3 gas atoms. All three nucleation site densities are seen to rise to a final value after a characteristic relaxation time of $\sim 2 \times 10^5$ s. The steady state values are given by:

$$\rho_p = \frac{\rho_o}{n_p^{1/3}} = \frac{4 \times 10^{15}}{n_p^{1/3}} \text{ cm}^{-3} \quad .$$

The calculations were made under the assumption that no embryos undergo thermal fluctuations to reach the critical nucleus size. Such "leakage" would, of course, invalidate the conclusions.

Figure 4 represents the activation barrier for nucleation as a function of the number of vacancies in the vacancy and gas embryo and clearly shows the effect of dislocations. The release of strain energy lowers the activation barrier by a factor of ~ 2, largely independent of embryo size. Spontaneous nucleation (which takes place when $\Delta G_k'/kT \lesssim 1$) occurs at a smaller embryo size than for matrix nucleation. The activation barrier is lowered even for nucleation in the absence of gas ($n_i = t_i = 0$).

A study of incubation time (Figure 5) shows it to be not strongly dependent on the embryo size and to have a value of $\sim 2 \times 10^5$ sec. In fact, we

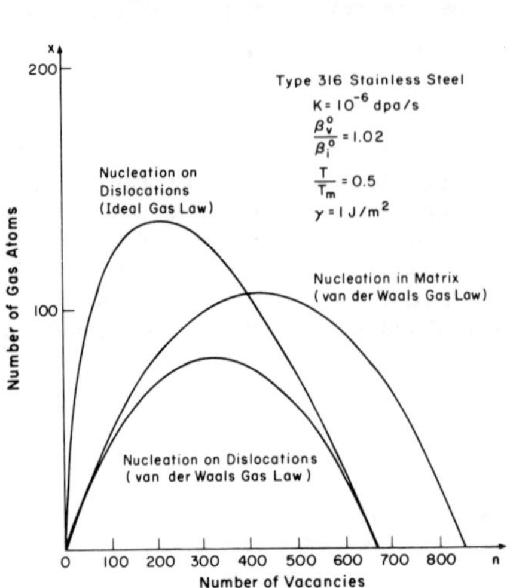

Figure 2 - Calculated $\dot{n}$ nodal lines for cavity nucleation on dislocations and in the matrix. The effect of the gas law assumed is also illustrated.

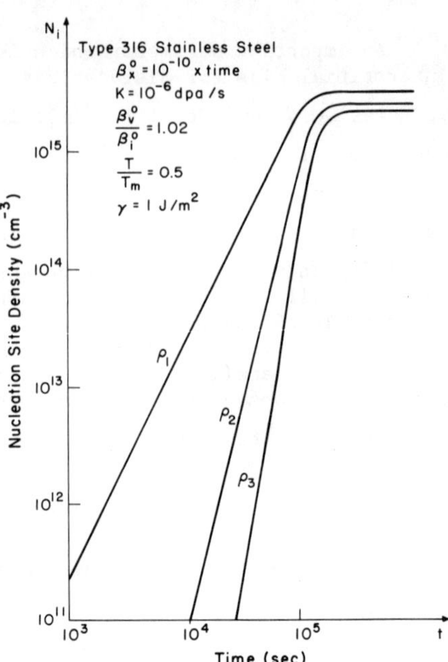

Figure 3 - Development of gas containing embryo densities with time. Embryos of 1,2, and 3 gas atoms are seen to increase to approximately the same steady state value at approximately the same time.

can take the Zeldovich factor as constant (meaning that the radius of curvature of $\Delta G'(n)$ around n_k is independent of the size of the critical nucleus) so that τ varies as $n_k^{1/3}$. Since the critical size varies in a range of 2, the incubation time varies only in a range of 1.3 . The release of strain energy increases the Zeldovich factor slightly and lowers the incubation time in the same proportion.

Figure 6 plots the nucleation rates on dislocations and in the matrix as a function of embryo size. For nucleation without gas atoms the rate is approximately 10^{-16} cm^{-3} sec^{-1} in the matrix, whereas that on dislocations is approximately 10^{-4} cm^{-3} sec^{-1}. For small embryos the nucleation rate is considerably higher for nucleation on dislocations than it is for nucleation in the matrix. Both nucleation rates saturate at approximately 10^{12} cavities per $cm^3 \cdot sec$.

Conclusions

1) The release of dislocation strain energy greatly reduces the activation barrier for nucleation on dislocations with or without the presence of inert gas.

2) Cavities on dislocations have a smaller critical size and thereby a higher probability for nucleation than do cavities in the matrix.

3) The incubation time is approximately the same for cavity nucleation on dislocations and in the matrix if the same gas arrival rates are assumed.

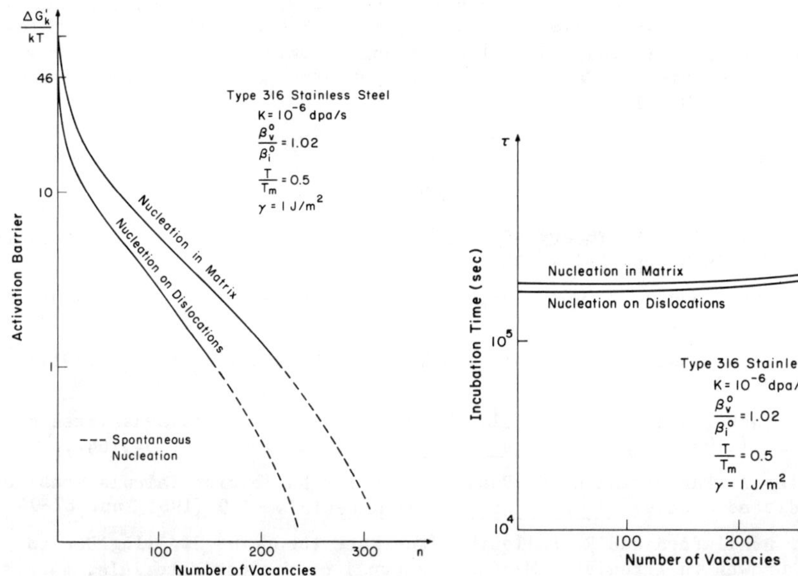

Figure 4 - Activation barrier for cavity nucleation vs. number of vacancies in the embryo. Release of dislocation strain energy by the cavity nucleus is seen to give a major reduction in the activation energy for cavity nucleation as compared to nucleation in the matrix.

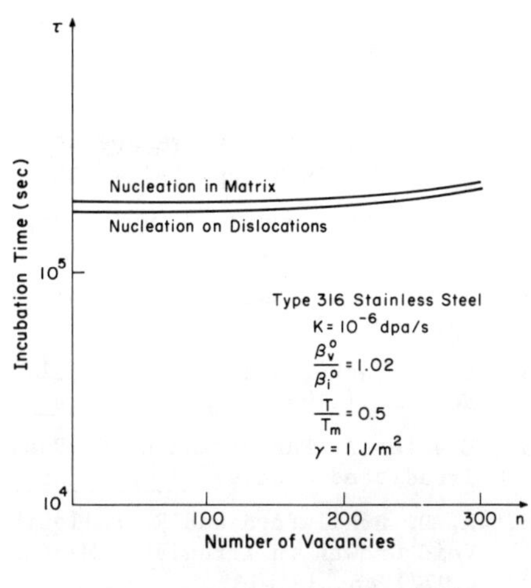

Figure 5 - Incubation time for nucleation on dislocation lines and in the matrix. The incubation time is largely independent of both nucleation mode and embryo size.

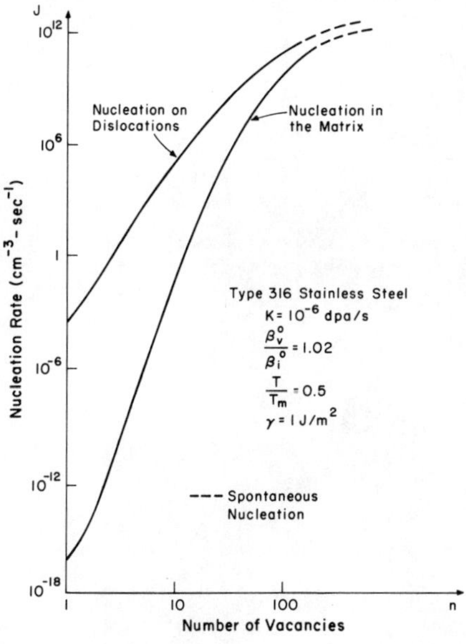

Figure 6 - Nucleation rate vs. size of the embryo for nucleation on dislocation lines and nucleation in the matrix.

Acknowledgment

This paper is based on work supported by the National Science Foundation under grant DMR-80-21244. Assistance by AMAX Corporation and the "Mission Scientifique à l'Ambassade de France aux U.S.A." is gratefully acknowledged by B. Bendriem. Calculations were performed at Information Processing Services of M.I.T.

References

1. K. C. Russell, "The Theory of Void Nucleation in Metals," Acta Metallurgica, 26 (1978) pp. 1615-1630.
2. J. Weertman and J. R. Weertman, Elementary Dislocation Theory; MacMillan, New York, 1964.
3. J. W. Cahn, "Nucleation on Dislocations," Acta Metallurgica, 5 (1957) pp. 169-172.
4. C. A. Parker and K. C. Russell, "Void Nucleation in Metals Assisted by Non-Ideal Inert Gas," Scripta Metallurgica, 15 (1981) pp. 643-647.
5. Charles A. Parker and K. C. Russell, "Cavity Nucleation Calculations for Irradiated Metals," Journal of Nuclear Materials, 119 (1983) pp. 82-94.
6. A. D. Brailsford and R. Bullough, "The Rate Theory of Swelling Due to Void Growth in Irradiated Metals," Journal of Nuclear Materials, 44 (1972) pp. 121-135.

subject index

Adhesive wear tests, 32
Advanced Test Reactor (ATR), 37
AES (Auger Electron
  Spectroscopy), 204
Aging, Iridium, 202
Alloying, corrosion, 5
Alloys
  A508-3, 173
  A533-B, 173
  AISI 316, 79, 141
  AMCR, 79
  Austenitic stainless, 73, 78
    111, 141, 239
  Co-base, Co-free, 15
  DOP 26, 202
  EP-838, 79
  Fe-Cr, 9, 10
  Fe-Cr-Mn, 71, 87
  Fe-Cr-Ni, 87, 111
  Fe-Mn, 87
  Fe-2 1/4Cr-1Mo, 10
  HT-9, 64
  INVAR, 103, 127
  Ir, 202
  Low cobalt, 17
  Martensitic steel, 63
  Modified 9 Cr-1Mo, 64
  NM-1, 83
  PCA, 159
  Pressure vessel steel, 225
  Refractory metals, 9
  Stellite, 16
  T91, 64
  Zircaloy, 35, 51
  Zr-5Be, 52
Aluminum, impurity in iridium, 202
Anisotropic creep, 42
Austenitic steel (see Alloys)

Basketweave microstructure, 51
Brazing, 51
Bubble nucleation, 240
Burgers vectors, 43
Burn period, 244, 246, 249, 251

CANDU PHW reactors, 35, 51
Carbon transfer, 3, 9
Cavitation (see voids, swelling)
Charged particle irradiation
  (see ion bombardment)
Charpy V notch ductility, 167
Chromium, effect on swelling, 91
Chromium equivalent, 78
Cobalt-base alloys (see Alloys)
  as hardfacing, 16, 17, 22, 31-33
Cobalt sources in nuclear plants, 16,
  18, 19, 21, 33
Cobalt, supply of, 16
Cold work, 39, 112
Compositional dependence, swelling, 87
Compositional oscillation, 111
Cooldown, 239, 244, 246-248, 251
Cooling rate, 51, 60
Copper clusters, 226
Copper, residual element, 168, 196
Corrosion, hardfacing alloys, 17
Corrosion, liquid metal, 3
Corrosion product formation, 7
Corrosion resistance, 80
Creep, 42, 67
Creep rupture strength, 63, 65
Critical void size, 243, 246, 247,
  253, 257, 266, 267

Decarburization, 9
Densification, 119
Density change, 91, 141
Depleted zones, effect on hardening,
  236, 239
Diffusion,
  enhanced, 230
  impurity, 205
Diffusion coefficient, vacancy, 102
Dimensional stability, 35
Dislocation density, 37
Dislocation loops, 126
Dislocations, cavity nucleation on, 257
Displacement cascades, 263
Displacement rate, effect on
  dissolution corrosion, 5, 7
Ductile to brittle transition,
  martensitic stainless steels, 67
  pressure vessel steels, 192
  temperature (DBTT), 192

Embrittlement, radiation, 167, 226

Energy dispersive x-ray analysis, 24, 114
Equilibrium diagram,
Fe-Ni-Cr system, 112, 130

Fabricability, 83
Fabrication, 37
Fast breeder reactor, 3
Fast induced radioactivity decay, 63, 68, 74, 105
Fast reactor, 87
Federal Republic of Germany,
pressure vessel steels, 174
Field ion microscopy, 226
Forgings,
pressure vessel steels, 171
Fracture properties,
pressure vessel steels, 167
Fracture toughness, 63, 65, 167
dynamic K_J, 167
France, pressure vessel steel, 168
Fuel sheathing, nuclear, 51
Fusion cycle, 246
Fusion reactors, 3, 77

Galling wear, 31
Gibbs free energy, 261
Grain boundaries,
Th stabilized in iridium, 201
Grain growth, effect of PuO_2, 201
Grain size, 35, 51, 53, 59
Growth, precipitate, 229, 235
Growth, void, 239

Hardening, radiation, 67, 225
Hardfacing, 17
Heat affected zones, 51, 52
Helium
effect on swelling in AISI 316, 157
embrittlement, 81
in cavities, 258
traps, 258
Heterogeneous nucleation, 240, 262
Homogeneous nucleation, 259, 265

Impurity
content, 51
in PuO_2, 204
reactions, 6
transfer, 9
Inert gas, 253
In-reactor deformation, 36
Interfacial dislocation, 258
Intergranular corrosion, 5
Intergranular stresses, 48
Intermetallics, Zr-Cr-Fe, 16
Interstitial bias, 265
Interstitial reactions, 6
Invar alloys, 103, 127
Ion bombardment, 99
Irradiation,
enhanced creep, 35, 36
growth, 35, 36, 39

Japan, pressure vessel steels, 174

Lattice parameter, Fe-Ni-Cr, 112
Lead-lithium, 4
Light water reactor, 226
Liquid metal compatibility, 81
Liquid metal corrosion, 3
Lithium, 4
Low activation, 63, 68, 73, 105

Manganese, effect on swelling, 97
Manganese stabilized, 77
Martensite, 51
Martensitic stainless steel, 63
Martensitic steel (see Alloys)
Mass transfer, 6, 10
Metallography, quantitative, 53, 54, 59
Microstructure, 37
Microvoids, 226

n, α reactions, 240
NaK, 4
National Research Universal Reactor (NRU), 37
Nickel,
effect on swelling, 91
equivalent, 78
replacement, 76
Nucleation,
cavities, 111, 239, 257
on dislocations, 266
rate, 260, 266, 267

Occupational radiation exposure, 16, 17, 25, 33
Ordering, 126

Oxidation of thorium, 201
Oxygen pressure over PuO_2, 216, 221

Parallel-plate microstructure, 51
Particles, second phase, 51, 52, 60
Pb-17 a/o Li, 4
Phosphorous
  additions to AISI 316, 162
  residual element, 168, 196
Physical metallurgy, 64
Plasma burn, 239
Point defect sinks, 241
Postirradiation deformation, 80
Potassium, 4
Precipitate ,
  hardening, 229, 235
  phases, 63, 69
Precipitation,
  radiation enhanced, 229, 235
Precracked Charpy-V notch ductility, 167
Pressure vessel steels, 167, 226

Radiation,
  occupational exposure to, 15, 16
  sources of, 15, 16
Radioactive waste, 74
Reactor
  EBR II, 142
  fast breeder, 63, 142
  fusion, 63, 73, 142
  light water, 37, 168
Refractory metals, 9
Residual elements,
  effect on embrittlement, 168

Saturation, swelling, 96
Schaeffler Diagram, 78
Segregation, solute, 126, 234, 236, 241
Silicon ,
  effect on swelling in AISI 316, 159
  in zircaloy-4, 52
SIMS (secondary ion mass spectrometry), 204
Sink strength, 241, 246, 264
Sodium, 4
Solute additions, 112
Solute segregation, 126, 229, 236, 241
Spectrometry, mass, 204
Spinodal decomposition, 126

Steel (see Alloys)
Stellite, 16
Strain energy, 257, 261, 262
Stress, applied,
  effect on swelling, 142
Stress relaxation, 35
Stress relief, 42
Supersaturation, 262
Surface energy, 245, 247, 251
Surface profilometry, 20, 24, 32
Swelling, 81, 87, 111, 141, 257

Temperature dependence,
  swelling, 87
Tensile properties,
  pressure vessel steels, 175
Texture, 37
Theory,
  growth of precipitates & depleted zones, 227
  hardening, 227
  nucleation, 257
Thorium, impurity in iridium, 202
Titanium, additions to AISI 316, 158
Tokamak, 239, 242
Toughness, 63, 67, 167
Transmission electron microscopy, 114
Tungsten, addition to iridium, 202

UBR, University at Buffalo Reactor, 168
Upper shelf energy, 192

Vacancy,
  diffusivity, 248
  loops, 226
  supersaturation, 242, 245, 246
Void ,
  dissolution, 239, 253
  growth, 239, 240, 244
  nucleation, 239, 257, 258
  number density, 239
  swelling, 81, 87, 111, 141

Waste disposal, improved, 74
Wear,
  measurement of, 18, 20, 24, 29, 32
  of cobalt-base alloys, 16, 18, 20, 24, 25, 29-31, 33
Welds, pressure vessel steels, 171
Widmanstatten structure, 51, 53, 59

Zeldovich factor, 234, 262, 266
Zircaloy-2, 35
Zircaloy-4, 51
Zirconium carbide, 52
Zirconium phosphide, 52, 60
Zr-Cr-Fe intermetallics, 52

Author Index

Bendriem, B., 257
Bloom, E. E., 73
Brager, H. R., 87, 141

Causey, A. R., 35
Christie, W. H., 201

DeVan, J. H., 3

Fidleris, V., 35

Garner, F. A., 87, 111, 141
Gelles, D. S., 63

Hawthorne, J. R., 167
Holt, R. A., 35

Klueh, R. L., 73

Northwood, D. O., 51

Ocken, H., 15

Pavone, D., 201
Primeau, S. J., 239

Quach, V., 51

Russell, K. C., 239, 257

Simonen, E. P., 225

Taylor, D. H., 201
Tortorelli, P. F., 3